Water in the Arab World
Perspectives and Prognoses

Edited by Peter Rogers and Peter Lydon

Papers from a Conference
Sponsored by the
Arab Fund for Economic and Social Development,
and Harvard University's
Division of Applied Sciences and Center for Middle Eastern Studies
at Harvard University
October 1–3, 1993

Published by the Division of Applied Sciences
Harvard University

Distributed by Harvard University Press

Library of Congress Cataloging-in Publication Data

Water in the Arab world : perspectives and prognoses / edited by Peter
Rogers and Peter Lydon
    p.   cm.
    "Papers from a conference sponsored by the Arab Fund for Economic
and Social Development, and Harvard University's Division of Applied
Sciences and Center for Middle Eastern Studies at Harvard University,
October 1–3, 1993."
    Includes bibliographical references (p. 18) and index.
    ISBN (invalid) 0-674-94789-9
    1. Water-supply -- Arab countries -- Congresses. 2. Water resources
development -- Arab countries -- Congresses. I. Rogers, Peter P., 1937–.
II. Lydon, Peter 1937–. III. Arab Fund for Economic and Social
Development. IV. Harvard University, Division of applied Science.
V. Harvard University. Center for Middle Eastern Studies.
    HD1698.5.W37   1994
    333.91 '0917' 4927-- dc20   94-37287
                                CIP

Printed in the United States of America.

Cover illustration: Satellite composite view of earth
© 1990 Tom Van Sant/The GeoSphere® Project, Santa Monica, CA

Table of Contents

Foreword i

Preface iii

Units of Measurement vi

Summary vii

1. The Water Problems of the Arab World:
    Management of Scarce Resources 1
    Abdul-Karim Sadik and Shawki Barghouti

2. Transboundary Water and the Challenge of International
    Cooperation in the Middle East 39
    John Waterbury

3. Overall Perspectives on Countries and Regions 65
    J. A. Allan

4. The Central Region: Problems and Perspectives 101
    Yahia Abdel Mageed

5. The Arab Mashrek: Hydrologic History,
    Problems and Perspectives 121
    Yahia Bakour and John Kolars

6. Water Resource Development in the Maghreb Countries 147
    Mohammed Jellali and Ali Jebali

7. Water in the Arabian Peninsula: Problems and Perspectives .. 171
    Jamil Al Alawi and Mohammed Abdulrazzak

8. Desalination, an Emergent Option 203
    *Taysir Dabbagh, Peter Sadler, Abdulaziz Al-Saqabi,
    and Mohamed Sadeqi*

9. Global Climate Change and Its Consequences for Water
   Availability in the Arab World 243
   F. A. Bazzaz

10. Conflict and Water Use in the Middle East 253
    Thomas Naff

11. The Agenda for the Next Thirty Years 285
    Peter Rogers

Biographical Notes 317

References 323

Index 341

Foreword

Economic and social development in the Arab countries depends, to a large degree, on the availability of good quality water. Most of the Arab region lies in the arid to semi-arid zone. Rainfall is low, variable and unpredictable in the majority of the area. Many of the Arab countries have a per capita water supply below the water poverty line of one thousand cubic meters per capita per annum. The population of the region is growing at a high rate. This, coupled with increased urbanization, improvement of standards of living and expansion of irrigation, has led to sharp increases in demand for water. The water supply-demand imbalance is aggravated further by the increase in water pollution both in scale and intensity. Inefficiency in water use has compounded this critical situation. The water resources prospects for the Arab region beyond the year 2000 seem at best gloomy. Water shortage will become a major constraint on development in many parts of the Arab world and could be the flash point of military conflicts in the future.

In view of the above the Arab Fund for Economic and Social Development has been actively involved in the issues of water in the Arab world. Recognizing the importance of water for all aspects of economic and social development, the Arab Fund has allocated more than 15 percent of its total lending to the water sector. The Fund is well aware of the disturbing outlook for water availability in the region and the need for immediate long-term solutions. Therefore, one of the urgent tasks of the Arab Fund is to help the Arab countries improve the performance of the water sector.

In 1986 the Arab Fund, jointly with the Kuwait Fund for Arab Economic Development and the Arab Center for the Study of Arid Zones and Dry Lands (ACSAD) sponsored a symposium to discuss water resources and water use in the Arab world. The symposium focused on the assessment of the water situation, future forecasts and problems facing the development of this sector. Valuable recommendations emerged from the symposium, including a recommendation to encourage cooperative programs in water resources development among Arab institutions and between these institutions and foreign and international organizations working in this field.

In 1992, Harvard University, one of the most prestigious institutions of learning, research and public service, was chosen as a cooperating institution to sponsor another symposium with the Arab Fund. Harvard's interest in international peace and prosperity was reflected in its acceptance to join the Arab Fund as a co-sponsor in this task. It brought together experts and senior officials in the water sector from Arab countries, regional and international organizations, academic communities, and independent specialists. Water issues in the Arab region were discussed and long-term strategies for the regional water sector were evaluated. Eleven commissioned papers on selected topics were presented at the symposium and were discussed at length and in depth by the participants.

The symposium was held at the Harvard campus October 1–3, 1993. Important recommendations were adopted that call for the improvement of the data base in the water sector, improvement of supply-demand management, formulation of long-term strategies, institution building and the intensification of research in this sector.

This book is the result of the efforts of many people to whom thanks are due. I want to thank the authors of the papers, commentators and all the participants. I also want to thank all the staff at Harvard University and the Arab Fund who contributed to the success of the symposium and to the editing and printing of this book.

Abdlatif Yousif Al-Hamad
Director General/Chairman of the Board of Directors
The Arab Fund for Economic and Social Development

Preface

The idea for a symposium on Water in the Arab World was first broached at a meeting in Boston among Abdlatif Al-Hamad, Richard Wilson, and Peter Rogers in November 1990. Despite gathering war clouds, Director General and Chairman of the Board of Directors of the Arab Fund Al-Hamad was convinced that after the Gulf War the issue of water would remain an irritant and possible cause of future conflict in the Middle East, and that this warranted close concern of the governments in the region. It was decided to hold a symposium sponsored by the Arab Fund at Harvard, under the auspices of the Divison of Applied Sciences and the Center for Middle Eastern Studies, expanding upon the 1986 conference on water resources and water use sponsored jointly by the Arab Fund, the Kuwait Fund for Arab Economic Development, and the Arab Center for the Study of Arid Zones and Dry Lands. The focus of the symposium was to be on the integration of the resource base with the economic, political, and social determinants of demand and supply. We wish to express our thanks to Mr. Al-Hamad for his support and encouragement, which began with those conversations and have continued through the publication of this book.

Between November 1990 and the actual meeting in Cambridge at the beginning of October 1993, a tremendous amount of preparatory work was carried out by a group at Harvard and by the Arab Fund staff in Bahrain and Kuwait. A steering committee of Ismail El-Zabri, Peter Rogers, Tony Allen, John Kolars, Thomas Naff, and John Waterbury identified topics and writers. Despite the dislocations caused by the war, Dr. El Zabri carried out most of the work enlisting participants and authors from the Arab countries. We are grateful to him for sharing his breadth of knowledge and contacts with us; without his help in all aspects the symposium could never have been held. We would also like to thank the Dean of the Division of Applied Sciences at Harvard, Professor Paul Martin, the Director of the Center for Middle Eastern Studies, Professor William Graham, and the Acting Director, Professor Edward Keenan, for releasing staff time and facilities, and for their encouragement to proceed with the symposium. Tom Mullins, Executive Director of the Center for Middle Eastern Studies, helped coordinate the Center's collaboration in the symposium.

Once the symposium was designed and dates chosen, logistical work began in earnest. An able team at Harvard carried it out. Susan Morrison kept track of the endless drafts and the extremely complicated travel and accommodation arrangements of participants coming from more than 20 countries, and she coordinated the work of the student assistants and organized the symposium facilities. While it is hard to single out individuals from a group effort, Radhika DeSilva made a tremendous contribution during the summer and fall of 1993. The other students who made the symposium arrangments work so smoothly were Jennifer Byrnes, Martha Crawford, Marwa Daoudy, Nagaraja Harshadeep, Lucia Lovinson-Golub, and Fiona Murray.

Editing a collection of papers from a conference is often a tedious and unrewarding process. We would like to report that editing this volume has been remakably free of the usual editorial blues. This is owed, first and foremost, to the collaboration of the authors. They cheerfully and speedily responded to our sometimes high-handed suggestions for their manuscripts. Secondly, one of us, Peter Lydon, accepted responsibility for coordinating all of the manuscripts and provided the first round of editorial suggestions; this avoided unnecessary duplication of coverage and inconsistency in style. Thirdly, the Arab Fund for Economic and Social Development was ever ready when logistic and other support was needed.

The work and dedication of the persons who chaired the sessions, commented upon the papers, and stimulated much of the discussions that followed is appreciated. The commentators were Fatima Al- Awadi, Nazli Choucri, Joseph Harrington, Jean Khouri, Miriam Lowi, Jeremiah Delli Priscoli, Ali Ahmed Attiga, Richard Sandbrook,Thomas Stauffer, Dale Whittington, and Richard Wilson.

Other participants who advanced the discussion, and thus contributed to shaping the papers found here, were Rami Abdulhadi, Fahd S. Al-Natour, Anjab Mohammed Ali Sajwani, Anthony Edwards, Farouk El-Baz, Sharif El-Musa, Ismail El Zabri, William Graham, Brian Grover, Robert Hunt, Omar Mohammed Joudeh, Edward Keenan, Salim W. Macksoud, Thomas D. Mullins, Qais G. Noaman, David Nygaard, Nemat Shafik and Adnan Shihab-Eldin.

The order of the chapters in this volume follows their sequence in the conference. The first day (chapters 1 to 3) considered overall perspectives.

The second day (chapters 4 to 8) was devoted to geographic subdivisions and to desalination, and the last day returned to the region-wide level and to issues of the future (chapters 9 to 11).

We would also like to thank Marc Kaufman for his able assistance and that of his staff at Desktop Publishing & Design Co. They turned many murky sketches into the finished drawings that are seen in this book, and have made the production part of finishing the manuscript almost invisible to us. Our appreciation goes as well to Sue Seymour, Executive Administrator of Harvard University Press, for her help in arranging to distribute this volume through the Press.

We believe that early and well-considered attention to water by the people and leaders of the Arab world can tame what otherwise could become a painful and divisive issue for these vital and growing societies. It is true that the positive adjustment and satisfaction of needs that can be gained by farsighted water management beginning now will be an almost invisible outcome, in contrast to the conflict, deprivation, and crisis that neglect of water problems could bring. Nonetheless, we hope that the gathering of minds and thoughts that the reader finds here will contribute to placing the region on the path of quiet progress in the management of its scarce water.

Peter Rogers
Professor
Division of Applied Sciences
Harvard University

Peter Lydon
Visiting Scholar
Institute of Governmental Studies
University of California, Berkeley

Units of Measurement

m^3	cubic meter
bcm	billion cubic meters; this measure of volume is one cubic kilometer of water. It equals .81 million acre feet.
mcm	million cubic meters; this measure of volume equals one thousandth of a bcm, or 810 acre feet.
cumec	cubic meter per second; a measure of flow, equal to 35.32 cubic feet per second.
hectare	2.47 acres. 100 hectares equal one square kilometer; 259 hectares equal one square mile.
lpd	liters per day, or liters per person per day. One thousand liters equal one cubic meter.

Summary

The fifty participants in the Conference on Water in the Arab World came from most of the twenty countries of the Arab region —from Arab, American, and British academic centers and from several governmental and international organizations. The conference heard eleven commissioned papers from participants, with prepared commentary on each, and gave considerable time to free discussion.

As they worked through voluminous data and views on the hydrology and water management of a populous but water-short region, participants in the seminar found general agreement among themselves on several recommendations.

In this summary, these are divided into three sets:

■ General principles to optimize water management in a situation of scarcity,

■ More specific measures that may apply to some but not all of the countries of a variegated region, and

■ Research, data exchange, and publishing activities to develop and disseminate full and current information for good decisions, and for public understanding of the region's water situation.

Background and the Problem

The Arab region has very low rainfall, and many of its parts have large unpredictable rainfall variations from year to year. Until now, the rising demand for water from Arab populations, which are growing at about three percent per year, has been met primarily by investments to increase the supplies of water available to cities and especially to agriculture. Water availability has been increased by building dams on surface water courses, drilling and pumping for underground water, desalination, and other applications of technology.

This era of finding new water and developing larger supplies at moderate cost is ending. No large rivers and natural lakes originate within the Arab region itself. The region's two largest river systems bring water from watersheds outside the Arab countries: the Tigris-Euphrates from eastern Turkey, and the Nile from Central Africa and Ethiopia. Turkey has a multi-billion dollar national project of dams, reservoirs and canals well underway to make greater use of the water of the Tigris and Euphrates, and the prospect in the middle or the long term is that the Nile upper riparians may also be able to manage and use more water within their countries. Thus, a long-term significant increase of water supply to the downstream Arab countries seems ruled out. The future may see less Tigris-Euphrates water than is now coming to Syria and Iraq, and less Nile water for Sudan and Egypt.

Within the Arab countries, most of the possible water resources have already been developed, and are now producing for human use virtually all the water that can be drawn from them. Many ancient deep aquifers are being mined, meaning that the present population of the region is taking waters that cannot be replaced for their descendants. Desalination is used, but it produces limited quantities of water at high cost. Reallocation of water between countries is extremely difficult to arrange. In broad conclusion, new and increased water supplies are going to be hard to find, and new water cannot be relied upon to make up large-scale insufficiencies, as it often has in the past.

But the water needs of the region are growing inexorably, with population that is both doubling every twenty-five years and also achieving rising standards of living, which mean much higher per capita water demand. Per capita water consumption in the Arab world has always been far below world averages, and a large part of the population is justifiably eager to emerge from the pattern of extremely low water use that is characteristic of underdevelopment in a dry region.

Two Needed General Responses

Thus powerfully rising demand is confronted with an annual supply that cannot easily be significantly increased. This is a description of a collision course, an impasse, or a policy crisis. To resolve it, the Arab Water symposium at Harvard agreed in recommending two responses at the level of general principle.

Give a more balanced attention to both demand management and supply augmentation

Primary attention must be balanced between trying to solve the problem through enlarging the supply of water, and an examination of the

F I G U R E   1

The Arab Middle East and Northern Africa showing the major surface water flows and the Nubian sandstone aquifer

(Consult Table 1 in Chapter 11 for names of rivers indicated here by number.)

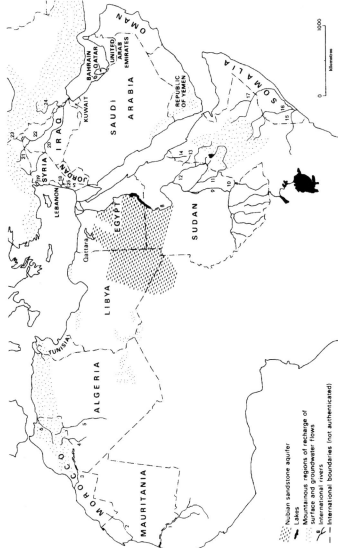

pattern of demand and of water use. As one participant wrote, if nothing is done to change the demand structure, with growth of population and rising standards of living the demand for water will increase so as to exceed any possible supply in the near future. Demand analysis and management must be practiced by governments, by the private sector, and by individual farmers and families in the general population.

Water development and water allocation must be more purposeful, conscious, and calculated

Both water supply and water consumption choices and decisions need to be made with increasing consciousness and purposefulness at all levels throughout the societies of the region. Each drop, or cubic meter, or cubic kilometer of water must be used so that it does the most good. This will probably mean both greater governmental attention to water, and subjecting water to market mechanisms of allocation. This is not an easy change for any society to make, since water has traditionally been thought of worldwide as a free natural good. Even in the dry Arab lands, this view of water is deeply embedded in the culture of leaders and of the people.

Shifting from Enlarging Water Supply to Analyzing Water Demand

We think of water shortage in terms of a dry faucet in a Cairo house or a Kuwaiti apartment, or of rural women walking many kilometers to bring drinking water to their villages, or of a factory shut down for lack of cooling or cleaning water. But all of these images risk misleading us about the problem. Water supplies for industrial and commercial activities, and for domestic use, including drinking water, are not genuinely threatened, largely because taken all together they use only about 15 percent of the region's water. Domestic and industrial users, moreover, generally can pay relatively high rates per unit of water to cover the costs of bringing water to them. Agriculture uses 80 percent to 90 percent of the water, and in this dry region, irrigation water takes the great majority of agriculture's overwhelming share in water demand.

The Arab region both imports food (in sharply rising quantities in the last decades) and exports food. Food transactions in both directions can be thought of as imports and exports of water. In this region a regulated water supply for irrigation is often an essential factor of production in agriculture. Because the amount of low-cost water is limited, agricultural decisions have important water use implications. In practice, agricultural decisions are water decisions, and vice versa.

Like nations around the world, Arab countries seek food self-sufficiency and food security, but the seminar participants believe that research is called for on the underpinnings of policies motivated by national food security. In what cases is domestically produced food more expensive than the same food would be if imported, and by how much? A realistic calculation, with careful scrutiny of subsidies, is needed. Objectively, what risks and vulnerabilities actually arise from imported food that do not apply to locally produced food? Can these risks be managed by means other than home production, such as diversified and competitive external sources? Are water-consuming food exports from an Arab country economically rational for the whole society, or only for the exporters?

Two exceptions to the general principle that little or no new water can be made available were raised at the conference: 1) Remote sensing indicates that there may be unexploited "fracture zone" aquifers in mountainous areas; research on these is recommended. 2) The conference recognized the Arab world's leading position in the application of desalination technology, and believed that in certain extremely dry and densely populated —essentially urban— locations, desalination can be the best or only alternative in a very constrained situation. Research on both technical aspects of desalination and on its economics is recommended below. However, the conference also recognized that desalination is intrinsically expensive, and not entirely free of environmental liabilities, especially in inland applications. Although having extremely low-cost energy available or having possibilities of pairing desalination with electricity generation can be favorable factors, for most of the situations in the Arab region it would be a mistake to be prematurely committed to desalination, since in general, conservation and demand management can meet wider needs at less expense.

Allocating water to demand more purposefully

Given water's scarcity in the Arab region, what are the "real" demands for water, and what are the water needs that can be met in other ways, or bypassed with the least damage? How can the claims of sustainability and equity in the management of the water resource be met, and the place of water in the region's historic culture and religion be respected?

The participants in the Arab water seminar at Harvard believed that at the national level, water production and use must be much more intensely managed and planned. With due care to avoid the dangers of bureaucratic control such as over-regulation and capture by special interests, explicit rationality and conservation must gradually replace outlooks based on tradition or vested interests. Each country must work seriously to reconcile development and environmental quality. Each country should formulate short- and long-term water strategies, and frame and enact legislation to permit well-debated strategies to be implemented.

The seminar participants also urge the augmentation or development of institutions and manpower specifically for water management. Water sector institutions must have more prestige and public attention and support. Such institutions must be built to do middle- and long-term planning, based on intensified coordination within countries with the managers of other sectors, notably agriculture. Strengthened water authorities should also carry out inter-country water coordination on a basin-wide or broader basis. Operating with wider social and political awareness and support than at present, the work of water experts and officials should be fully integrated with general national economic planning.

Water as an Economic Good

In modern circumstances, money or other resources must be spent to supply water to a home, a factory, or a farm. The resources may be public or private, of course, and water can come in a variety of qualities, timings, and quantities. The tasks that water is used for can often be accomplished with less water or by other means, while at the same time a given amount of water can be used for a variety of purposes. Often a choice must be made among the possible uses for water since the amount of water available is not enough to meet all uses. This increasingly will be the case. All the above may be summarized by saying that water is an economic good. When we talk about using it purposefully, we mean allocating it rationally and fairly among people and uses, in order to serve explicit goals such as equity, prosperity, autonomy, and sustainability. For this, it is indispensable that Arab governments and societies gradually and progressively recognize water's economic character.

A basic principle of economics applies in this situation. If a good is free, people in a position to do so will demand it without limit. Demand will be infinite. But, if a specific cost is attached to a good, that will change both the demand for it, and the supply. A reasonable price for water will make consumers think about how much they wish to consume, and will tend to restrain their demand, thus freeing certain quantities of water for other uses. Tunisia and Morocco, for example, have found that incrementally raised water prices encourage consumers to apply water to only its best and most productive uses. A higher price will also tend to make available greater supplies of the commodity (a sufficiently high price for water in Arab countries, for example, would bring forth essentially unlimited supplies of expensive desalinated water).

The seminar participants, therefore, stressed that if using water is gradually linked more directly to meeting a portion of its cost of supply, water use decisions throughout society, down to the level of individual farm and family consumers, will be more rational, and more cautiously respectful of the basic limitations of supply in the region. There need be no contradiction between this principle and the equity requirement that water necessary for health and nutrition be available to all members of the population at affordable cost.

More specific measures

■ To frame and carry out the integrated national water strategies recommended above for the region, a restructuring of the water authority in each society will be needed. Cooperative and complementary roles must be developed and enacted into law for the political level of decision-making, for technically qualified water managers and for water users. Consumers should be encouraged to organize into groups to articulate their needs and desires, to deal responsibly with official structures, and possibly to take upon themselves tasks in the final distribution of water.

■ To perform on a long-term basis the large tasks of national adaptation that lie ahead, major resources must be committed to capacity building in the water planning and management sector. National hydrological technical and administrative institutions need to be strengthened, or created, with provision for the recruitment and training of high quality manpower and its adequate compensation throughout a career.

■ Modern computerized data processing systems need to be installed, with the training and manpower development programs that they imply, for water data base development, for water management, and for timely dissemination of information. This will also support and facilitate the research agenda discussed below.

■ Water laws or general planning legislation should require the coordination of activities among different organizations within a government or country. In particular, water authorities should participate fully in the appraisal of agricultural policies, since these depend for their success on water, and agricultural decisions reciprocally have immense effects on water availability.

■ There should be regularized and institutionalized forums for consultation among the water authorities of neighboring and co-basin countries. These arrangements for consultation should reach outside the Arab region when this is called for by the geography of particular basins. Such consultations can be the vehicle for data-sharing, and can initiate joint model building, which often generates instructive and useful exchanges.

■ Under national water strategies, the drilling of water wells and the volume of water pumped from them should be reported to national water authorities. Responsible authorities should be ready to regulate the use of aquifers that cross international boundaries, or that are above their sustainable yield, or close to that point. Consideration should be given to levying charges according to the volume of groundwater extracted.

■ The seminar members recommend the most serious, country-by-country consideration of introducing cost recovery for water use in situations where it is not practiced, or increasing it where it is in use. Probably the most effective and most self-executing strategy for demand management, and the one with the greatest promise to shrink pressure on a scarce resource, is to link water use with support of the costs of providing water through a system of fees. Such effects were seen, for example, in the consumer response to the energy revaluation of the seventies. Water fees for cost recovery are practiced within the Arab world in Morocco and Tunisia and elsewhere, but their adoption or augmentation by other countries, with the requisite installation of metering and accounting systems in the different parts of the water sector, still remains a major task. It can expect to encounter large scale public and political resistance. Water charges would certainly require wide public debate and careful public deliberation to achieve sufficient social and political endorsement to begin, or to be increased significantly. Charges, and the apparatus to implement them, could be introduced or augmented gradually, starting with fees that would recover only a fraction of water delivery costs at the beginning. Even partial cost recovery is however the principal means to spread the habit of very conscious water use throughout every level of society, to provide to every user a logic and incentive to conserve water, to hold use to sustainable levels, and to mobilize the necessary capital to make water available equitably to all parts of the population. Cost recovery repre-

F I G U R E   2

Rainfall in the Middle East and North Africa

(discussed in Chapter 11)

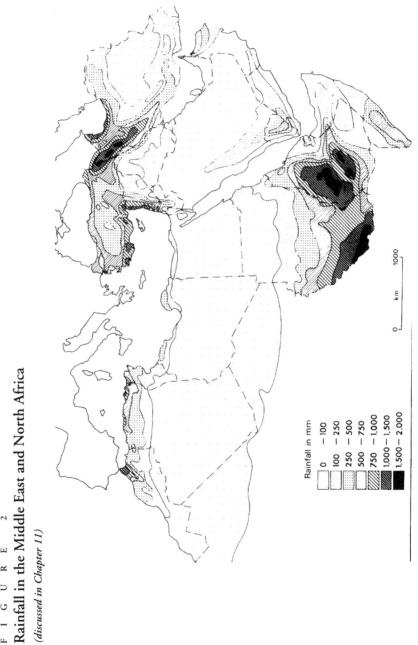

Rainfall in mm

0 — 100	
100 — 250	
250 — 500	
500 — 750	
750 — 1,000	
1,000 — 1,500	
1,500 — 2,000	

0   km   1000

sents open recognition of water as an economic good, and will encourage the application of water to its most productive uses.

- Making water entitlements tradable could prove helpful.

- In allocating water, conflicts of interest and disputes inevitably arise. Local allocation and local dispute resolution should be based on an understanding of traditional conflict management approaches.

- There should be more use of treated effluent in applications that are safe and that may relieve pressure to use non-renewable or very expensive alternative sources.

- An ethos and practice of water conservation in all water-using sectors is of great importance, and must be supported by the national water strategy. Funding agencies, in supporting projects, should allocate a portion of their investment to measures to conserve water. Educational campaigns and school curricula to educate the public on a country's water situation, and the steps necessary to deal with it, can make a major long-range contribution.

- It is recognized that policies inspired by sound economic principles alone will meet with strong opposition from individuals, constituencies, and officials who will be disadvantaged by them. An important element of any study of policy should be a review by local scientists, and by local officials and legislators, of the political economy of water and the measures and incentives needed to enable the installation of sound water allocation and management policies.

Recommendations for research and publishing

Symposium members laid great stress on the need for economic and hydrological research, data exchange, and circulation of information for debate and discussion. The following topics were particularly mentioned as needing attention:

- Following on the recognition of water as an economic good, research is needed to establish better the actual economic value and productivity of water in different places and applications. Different forms of irrigation, different crops, different treatment and distribution systems have different cost/benefit profiles, which must be worked out. Estimates from such analysis must be disseminated for debate, and to increase the public's awareness that various water uses have specific costs and returns.

- The willingness and capacity of different parts of the public to pay for water should be examined by economic and cultural research. Making water subject to market forces is intended to make the use of water more flexible and responsive to real needs, not to limit access to water for people who need it. Subsidies may well be in order for parts of the population with limited capacity to pay. For example, in some cases poor people are now buying domestic water from street vendors at very high prices that could well be reduced by improved distribution and by targeted subsidies.

- Given concerns about food security, research should investigate whether or not food is reliably available on the world market at economic prices, and the policy implications of the findings from this inquiry should then be assessed. Can the exploitation of a food-importing country by food surplus nations be excluded, or is it a real risk? If it is a danger, can it be ameliorated by measures such as procurement from multiple sources? Research should also clarify the ability of Arab countries to pay for imports through exports in which they have a robust comparative advantage. Results of such studies should receive dissemination; the public needs to be more aware of the interchangeability of food and water through international trade, and the degree of risk, if there is a risk associated with it.

- A necessary preliminary to the adoption of sound water management policies, and especially policies based on demand management principles, is study of the interests of the communities dependent on agriculture and especially of those government that have a deep and long-lasting involvement in water management. Until these groups can feel that they have a secure future they will strenuously impede the adoption of policies that they will perceive to be to their economic and social detriment.

- If a shift of water away from agriculture is judged to be needed, studies should clarify the consequences and practical aspects of such a change. What are the suitable incentives, the operational steps, the tradeoffs and necessary mitigations, and what will be the role of stakeholders in the decision-making?

- Broad, multi-country recommendations such as those presented in this summary need testing through case studies and other forms of close empirical examination. Detailed empirical work should form a substantial part of a national or regional research agenda.

■ Researchers should build alternative water future scenarios. When done cooperatively between two or more countries, these would be joint interactive models of regional land- and water-use systems.

■ Desalination needs further research, including public sector–supported research, and research on desalination processes done jointly by producers of the equipment in industrialized countries and those with the experience of using and maintaining it in arid countries. Close study of the economics of desalination in comparison with alternatives such as long-distance transport of water, is needed. Such research should be attentive to the long-term and to the overall water-use cycle when desalination is employed, and not limit itself to the desalting process alone.

■ In water policy research and decision-making, each full basin should be considered. The effects of actions or inaction in the upper basin upon those living lower down on a river system, perhaps in another country, must be taken into account.

■ There should be continuing study of groundwater, particularly of trans-border aquifers, to encourage monitoring them for cooperative management.

■ Meteorological studies, with improvement and exchange of meteorological data, should focus on annual rainfall variability, and on early awareness of potential long-term climate change.

■ Administrative and water management studies should compare what is being done in one sub-region with what is needed in another. For example, the Mashrek should be more aware of what strategies are successful in the Mahgreb.

■ A regional water data center, or Regional Information Clearing House (RICH), is needed to collect and organize data, and to facilitate data access and sharing among Arab countries.

■ A reappraisal of the role of rain-fed agriculture, including the viability of supplemental irrigation, in the whole water resource picture is needed, aimed at maximizing the rainfed contribution to food production and productive employment.

Conclusion

The seminar participants believed that some published discussions of the water problem in the Arab countries have exaggerated immediate dangers. A majority of participants did not foresee, for example, military conflict over water in the next twenty years. The seminar participants did believe that with early, prudent, and vigorous demand management responses by Arab societies under the headings of 1) explicit strategy formation, legislation, and implementation, 2) institution building, and 3) research, the clearly looming policy crisis in water in the Arab world can be kept from turning into a set of genuine national resource crises. Such resource crises could only have heavily damaging effects on the economies, the lives, and the living standards of Arab populations, and on the possibility of cooperative relationships among Arab countries.

1

The Water Problems of the Arab World: Management of Scarce Resources*

Abdul-Karim Sadik
Shawki Barghouti

Introduction

The water situation in the Arab world is attracting increasing attention. It is serious, the challenges are for the long run, and the proposed solutions require regional cooperation and coordinated action plans. Different countries in the region confront different problems with their water sectors. Some countries have too little water, or no scope for additional supplies without massive technological and financial inputs. Others have adequate supplies, at least for the short run, but manage their resources poorly. The common features of the water sector in the Arab world include:

■ The resources are scarce and expensive to exploit.

■ The problem extends beyond the national political borders, but the technical basis for regional cooperation is not well established nor politically endorsed.

■ Municipal and industrial water requirements are increasing sharply, as the population is growing at about 3 percent annually, and rising incomes bring about increased demands for water. The per capita consumption rates in the Arab countries are among the lowest in the world and they are likely to double and triple over the next 20 years. Municipal and industrial water as a proportion of total water use is expected to double in the next two decades. These increases in demand for domestic and industrial purposes will require a revaluation of the present allocation to agriculture and irrigation.

---

* The views expressed in this paper are solely those of the authors.

■ For the foreseeable future, a realistic approach to the water problem in the Arab region is best limited to improving management and allocation practices, and upgrading and modernizing water delivery systems.

A prospective shift in allocation from irrigation to domestic water use requires careful analysis of the various factors affecting the improvement and development of the water sector (resource management) and the efficient utilization of the water resource (demand management).

The paper will review the water balance (supply and demand), and the evolving relationship between water and economic development, especially agriculture. We will assess the political and economic costs of the water problem and consider the problem's serious implications for the social and formal institutions that are involved in water and in the economic development of the region.

The Land and Water Resource Base

Unexploited arable land and reliable water resources have been exhausted in most of the Arab world. The region now confronts two major challenges. First, high population growth is forcing expansion of cultivation onto marginal land and engulfing prime agricultural land in accelerating urbanization. Second, the low return on agricultural enterprises in rainfed farming discourages investment in this activity and increases pressure to expand irrigated agriculture. The conflicting and competing demands for land and water in the Arab world hinder effective management of the scarce water resources in the region.

The population of the Arab region is about 3 percent of the world's population, but it receives from nature only 1 percent of the world's renewable water resources. Population growth is likely to reduce the per capita availability of water in the Arab countries by about 50 percent by the year 2025. For Arab countries in the Gulf annual per capita water resources are extremely low, ranging between 107 m^3 in Kuwait and 148 m^3 in Saudi Arabia. The equivalent availability in Egypt is 1120 m^3 and in Jordan is 368 m^3 (World Resources Institute 1992). The per capita availability of water is changing annually as population increases without commensurate increase in renewable water resources. A World Bank Report (1993) states that "within one's lifetime, annual average per capita renewable supplies — excluding so-called 'fossil' aquifers in the Arab region — will have fallen by about 80 percent, from 3,430 m^3 per capita (in 1960) to 667 m^3 (in 2025)." The World Resources Institute (1992) identifies the Middle East as the region where water shortages have reached crisis proportion, and have become a prominent political issue, especially along the international

river basins of the Tigris and Euphrates, the Jordan and the Nile. In light
of the predominant share of water that is used for irrigation, we wish to
stress that the water crises cannot be objectively analyzed nor adequately
addressed without a thorough consideration of agriculture in the region.

Water Allocation: Current Trend

World Resources Institute (1992) indicates that annual renewable
water resources in the Arab Region average about 350 billion cubic meters
(bcm). Of this, about 125 bcm, or 35 percent are provided by river flows
from outside the region: 56 bcm by the Nile, 28 bcm by the Euphrates, and
38 bcm by the Tigris and its tributaries. Besides renewable surface water
and groundwater, there are substantial non-renewable groundwater resources and countries in the region have varying access to brackish water
and unlimited seawater.

Irrigated agriculture receives the lion's share of water resources in the
Arab world. Table 1 provides percentages of water use by major categories

T A B L E   1

Sectoral Water Allocation in the Arab World *(percentage shares)*

	Domestic	Industry	Agriculture
Algeria	22	4	74
Bahrain	60	36	4
Djibouti	28	21	51
Egypt	7	5	88
Iraq	3	5	92
Jordan	29	6	65
Kuwait	64	32	4
Lebanon	11	4	85
Libya	15	10	75
Mauritania	12	4	84
Morocco	6	3	91
Oman	3	3	94
Qatar	36	26	38
Saudi Arabia	45	8	47
Somalia	3	0	97
Sudan	1	0	99
Syria	7	10	83
Tunisia	13	7	80
UAE	11	9	80
Yemen	4	2	94

Source: World Resources Institute 1992.

in the Arab countries. The share of agriculture can be as high as 99 percent, as it is in Sudan. In all of the other Arab countries (with the exception of Bahrain, Kuwait, Qatar, and Saudi Arabia, which rely on non-conventional sources of water), agriculture's share exceeds 50 percent; and for twelve countries it is 80 percent or over.

Based on these sectoral withdrawal percentages for individual countries, region-wide averages of water use in the Arab countries are: domestic (6.9 percent), industrial (5.1 percent), and agricultural (88 percent). It is estimated that 163 bcm of water are used annually for agriculture and industry (Arab Fund 1992). Total water uses, therefore, amount to about 174 bcm, and are allocated as shown in Table 2.

The above figures show that current water consumption per capita of 792 m^3 is equivalent to 54 percent of renewable water available per person (1,460 m^3). These figures also indicate that average domestic water use in the Arab world is about 140 liters per person daily (lpd), and that irrigation consumes, on average, about 14,000 m^3 per hectare (ha), based on a total irrigated area of about 11 million ha (FAO 1991).

Domestic and industrial uses

At current population growth rates, and current per capita average consumption, domestic water use is projected to reach about 13 bcm by the year 2000. However, domestic water use is still very low in many parts of the Arab world, with nearly one-third of the total Arab population still having no access to safe water (Arab Fund 1992). If only a minimum per capita average of 200 lpd is used in calculation, total domestic water use, at current population growth rates, would exceed 21 bcm by the year 2000, and 43 bcm by the year 2025. Industrial use, on the other hand, at current per capita average and current population growth rates, would reach a total of about 11 bcm by the year 2000, and about 22 bcm by the year 2025. At a rate of increase of industrial use at 3 percent per year, which is the expected growth in the industrial sector, total industrial demand, at

TABLE   2

Arab Region, Total Water Use by Sector

Sector	Percent (%)	Volume (bcm)	Per Capita Use (m^3)
Domestic	6.9	11.24	51.10
Industrial	5.1	8.31	36.83
Agriculture	88.0	154.68	703.10
Totals	100.0	174.23	791.95

Table compiled by authors based on data available in *Agriculture & Development*, a periodical of the Arab Organization for Agricultural Development, 1986 and 1993, Khartoum.

current population growth rates, would amount to about 14 bcm in the year 2000, and about 37 bcm in the year 2025 (see Table 3).

If population growth remains at nearly current rates during the next decade, total domestic and industrial demand is likely to reach 24 bcm by the year 2000. Domestic and industrial uses are projected as shown below in Table 3, but it may be noted that moderated population growth rates (Bos et al 1992) could reduce total domestic and industrial demand by about 9 bcm by 2000 or 16 bcm by the year 2025.

Is expansion of irrigation possible?

Water use for agriculture is significant in 14 Arab countries where the irrigation sector consumes about 154 bcm to serve about 10 million ha. About 50 percent of this total is accounted for by nine countries (see Table 4). Iraq faces sharp reductions due to upstream development on the Euphrates and, to a lesser extent, the Tigris. Egypt and Sudan have tried, with limited success, to expand areas under irrigation, and both countries appear at the limit of their horizontal development of agricultural land. Egypt's attempts to utilize the return flows downstream are an issue to which we will return later. Morocco (as well as Algeria and Tunisia) have severe regional deficits and expansion of irrigation may require expensive water transfer to dry regions. Withdrawals in the remaining Arab countries, especially Jordan, Libya, Saudi Arabia, the Gulf States, and Yemen already exceed renewable supplies. Expansion of irrigation can only be discussed in the Arab countries when urgent conservation measures have been introduced including modern water scheduling and sharing operations, modernization of irrigation networks and on-farm systems, use of less water-intensive crops, and restructuring of the agricultural sector. We will address this issue within a strategy aimed at improving management of prime agricultural lands and maximizing productivity of major food crops in rainfed areas.

T A B L E   3

Projected Non-Agricultural Water Use in the Arab Region *(bcm)*

	1990	2000		2025	
Domestic	11	13 (a)	21 (b)		43 (b)
Industrial	8	11 (a)	14 (c)	22 (a)	37 (a&c)
Total	19	24	35	65	(80) (d)

(a)   At current population growth rates and current per capita average consumption.
(b)   Assuming per capita consumption at 200 lpd.
(c)   Assuming 3% increase for industrial use.
(d)   Total assuming 3% in industrial use and 200 lpd.

Table compiled by authors based on data from water studies by staff of the Arab Fund, 1986–1993.

The region's prime agricultural land, defined here as land receiving upward of 350 mm of rainfall annually with suitable soils for crop production, is less than 20 percent of total land in the region. While these lands do not have severe inherent limitations to crop production, there is a limit to the abuse and exploitation that any soil can withstand. Most of these lands are intensively cultivated, and farmers, on average, maintain a winter/summer rotation that allows for close to 80 percent cropping intensity in average rainfall seasons. High pressure on the land has been associated with this high intensity and indicates that farmers allow only about 20 percent of their land to rest annually in order to regenerate its moisture, nutrients and fertility. (Productive regions in the United States, Canada, and Australia, on the other hand, can afford to leave 50 percent of their land fallow, which results in higher yields and more productive crop rotation.) The growing season in the Arab rainfed lands is about 180 days, which allows the major food crops, especially wheat, barley, and pulses, to be grown. Yields of these crops are relatively high considering the soil and rainfall limitations, but do not usually exceed 50 percent of those achieved under irrigation (see Table 5).

FAO production yearbooks over the last two decades indicate almost no expansion of arable lands in the Arab world during this period. This situation is alarming, indeed, especially since the population in the region continues to grow at a relatively high rate of about 3 percent annually.

Table 6 illustrates the increasing pressure on arable land; in a matter of only 15 years the competition index (number of people per hectare) has increased over 50 percent in countries such as Egypt, Jordan, Syria, and Yemen. As noted, the expansion of arable land in the Arab region has been

---

T A B L E   4

Water Allocation and Availability for Irrigation in Nine Arab Countries (*bcm*)

	1990	2000	2025
Water Resources	235	235	235
Water Use:			
Domestic	4	13	23
Industrial	3	6	14
Agric'l (curr)	80	80	80
Water Use Total	87	99	117
Balance	148	136	118

Source: World Resources 1993, and Annual Production Yearbook of the FAO, 1991 and 1992. The nine countries are Egypt, Jordan, Saudi Arabia, Syria, Morocco, Iraq, Sudan, Tunisia, and Algeria.

T A B L E   5
Yields of Selected Crops and Productivity Characteristics *(ton/ha)*

	Irrigation	Rainfed Land	
		Prime *>300 mm*	*Marginal* *<300 mm*
Wheat	4.42	2.15	0.65
Barley	2.85	1.65	0.93
Sorghum	4.20	1.90	0.72
Legumes	3.80	2.70	0.40
Growing Season	All Year	180 days	110 days
Cropping intensity	200%+	90%	60%

Source: Shawki Barghouti and John Hayward (1992): Land and Water Resources in the Middle East and North Africa: Issues and Challenges, EDI/ICARDA, 1992.

T A B L E   6
Population Pressure on Arable Land in Selected Arab Countries *(Population in thousands, land in thousand hectares)*

	1974	1979	1984	1989
Egypt				
Population	36,289	40,875	46,511	51,233
Arable Land	2,707	2,304	2,310	2,310
Competition Index (pop/ha)	12.4	17.7	20.1	22.2
Jordan				
Population	1,850	2,250	2,712	3,167
Arable Land	290	297	304	310
Competition Index (pop/ha)	6.4	7.6	8.9	10.2
Lebanon				
Population	2,767	2,669	2,668	2,678
Arable Land	237	220	210	208
Competition Index (pop/ha)	11.7	12.1	12.7	12.9
Syria				
Population	7,438	8,800	10,458	12,082
Arable Land	5,682	5,230	5,104	4,889
Competition Index (pop/ha)	1.3	1.7	2.0	2.5
Yemen				
Population	6,991	8,219	9,758	11,272
Arable Land	1,355	1,366	1,370	1,376
Competition Index (pop/ha)	5.2	6.0	7.1	8.2

Table compiled by authors.

meager in the last two decades. From 1970 to 1987, Iraq, Jordan, Libya, Morocco, Saudi Arabia, Syria, and Tunisia added about one million ha under irrigation (about 12 percent of total irrigated area). No significant expansion in the rainfed areas has been recorded during the period. There is little scope for expansion of arable land in the Arab world, since much of the land being farmed now has thin soils, rainfall is limited and highly variable, and the population is being squeezed to face greater farming challenges. Because of the increasing inadequacy of rainfed farming to meet the region's growing needs for food, several Arab countries have turned to irrigation. As Table 7 shows, about 20 percent of the total cropped area in the Arab world is under irrigation.

Water and agriculture

In most Arab countries the large allocation of water to agriculture has been driven by a strategic aspiration to attain food self-sufficiency. The Arab region imports more than 50 percent of its food requirements now, and according to recent studies by the Arab Organization for Agriculture Development (AOAD), this gap is likely to double in the next two decades

T A B L E   7

Land and Water Utilization in Certain Arab Countries
(land in thousand hectares)

	Total Cropped Area	Cropped Area (ha/cap)	Irrigated Area	% of cropped area irrigated
Egypt	2,585	0.05	2,559	99
Libya	2,150	0.47	129	6
Mauritania	199	0.10	24	12
Morocco	9,241	0.37	1,300	14
Jordan	450	0.11	49	12
Syria	5,503	0.44	660	12
Yemen	1,480	0.15	266	18
Saudi Arabia	1,185	0.08	420	36
Tunisia	4,700	0.47	282	6
Iraq	5,450	0.29	2,300	42
UAE	39	0.02	5	13
Sudan	12,510	0.50	1,900	15
Lebanon	301	0.11	88	29
Algeria	7,605	0.30	304	4
TOTAL	53,398		10,286	20

Table compiled by authors from World Bank and FAO agricultural sector studies and reports, 1980–1992.

TABLE   8

Percentage of Self-Sufficiency Major Food Commodities in the Arab World

	1970	1980/85	1986/90
Wheat	60.5	43.2	48.5
Other Cereals	80.1	49.5	50.2
Sugar	34.1	29.4	33.8
Oil Seeds	58.5	37.7	34.3
Meat	95.5	71.8	78.9

Source: *Annual Report: Agricultural Statistics*, Arab Organization for Agricultural Development (AOAD), Vol. 2, section 11, 1992.

(Table 8). The Arab countries' total demand for wheat exceeded total production by about 22 million tons in 1990. The wheat imports of Egypt alone amounted to 6.6 million tons that year. Egypt's attempts to increase cereal production in the new lands failed because this type of farming is not profitable under irrigation in the new lands, and there is not enough water to satisfy the needs of cereal crops there.

The principal irrigation schemes in the region were originally constructed to provide additional staple food supplies, mainly wheat (except in Sudan, where irrigation works were built to grow cotton). During the 1980s, staple world commodity prices declined and returns to irrigation projects diminished. As a result, farmers shifted to growing vegetables, fruits, cotton, and oilseeds. Governments have been reluctant to support this shift on the grounds that such changes may endanger national food security. State intervention to increase staple food production on irrigated lands has taken the form of administrative controls on cropping choices rather than market incentives, leading to a deterioration of incentives in the sector in the light of unhelpful pricing policies and low returns to the farmer in many countries in the region (Egypt, Jordan, Syria). This in turn is reflected in the poverty of management of irrigated land, where field efficiency in the traditional sector (gravity irrigation for staple food crops) is low and cropping intensity of irrigated areas averages 140 percent (subject to availability of water), compared with 240 percent in East Asia. Although water is scarce and expensive, low-value crops and water-demanding production systems have been encouraged. Poor water policies and heavy government intervention in cropping patterns favored crops considered strategic by governments, such as grains and sugar for local consumption, and cotton for export, at the expense of other less water-demanding crops with higher market value.

While the countries in the region have expanded their investment in irrigation, the agricultural sector itself has been facing three important challenges. The first is the declining contribution of agriculture to the national economy (Table 9). Farmers' incomes are lower than urban incomes. The non-agricultural sectors are drawing resources away from agriculture, and low prices and incomes are "pushing" farmers into urban jobs. Many farmers are diversifying their incomes by seeking part-time employment outside the farming sector. Most successful countries use this structured transformation for more efficient resource allocation and better income distribution. Prevailing pricing and incentive policies in the region for both water and agricultural commodities have not encouraged smooth and timely diversification of the agricultural sector. Agricultural policies in Egypt have favored cotton, sugar, and food grains. Agricultural policies in North Africa, Syria, and Jordan have also provided incentives for food grains. The return on investment (especially for irrigation, and operations and maintenance of irrigation systems) for these crops is well below that obtained from growing fruits, vegetables, and industrial crops.

The second challenge is the slow adoption in the Arab region of modern agricultural production technology. The rapid development of advanced technology in agriculture has enabled farmers in several regions of the world to increase their output per unit of input (labor, water, seeds, etc.) and provided them with new production options and more flexible

---

T A B L E   9

Percentage of Agriculture in Gross Domestic Product (GDP) and Employment in Selected Countries *(1965, 1988 and 1990)*

	Percent Share of Agriculture in GDP			Percent Share of Labor in Agriculture		
Country	1965	1988	1990	1965	1988	1990
Egypt	29	20	17	55	42	35
Syria	25	22	30	52	25	24
Morocco	22	18	16	61	40	38
Tunisia	20	17	14	49	32	30
Jordan	18	8	7	37	6	6
Algeria	13	8	10	57	24	24
Sudan	49	35	30	82	71	70
Yemen	52	22	20	73	64	62
Mauritania	30	34	24	89	69	68

Source: World Development Report – 1990 (World Bank).

farming systems. High-yielding varieties of wheat and rice largely contributed to the doubling of yields in several Asian countries (Barghouti 1990).
The Arab world is yet to fully employ these technologies, but since
horizontal expansion of its agriculture is no longer possible, production
growth must come from yield increases and increased cropping intensity.
The basic technologies responsible for output increases are improved plant
varieties and animal breeds, agrochemicals, and irrigation systems.

Technological improvements in irrigation systems have also expanded
production opportunities. Whereas traditional irrigation technologies
(furrow, border, and flood irrigation), which deliver water to plants by
gravity, usually result in substantial water losses and poor uniformity in
water distribution, modern irrigation technologies (particularly sprinkler
and drip irrigation) are more water efficient. They open greater opportunities to cultivate sandy and rocky soils with low water-holding capacity,
and to farm low quality land and steep slopes. The transition to modern
irrigation has also enabled regions facing limited water supplies to shift
from low value crops with high water requirements, such as cereals, to high
value crops with lower water needs such as fruits, vegetables, and oilseed.
New irrigation methods allow the use of low-quality (e.g., highly saline)
water in regions with high temperatures and high evaporation rates.

These technological advances, however, have been a double-edged
sword for the Arab region. The changes in both production and irrigation
have increased output and created strong competitive markets at the world
level; in comparison with world price levels, cereal production in the Arab
region is now expensive. Arab farmers cannot justify expensive investment
in irrigation or in diverting water to produce crops (notably sugar and
cotton, as well as cereals) that are available on the world markets at prices
lower than production costs in the Arab world. Under these disadvantageous conditions, Arab farmers have had to search for new crops, new
technology, and new production systems to sustain their farming enterprises. Thus, improved technologies have not always brought with them
stability and improved incomes.

The third important challenge is the increasing awareness of the
negative impact of current intensive production practices on the environment, and concern about these effects. Three major problems of land
degradation face the water sector: silting, waterlogging, and salination.

The upland watersheds of most rivers in the Arab world have been
deforested and overgrazed. Erosion from seasonal torrential rains have
stripped off the soil of the upland areas in the region and have poured it into
the streams, which in turn have carried it as suspended sediments hundred

of miles downstream. The silt settles in channels and fields, and thus clogs the irrigation works.

In the irrigated areas of Iraq, the scourges of waterlogging and salination are as prevalent as ever. The irrigation water from the twin rivers, Tigris and Euphrates, contains a certain concentration of salt, which accumulates in the top soil as a consequence of evaporation in the hot weather prevailing in the area. According to some reports the total land under irrigation shrank from about 7.5 million hectares in the early 1970s to about 6 million in the 1980s. Despite its great agricultural potential, Iraq in the last decade became a net importer of grain to feed its population of 20 million.

Similar problems are rapidly arising in Egypt as a result of the increased intensity of irrigation made possible by the construction of the Aswan high dam. The dam makes it possible to maintain a nearly constant year-round water level in the Nile. This allows easy pumping of irrigation water throughout the year, but it also raises the water table and impedes drainage, so Egypt is now experiencing the maladies of waterlogging and salination to which it had for so long seemed immune. Agricultural drainage water in upper Egypt flows back into the Nile River, aggravating the already growing salination problems caused by the rising water table.

Policies and Government Actions

While each Arab country is attempting to tackle its situation according to its local needs, all face the imperative of formulating consistent national water policies as a matter of national priority, and addressing the allocation of water to agriculture. In almost all countries irrigation is the main user of water, and vigorous constituencies pay close and active attention to issues affecting it, such as those related to crop production, efficiency, modernization of systems, pricing and allocation of water, and protection of the environment from the serious consequences of the continual intervention in natural hydrological cycles.

Strategies and policies are not always consistent from country to country, or even within countries, nor should they be, since the situations vary greatly. Irrigation systems vary from traditional farmer-built spate irrigation schemes in Yemen, to basin flooding in Egypt, to modern drip irrigation systems used in greenhouses for fruits and vegetables in Tunisia, Jordan, and Morocco. Water projects range from earth dams in eastern Jordan, to a modern irrigation network in Doukkala in Morocco, to large, sophisticated and multipurpose dams in Egypt and the Jordan Valley, to a small reservoir to provide water to the main cities in Algeria. What is missing is an integrated water plan which makes effective use of a river basin and its local resources and takes account of different factor prices and constraints in each country.

In some aspects, the region has had great success in water management, and it has supported the adoption of modern technology in canal conveyance systems and on-farm water distribution, as is illustrated by Doukkala and new lands development in Egypt. Nevertheless, the water problem is intensifying. The growing scarcity of water resources in the region is caused by three major factors: rapid population growth, deteriorating water quality, and the rising cost of new water projects. Let us look at two country experiences.

Egypt and Jordan provide vivid and contrasting examples of the severity of water shortages and of the agricultural sector's response. Jordan, which receives an annual average of 250 to 325 mm of rainfall, consumes 110 percent of its water stock annually and has accumulated a deficit equal to one full year's supply through overpumping shallow aquifers. More efficient use of water is Jordan's top policy and investment priority. Egypt, which is completely dependent on the flow of the Nile River, experienced major crises during the droughts of the 1980s. Its program to expand agricultural land on both sides of the Nile Delta is being threatened by water shortages. A new land development program is proposed to make use of recycled irrigation water, which is likely to be more saline after its original use. Using poor-quality water may threaten the long-term sustainability of the newly reclaimed lands. Where the contribution of agriculture to the national economy is declining, farmers tend to introduce export-oriented high value crops to increase their income. This situation is evident in Egypt and Jordan where new modern irrigation systems have become the main investment to commercialize the agricultural sector, enabling it to face increasing competition from other sectors in the economy. A further reason to proceed with commercialization and diversification of the agricultural sector is to absorb rural labor (to discourage migration to already crowded urban centers), and to increase the return to scarce land and water resources.

Egypt

Egyptian water management, reliant on the Nile for 95 percent of its supply, is dominated by agricultural consumption, which is served by a rigid water distribution system with more than 1000 km of main canals and 30,000 km of secondary ones. A second focus is the continuous threat of Mediterranean salt-water intrusion into the Nile delta. Over 95 percent of agricultural production is derived from irrigated land. Egypt's population grows at an annual average of 2.7 percent, adding about one million people every ten months. The area of arable land per capita is dismally low; by 1991 there were 24 people to every hectare of arable land. Attempts to expand agricultural land have been difficult and expensive. Of the 3 million hectares of cultivated land in Egypt, about 263,000 hectares were reclaimed on both sides of the Nile Delta in the last 40 years. There are plans to add about 400,000 hectares of new lands, but water scarcity is challenging this endeavor.

During the 1980s, in an effort to increase water and land use efficiency, the government launched a major policy shift regarding land use in the reclaimed area. It realized that the overall performance of state farms was inefficient, and that they were unable to adopt quickly new farming practices that could adjust to arid and desert conditions. Changes in land use patterns in the new lands were made necessary because the area was likely to consume 10 percent of all the water available in the country. The need for efficient utilization of water has urged the government of Egypt to establish responsive systems in land property rights, water use and management, and adoption of modern production technology. Modern irrigation has allowed the new lands to improve agricultural production and shifted cropping patterns toward high-value crops, including fruits, vegetables, and oilseed.

There is continued pressure to change traditional cropping patterns in irrigation systems in the old lands in the Nile Valley that cause large water losses; for example, 25 percent to 30 percent of irrigation water supply is allocated to rice and sugarcane (Abu Zeid and Rady 1992). These crops seem profitable from the farmer's perspective but are expensive from a social perspective in view of the free water supplied to the "old lands" farmers while their counterparts in the new lands pay substantial sums for investment and delivery of water.

Recent World Bank studies, supported by independent studies carried out by Abu Zeid and Rady (1992), confirm that improving the present land use and water delivery systems faces serious socio-economic problems including:

■ Fragmented lands and small holdings impede the establishment of efficient irrigation methods.

■ The present water rotation system is unsuitable for modern technologies, such as sprinkler or drip irrigation, which usually require a continuous supply of water.

■ Accurate water control under the present irrigation regime is difficult, especially with water demands of different frequency and quantity for different crops.

These studies proposed several steps to address this situation, including:

■ Increasing the efficiency of the present irrigation system in old lands and developing modern irrigation methods in new lands that would allow planning a reliable irrigation schedule for different crops.

■ Improving drainage networks and reducing losses.

■ Recovering the cost of irrigation development and operation from farmers.

Environmental dimensions are also part of the water situation in Egypt. Because water is the major constraint to further expansion of agricultural areas, treated wastewater has become a new source of irrigation water. Steps have been taken to establish pilot projects on the use of treated wastewater in agriculture and to address several environmental issues associated with improved water management. For example, agricultural drainage water in upper Egypt is discharged back into the Nile River. This affects the salinity of the main river downstream and in the delta. Strict policies have been imposed regarding the release of Nile water downstream of the Aswan High Dam to protect water quality in the river for downstream users. The quality of water in the river (salinity, pollution) decreases gradually toward the delta and coastal plains. A potential decrease in drainage water quantity and an increase in its salinity will occur when irrigation efficiency is improved both in the conveying system and at the farm level. An immediate environmental challenge will be faced in the coastal zone where the increasing reliance on drainage water for irrigation could create serious salinity problems.

The likely challenges to the sustainability of water resources in Egypt include salinity, waterlogging, and the declining quality of fresh water as a result of continuous discharge into the Nile system of usually untreated wastewater from domestic and industrial use (World Bank 1990). Also likely to aggravate pollution is the increased use of chemical fertilizers, especially nitrogen, phosphorus, and potash. This use has increased fourfold in the last two decades, in response to the Aswan High Dam's reduction of the flow of silt downstream. The use of herbicides, especially acrolein to control submerged weeds in canals and ametryn to control water hyacinths in drains, has caused serious environmental hazards and is likely to be discontinued. This problem highlights another challenge: new technologies may have adverse environmental impacts and threaten the sustainability of the system.

Jordan

Jordan has an arid climate, and its basic quantity of available water is far lower than Egypt's. Only 5 percent of its land area receives sufficient rainfall to support cultivation. Less than 10 percent of its agricultural land is irrigated but that fraction accounts for about 20 percent of the value of exports. Jordan's surface water resources, especially in the Jordan Valley, have been extensively developed despite high investment costs. Irrigation in the Jordan Valley consumes about 65 percent of the country's total utilizable surface water. Irrigation efficiency varies from 70 percent for the lands irrigated through direct pressure pipes to 38 percent for surface irrigation. The severe water shortage in the valley restricts cropping intensity to 115 percent (which next to Sudan, is among the lowest ratios for irrigated lands worldwide). An increasing proportion of irrigation water consists of return treated sewage flows, expected to rise to about 35 percent of total irrigation water in the next two decades.

Over the years, Jordan's priority objective has been to expand its irrigated areas; intensification of cropping has been a secondary goal. As the serious water outlook became obvious to decision makers and planners, government policies have given higher priority to intensified land use and are gradually curtailing irrigation expansion in the Jordan Valley. This shift in policy is long overdue. Because of water shortages, farmers were unable to grow seasonal crops on all their land. They utilized less than 100 percent of irrigated land annually; part of the irrigated land was left fallow especially during the summer season when water is scarce. With the introduction of new water delivery systems (pressurized pipes for drip and sprinklers) some farmers were able to grow two crops annually, and cropping intensity in the Jordan Valley increased continuously over the 1973/74–1984/85 period, reaching 130 percent in 1985. With such low cropping intensity, irrigation becomes profitable only when farmers shift cropping patterns toward high value crops (vegetables and fruits). (In contrast, cropping intensity in Egypt is 230 percent and in Punjab, India, 210 percent.)

The basic issue is the relatively poor water resource. There are employment and equity benefits to be gained from expanding the irrigable land areas in the Jordan Valley, but the cost of increasing production using this option are high relative to the cost of intensifying land use. The latest expansion of irrigated areas has cost about US$6,000/ha in public expenditure plus private sector on-farm investment of about $3,000/ha. Where totally new systems are proposed (i.e., in the southern parts of the Jordan Valley), the development costs reach about $20,000 per hectare (World Bank 1990). With the increasing water shortages, further investments in land expansion may not be justified; indeed, they should be thoroughly evaluated.

What is the cost of irrigation water in Jordan? It is difficult to put a price on it. In some countries, farmers are charged according to the area they cultivate, or the crops they grow or the volume of water delivered to the field. In Jordan, farmers pay the cost of operation and maintenance (O and M), although the O and M cost per m^3 of water can be erratic due to irregular supplies. O and M costs in the Jordan Valley fluctuated from $.03/$m^3$ in 1985 to $.05/$m^3$ in 1991, largely because the volume delivered to the farmers declined from 143 mcm in 1985 to 108 mcm in 1991. Since 65 percent of the O and M costs are for salaries, the cost of O and M per m^3 is likely to decrease wherever more water is made available through the East Ghor Main Canal. The supply of water in this canal fluctuates according to the season (high in winter and low in summer) and is subject to the level of cooperation between Israeli and Jordanian authorities.

While the per unit cost of O and M may decrease as the volume of delivered water increases, the marginal cost of water, however, will increase because it is a function of the cost of developing future supplies. In this case, a future water resource for Jordan is the Al-Wahda Dam. A recent analysis of the cost of this dam indicates that the marginal cost of water development is $0.34/$m^3$. Adding the O and M cost would bring the marginal cost to about $0.44/$m^3$ (subject to the amount of water diverted to the Jordan Valley).

With the construction of the Al-Wahda Dam, Jordan will have exhausted all existing water resources. The cost of new water thereafter would be the cost of seawater desalination, which according to figures available from a recent study (Berkoff 1994), would probably be about US$1.92/$m^3$. Under prevailing crop production technology, prices, and water use efficiency, agricultural use of desalinated water would not be justified at such a cost. Irrigated agriculture will need to be restructured and will need to rely on recycled and marginal quality water. If full water charges are applied to agricultural crops, major shifts would have to be made in cropping patterns and irrigation technology. A recent analysis of the net return to water in the Jordan Valley indicates that under present management conditions, many of the higher value crops would remain profitable even with a full cost recovery of water charges.

The economic value of water is difficult to estimate, as international prices for most products that depend on irrigation are undefined, due to seasonal price variability and differentiated products and quality. Despite this uncertainty, it is clear that irrigation water for highly efficient agriculture, using modern technology for the production of high value crops and for water delivery at the system and farm level, could be profitable. It is equally clear that in the case of Jordan, most traditional crops (cereals and vegetables) are not profitable, especially if full cost recovery of water is imposed.

The government of Jordan has initiated several activities to deal with this situation:

■ increasing the proportion of the irrigation water that consists of return sewage flows;

■ converting irrigation systems in the Jordan Valley from surface to pressure pipe systems;

■ introducing automated water management control systems at the farm level;

■ imposing area restrictions on low-value and high-water-consuming crops;

■ selling a fixed volume of water to farmers, determined annually according to water availability;

■ strengthening research and extension on alternative crops, especially those with low water requirements;

■ designing a practical and objective allocation system using long-run management and cost pricing of water to help guide the allocation of water to irrigation against competing demands, and, within agriculture, to help determine optimal cropping patterns;

■ developing water harvesting practices in major watershed areas, especially in the Zarqa Region;

■ establishing a regional cooperation network for hydrology and water resource management (further discussed in the next section) (World Bank 1990b).

Water Management

In the previous section we highlighted the resource problems of the water sector in two countries, emphasizing that scarcity of supply can be compounded by problems of management and ineffective utilization. Let us look at a sample of management problems region-wide.

Supply management

The precarious water situation dictates that the countries of the region devise new water management policies in order to deal appropriately with both water quantity and quality in an integrated fashion. The experience in water management in the region is mixed. The emphasis on supply management has, over the years, given high priority to locating, developing, and managing new resources. Demand management and improvement of patterns of water use received less attention. Also, the region suffers from planning that inadequately integrates environmental concerns, and, at the same time, does not sufficiently provide the policy tools and analytical framework for choosing among alternative uses.

As new water sources become increasingly inaccessible, efficient supply management will require careful analysis of new investment and non-conventional sources. The region already accounts for about 60 percent of total world desalination capacity. Also needed is the upgrading of the management tools applied to the high seasonal and year-to-year variability of water supplies in the region. The stream flow of main rivers in the region varies markedly during the year in response to rainfall/runoff patterns. For example, low flows on the Tigris and Euphrates have been recorded at less than one third of the average annual flows and on the Jordan at less than one half. Management of the variability of water supplies requires a sensitive approach to pricing water under high risk conditions. Cross-sectoral comparison of water's opportunity cost becomes increasingly important during drought seasons, but is very difficult to estimate. The countries of the region need to develop systematic contingency planning to ensure that response to drought on national or locational basis minimizes adverse effects. The experience of the region is alarmingly limited in systematically distributing the risk of drought among different users in a planned and equitable manner.

With the diminishing opportunities to increase water supply, and the escalating costs of projects to augment supply, governments must give far greater emphasis to demand management. This approach requires that direct measures be devised, including regulations, technology, pricing, marketing incentives, and public education. Since irrigation is the main user of water in most of the Arab countries (except for Bahrain, Qatar, Kuwait, and Saudi Arabia), special attention is needed to address its management deficiencies. Some argue that relatively small water transfers from agriculture would substantially increase water's availability to other sectors. For example, a 10 percent transfer from irrigation in Morocco could almost double water supplies to the domestic sector, and a similar transfer in Jordan could increase water supplies for domestic use by more than 40 percent. The discussion of water transfer from agriculture to other sectors can be analyzed within a framework that addresses the efficiency of irrigation systems.

Xie et al. (1993) assembled figures on irrigation water use efficiencies at various levels of an irrigation system (conveyance, distribution, and field application) for a number of developing and developed countries. For the Arab countries included in their review (Jordan, Morocco, West Bank and Gaza, Egypt, Syria, and Yemen), the efficiencies at the network level ranged between 50 percent for Yemen and 75 percent for Jordan. Morocco attained 70–74 percent; the West Bank and Gaza, 74 percent; Egypt, 67 percent; and Syria, 60 percent.

More important than the network efficiency measure, however, is the overall efficiency level that encompasses losses from both the irrigation network (conveyance and distribution) and field applications. Xie et al. (1993) also provide measures of overall efficiency for 12 developing and developed countries. The average for the developing countries was 30 percent. The four Arab countries included in this survey show overall efficiency levels ranging between 20 percent for Yemen (because of the prevailing spate irrigation system) and 53 percent for Jordan, with Morocco achieving 42–49 percent, and Syria, 30 percent.

These comparisons suggest that opportunities exist for various Arab countries to save water through improving irrigation efficiency. Given the relatively low irrigation efficiency levels, modernizing conveyance, distribution, and on-farm application of irrigation water would potentially conserve substantial amounts of water. For Egypt, Abu Zeid and Rady (1992) estimate that between 10 and 15 percent of irrigation water could be saved by optimizing water use.

Similarly, Syria's irrigation systems experience considerable losses. According to Bakour (1992), the efficiency of irrigation networks is 60 percent, and water loss in agriculture is 50 percent. He recognizes the importance of modern irrigation techniques (which are used on only one percent of the country's total irrigated area), and estimates that the application of drip irrigation could reduce water consumption by 45 percent. In addition as much as 20 percent of irrigation water can be saved by increasing the network efficiency to 75 percent, which would require overcoming many difficulties and modernizing the existing irrigation system.

Low irrigation efficiencies can be attributed, in addition, to technical problems of conveyance, distribution, and on-farm application, and to poor maintenance of irrigation structures, often caused by inadequate resources for operation and maintenance. For example, investment in irrigation in Sudan has declined by nearly 40 percent since 1982, leading to poor maintenance of irrigation systems (Zaki 1992).

Financing

Upgrading the efficiency of existing water supply systems, both irrigation and municipal, and undertaking new investments to meet the rising demand for water will require funds that are not easily available under current meager financial resources and stiff competition for budgetary allocations. In almost all the Arab countries where major rehabilitation or modernization of water supply systems are needed, financing is a principal constraint. The share of agriculture and irrigation in public investment has declined over the last two decades, especially in the countries where the agricultural sector can still play a significant role in the economy (Table 10). Investment for new projects is becoming increasingly expensive because prime land is not available and easy access to water has reached the limit. For example, chronic shortages of groundwater in Amman led to the use of surface water resources, which raised the average incremental cost from an estimated \$0.41 per m^3 to \$1.33 per m^3 (World Bank 1993). According to Elahi (1992), the Sudanese system needs massive rehabilitation to restore productivity to levels prevailing in the 1970s.

It will be noted that the percentage share of Agriculture/Irrigation in public investment declined from 14 percent in 1970–75 to 9 percent in 1981–85.

TABLE   10

Public Investment in the Arab World *(Current US$ billions)*

	1970-75	1976-80	1981-86
Total Investment	56.00	283.60	683.50
Agriculture/Irrigation	7.80	25.70	63.80
Industry	11.90	64.00	140.40
Transport	12.10	46.00	94.30
Construction	8.10	37.00	90.00

Source: Salem 1993.

Possible Solutions

Comprehensive National Policies and Regional Cooperation

Complex water problems cannot be resolved by piecemeal or fragmented actions. Delivery of water in the right volume, with suitable quality, at the right time and at the right place, requires an integrated and comprehensive approach to the management of water resources. A system must be designed on the basis of a country's national objectives, in a framework of regional cooperation supported by a reliable regional data base, buttressed by suitable economic policies, as well as a legal and regulatory framework conducive to water conservation and augmentation.

Regional and national hydrological information is an essential requirement for sound management of water resources. Planners need up-to-date data on quantity, location, and quality of renewable water available from surface and ground sources. The Arab countries have made considerable progress in this regard; nevertheless, they have not yet accomplished full identification and assessment of the resources. Khouri et al. (1986) highlight the geological, hydrological, climatic, and economic constraints on accurate assessment of groundwater in the Arab region, and point to the need for further studies and investigations of techniques that can be adapted and applied under arid conditions. Measurement of surface water is usually easier than that of groundwater, yet despite the limited number of river basins with continuous flow (not more than 50 in the whole Arab region), adequate knowledge about the full potential of these rivers is lacking because meteorological networks are weak or non-existent, except for the Nile, Jordan, and some measurement stations for the Tigris and the Euphrates (Khouri et al. 1986). Adequate planning and project preparation requires more and more detailed data to evaluate all aspects that may affect a project and all possible effects of a project. A World Bank report (1993) indicates that such data are often inadequate or unavailable because the capacity of hydrological agencies is, in many countries in the region, hampered by a lack of funds, outdated equipment, low salaries, and insufficient trained staff. As a result, data are poorly recorded, data transmission from measuring station to agency centers is unreliable, data processing is inadequate, and data publication often far behind schedule.

Recently, Algeria, Morocco, and Tunisia requested assistance from the World Bank to upgrade the national hydrological information networks. Morocco is consolidating its water policy based on basin planning that requires detailed data to verify program and projects within the hydrological system. In Jordan, old national and regional plans and data are being updated with United Nations Development Program (UNDP) assistance. Tunisia is upgrading its hydrological network in order to better plan and manage the distribution system that the country has established to integrate water management over a substantial part of the national territory.

An integrated approach

Integrating the water sector with the overall planning and management of the economy is essential to adequately handle diverse and complex water supply and demand issues. Handling water investments and water uses in response to individual sub-sector demands is an increasingly inefficient and unsustainable practice. Generally, priorities for water allocations are accorded to domestic use, agriculture, and industry, in descending order. Allocation among categories, however, is a political decision (Frederiksen 1993) reflecting the broader social, economic, and environmental objectives that water, as a national resource, should properly serve. For proper alignment with such broad objectives, it is imperative that the main responsibilities for water resources management lie with governments. The broader objectives can best receive their due from government institutions capable of undertaking the various functions of water development and management within a framework of appropriate laws, policies, regulations, and procedures.

Some Arab countries have recently moved toward coordination of the activities of various water organizations and have introduced modifications to the laws, regulations, and procedures governing water administration and use. Algeria, Egypt, and Jordan have, in the recent past, begun operations to support sector planning efforts (Rajagopalan 1992). Morocco, however, has undertaken successive initiatives since 1962 to develop coherent policies and capable institutions to promote its irrigation sub-sector, including comprehensive legislation, the bulk of which was issued in 1969, to promote the rational use of water, land, and human resources in the irrigated schemes (Darghouth 1992).

Abu Taleb et al. (1992) describe the institutional arrangements under which the water sector in Jordan was managed before 1984 and the modifications introduced in 1984 and in 1987 together with water laws to enhance the allocation of water by government. The underlying premise was that water is a public good to be used, with consideration of equity, for the development of the country's various regions. The government recognized that with demand pressures escalating and water imbalances beginning to appear, the water sector would need a well defined and integrated approach.

In this connection, much needs to be done in Sudan. Zaki (1992) recognizes that coordination among irrigation water institutions is vital, and that future policies will be oriented toward linking all national and regional water organizations through the creation of a coordinating body for the water sector.

Elsewhere in the Arab world, including Syria, institutional arrangements, water laws, regulations, and procedures have been modified at various points in time with varying levels of coordination to better manage water resources; nevertheless, more efforts are needed for the adoption of comprehensive and integrated approaches based on country-specific conditions and problems.

Cost recovery and system turnover

Water's value and uses have always been heavily influenced by social, economic, and political factors. Historically, water for various uses was provided either free or for less than its cost. At present, a wide spectrum of water charge levels relative to actual costs prevails in various countries; some impose volumetric charges, others charge according to area irrigated or crops cultivated. Water rates are also diverse, reflecting social, economic, and political interests. There is neither a universal policy for water cost recovery nor a suitable and broadly accepted criterion to reflect the diverse issues involved in water use.

Cost recovery is not a purely financial policy. With the increasing scarcity of water and rising demands, cost recovery has emerged as an important conservation policy. It can achieve a two-fold objective of collecting revenues to reduce the burden on government budgets, and at the same time creating incentives for consumers to save water.

Water charges should take into account the costs of operation and maintenance and of capital. Provision of water requires facilities and services for which, normally, capital expenditures (investments), and operation and maintenance expenses are incurred, in addition to administrative and financial expenses (interest in the case of borrowed funds to finance investments). This breakdown of costs leads to another question related to cost recovery: are all costs to be recovered, or only some of them? In each case what is the rationale, or the justification, or the underlying premise?

Theoretically, there are several cost recovery concepts: service cost, opportunity cost, marginal cost, average cost, ability to pay, and market-determined cost. The selection of a pricing mechanism is influenced by a number of factors such as sectoral use, level of subsidies, conservation incentives, equity, ability to pay, and poverty alleviation. The dual objective of generating income and encouraging water efficiency and conservation through a cost recovery policy inevitably requires designing innovative informal institutions with strong beneficiary participation.

In the developing countries, governments seldom recover the full costs of providing water services. Most irrigation, some large urban water supply systems, and all waste treatment are subsidized (Frederiksen 1993). Subsidies, particularly for irrigation, historically have served as incentives to agricultural production and to encourage settlements in rural areas, as in the United States Reclamation Act of 1902 (Teerink and Nakashima 1993). Such objectives are now found in many developing countries, though many countries are moving toward reducing or eliminating such subsidies. Fiscal and budgetary constraints, and increasing water scarcity have contributed to the new orientation in favor of an upward adjustment of water charges.

In the Arab countries, water costs are only partially recovered. In Syria, irrigation water charges do not include the capital costs of water supply and they cover only a small portion of the current costs of water supply. The amount of irrigated area, and not the quantity of water used, is taken as the basis for charges. Extraction of groundwater, regardless of volume or depth, is exempted from charge. However, domestic water charges are based on volume blocks of monthly consumption (Bakour 1992).

In Egypt, cost recovery and water pricing issues are complex. Abu Zeid and Rady (1992) highlight the social, religious, and political issues involved and point to the need for addressing them before adopting water pricing and cost recovery policies. They indicate that irrigation water is not charged, but that part of the irrigation water costs are recovered through land taxes. Despite the serious work recently initiated on water pricing, Abu Zeid and Rady foresee that considerable resistance can be expected from farmers to the implementation of water pricing policies. They emphasize the importance of consultation with farmers on these issues.

In Sudan, joint charges are applied on water and land together, and are collected from farmers by deducting them from their product sales (Zaki 1992). These charges evidently only partially cover water O and M costs, since Zaki calls for gradual replacement of the flat water rate system by metering devices, to achieve conservation and to generate income to cover O and M.

However, some Arab countries — particularly Jordan, Morocco, and Tunisia — have recently addressed water cost recovery issues with a view to recovering the maximum possible of the costs of providing water services.

Jordan applies a fixed rate of $0.02 per M³ for irrigation water. For domestic use, water is charged progressively under a block structure which differentiates among localities, with higher rates for the capital city, Amman, lower rates for suburban areas outside Amman, and further rate reductions for consumers in the Jordan Valley (Abu Taleb et al. 1992).

Morocco's water cost recovery policy was substantially adjusted during the 1980s. The applicable charges for irrigation water in 1985 amounted to a full recovery of O and M costs and up to 40 percent of irrigation-related investment costs (Darghouth 1992). Doukkala, one of the Regional Agricultural Development Authorities (ORMVA) has attained financial autonomy through increased water charges (Tuijl 1989).

The Arab region should reduce the heavy role of the public irrigation department in the daily management of irrigation. Several countries have successfully turned O and M over to the users. The region should assess this option in addressing cost recovery.

It must be remembered that opportunity cost pricing is not applied in the developed countries, nor is marginal cost pricing found in the water sector (Frederiksen 1993). Water experts and economists argue that water markets are important tools in adjusting water allocations under scarce conditions. These experts argue that allocation problems can be addressed by allowing free trade in water use rights because it may ensure the supply of water to high-value uses in urban areas and provide incentives for more efficient use in agriculture. Market forces may not be relevant, however, when addressing transaction cost, inter-basin planning, environmental cost of water development, and cross-generational conservation and development of the sector. A prerequisite for the functioning of a market price mechanism is ownership of water. Local water markets almost invariably emerge when private interests or individuals control particular water supplies or assets, and can deal directly with their consumers. Examples include the sale of irrigation and domestic water from individually-owned tubewells or pumps, and the provision of domestic water by private-sale tanker in urban areas poorly served by public supply. But, it is unrealistic to expect that a general reallocation between sectors or improvements in water quality can be effected through the market, at least for the foreseeable future. In a few places (e.g. Australia and western United States) market mechanisms complement public allocation policy. Similar mechanisms in the Arab countries can only be established when several problems are resolved: it is necessary to strengthen legal, institutional, and financial water arrangements, and to improve the social and economic conditions of poor regions that are likely to be affected by market transfer of water.

Based on his examination of entities in a number of developed and developing countries, Frederiksen (1993) found that all entities (public and private) follow the worldwide practice of pricing water based on the actual cost of water service, less subsidies; and that charge mechanisms are used for no purpose other than recovery of investment and O and M costs, and to encourage conservation through modification of the internal rate structure. Full cost recovery from water consumers has rarely been achieved by developing countries, nor have all developed countries reached the stage of full recovery of water service costs, Frederiksen observes. Irrigation water costs are generally only partially recovered and, in developing countries recoveries are often very low. While higher prices create incentives to avoid excessive use and to conserve, the specifics of the rate structure are important for any pricing mechanism, especially when social and poverty alleviation concerns have to be addressed.

In World Bank support of water supply and sanitation (WSS) projects in developing countries, the loan agreements contain covenants to ensure appropriate water charges in line with the Bank's policy that the recovery from beneficiaries should be sufficient to meet O and M and to achieve a fair return on capital investments. A review carried out by the Bank's Operations and Evaluation Department (OED) in 1988 found that cost recovery was rated as unsatisfactory in 80 of the 114 completed irrigation projects, and that financial covenants were not fulfilled in 78 percent of the country recipients of WSS loans (World Bank 1993). The reasons given by the Bank's OED for the unsatisfactory cost recovery results include poor maintenance of irrigation and WSS projects, and the reluctance of many governments to collect fees. In irrigation, in particular, inadequate O and M, low quality of service, and low reliability of water delivery contributed to the inadequacy and non-fulfillment of cost recovery targets.

The prospects for achieving full recovery of water cost do not seem promising in the short to medium term. In view of the present low levels of cost recovery, and the political difficulty in adopting opportunity cost pricing immediately, the World Bank (1993) considers a water pricing method that would ensure financial autonomy as a good starting point. However, the Bank also recognizes country differences in water require- ments, endowments, institutional capacities, poverty profiles, and environmental problems, and maintains that reforms and the time frame for their implementation should be on a case-by-case basis.

Regardless of when in the future it is possible for governments to ensure financial autonomy for their water operations, in the meantime there is an immediate and urgent need in many cases to provide funding for O and M and the rehabilitation of supply systems, especially irrigation. Unless reliable water is provided in appropriate quantities at the right time, farmers will remain reluctant to pay water charges, especially in the absence of strong incentives. Regardless of whether charges are high enough to cover O and M, funds should be provided to sustain the operation of the facilities adequately. Farmers' willingness to pay would be enhanced by providing them not only with reliable water supplies, but also with technical support to promote the profitability of their agricultural activities.

Conservation

Water can be conserved through better on-farm and system management. Authorities should address reducing distribution losses, recycling wastewater, introducing appropriate cost recovery levels, adopting water-efficient technologies, changing cropping patterns, and increasing public awareness.

Water losses in irrigation and municipal uses identified earlier in certain Arab countries are substantial, and can be reduced significantly by modifications to the existing water conveyance, distribution, and on-farm inlets that cause most of the losses. In certain cases where the systems have deteriorated considerably because of lack of adequate maintenance, rehabilitation may be needed. New irrigation projects should give sufficient attention to construction standards conducive to efficient water delivery and to adequate operation and maintenance of the facilities. Irrigation systems currently being modernized to serve the new lands in Egypt incorporate modern designs that address water recycling and quality.

The quest to conserve irrigation water should address the technological problems causing losses. Seepage, leakage, percolation, evaporation, and unproductive evapotranspiration are the main causes for water losses in water distribution and application. In arid climates, the efficiency of irrigation networks, as discussed previously, can be significantly improved through well-operated lined or piped-conveyance canal systems and the application of modern on-farm irrigation methods. The use of low-pressure buried pipes in conveyance systems, as in Cyprus, for example, can produce a high network efficiency of 95 percent (Tuijl 1993).

Modern on-farm irrigation methods, such as sprinkler and drip irrigation, can reduce water losses considerably. Efficiency rates of 70 percent for Cyprus and Jordan, and 67 percent for Morocco, are attributed mainly to the application of sprinkler and drip irrigation.

Drip irrigation can increase agricultural productivity, reduce water consumption, or do both. The application of drip irrigation in Jordan to 60 percent of the total irrigated area in the Jordan Valley increased average yields for vegetables, and those for fruits more than doubled (Abu Taleb et al. 1992). In Syria, drip irrigation techniques are applied on areas not exceeding 1 percent of the total irrigated area, but they have a potential for reducing water consumption by 45 percent, while sprinkler techniques could reduce it by 20 percent (Bakour 1992). In other Arab countries, the use of modern irrigation techniques is either non-existent or very limited. In Egypt, sprinkler irrigation does not exceed 21 percent of the total irrigated area (Abu Zeid and Rady 1992). In Morocco it covers only 16 percent, and 11 percent in Tunisia (Tuijl 1993).

In addition to their potential for increasing water use efficiency, drip and sprinkler irrigation technologies have also provided opportunities to farm low-quality lands and cultivate sandy and rocky soils, and have enabled countries with limited water resources to change cropping patterns by shifting from high-water consuming and low-value crops to low-water consuming and high-value products (Barghouti 1992).

Under current prevailing technologies, prospects for reducing evapotranspiration losses from crops are very limited, but some country experiences show that large savings can be achieved by controlling evaporation losses from the water surface of small system reservoirs (Xie et al. 1993). However, such losses within the system are usually very small compared with on-farm evaporation, where losses can be at least 10 times greater (Chitale 1992).

Existing cropping patterns contribute to heavy water consumption. Cereal crops generally consume more water than other crops, and cereals, especially rice, dominate agricultural production in Egypt. Wheat, alone, accounts for 35 percent of total agricultural production in Morocco (Darghouth 1992), and wheat is also a major crop in Sudan. Water scarcity, however, is bringing about increased awareness of the cropping pattern issue. Abu Zeid and Rady (1992) note that rice and sugarcane crops in Egypt are allocated 25 percent to 30 percent of irrigation water. Free water supplied to farmers make the crops profitable to them, but expensive from a social perspective. Abu Zeid and Rady indicate that a strategy to change cropping patterns will be important for future water allocation.

Cropping patterns in Sudan lead to economically doubtful uses of water. Economic analysis there estimated that the domestic resource cost for cotton to earn one unit of foreign exchange is 0.37, compared with 0.55 for wheat (Zaki 1992). In Egypt, Sudan, and most of the Arab countries, cropping patterns are heavily influenced by national policies seeking to achieve food self-sufficiency.

Voluntary change in cropping patterns can be expected in the future. As countries embark on economic reform and structural adjustment (Morocco and Tunisia launched their reform programs during the 1980s, and Egypt's economic reform program is underway), including private sector development, privatization, and trade and price liberalization, the new environment will facilitate increased private sector participation in economic activities. The impact of this liberalization on agriculture could be far reaching. This, together with new government policies to lift controls on agricultural crops will induce a shift, albeit slow, towards more profitable crops. Morocco has recently freed non-industrial crops from previously mandatory cropping patterns (Tuijl 1993). When farmers are provided with options they will adjust over time towards more profitable crops, in which they have a comparative advantage based on agro-ecological and economic factors (Barghouti 1992).

Loss reduction is also possible in municipal utilities. Adequate maintenance, rehabilitation of old and poorly maintained distribution networks, and fixing leaks can greatly contribute to conserving water. Khadam et al. (1991) show that the potable water network in five locations of Al-Riyadh City in Saudi Arabia reduced water losses significantly by repairing all major leaks, and taking other steps to reduce leakage. Conservation can also be pursued by introducing household water-saving devices and providing incentives to use less water for the same services through a pricing mechanism with an appropriate structure and rate level.

Non-Conventional Sources

Growing water shortages will also inevitably lead to greater attention to non-conventional water sources, such as wastewater reuse, desalination of seawater, rainwater harvesting, and cloud-seeding.

Xie et al. (1992) cite several examples in China, Chile, and Mexico, where extensive agricultural areas around major cities use wastewater for irrigation; it is also an important source for industrial use in Beijing, China. In Egypt, treated sewage water now amounts to only 200 mcm, around 0.36 percent of total water resources. However, Abu Zeid and Rady (1992) indicate that under the most feasible scenario for future water supply and demand, treated sewage could be raised to a level of 2 bcm. In Jordan, about 40 mcm of effluent are produced annually, with prospects of reaching 70 mcm in the year 2000 (Abu Taleb et al. 1992).

Serious consideration is being given in Syria, Jordan and Yemen to water harvesting projects. These are usually very small local dams in the beds of small or even intermittent streams (wadis) to capture water from rare but intense rainfalls. In addition to providing more irrigation water to local users, these are expected to ease some of the pressure on the overdraft of groundwater (Al-Weshah 1992). Nevertheless, the reliability of water harvesting will be constrained due to the direct dependency of this source on unpredictable rainfall, and because of high evaporation losses.

As a non-conventional source of water, desalination of seawater is technically a sound alternative, and areas with water shortages are turning increasingly to desalination plants to supplement water resources. Globally, there are about 8,000 of these plants each producing over 100 m^3 per day, in about 120 countries. The Middle East accounts for nearly two-thirds of the world's installed desalination capacity, with much of it found in the Arabian Peninsula (Keenan 1992). Saudi Arabia is a world leader in desalination and holds about 27 percent of the estimated 14 million m^3 of global capacity. In water-scarce countries with available energy supplies, such as Saudi Arabia and Kuwait, desalination is a main source of urban water supply.

The high cost of desalting water is the main reason for its limited application so far. Desalination is still three to four times more expensive than conventional sources of fresh water, costing US $0.40 to $0.60 per cubic meter when brackish water is used, and $1.05 to $1.60 per cubic meter when seawater is desalinated. Data cited by Frederick (1993) for four proposed desalination projects in California show that the cost of desalination in 1990, using reverse osmosis plants, ranged between $1.32 and $1.87 per m^3, depending on plant capacity, intake salinity, and product quality. Countries in the Middle East and elsewhere will seriously consider desalination as conventional water sources become scarcer and more costly. In Jordan, for example, the cost of developing marginal sources of surface water is estimated at $6 to $7 per m^3 (Abu Taleb et al. 1992). But for this country the practicality of desalination is called into question since it has access to the sea only at Aqaba on the Red Sea, about 400 km. south of the consumption centered on Amman (Al-Weshah 1992).

Notwithstanding differences in desalination techniques and their associated costs, technological advance, albeit slow, has contributed to reducing desalination costs. For most situations involving desalination of seawater or brackish water there is increasing evidence of the competitiveness of reverse osmosis due to a significant reduction in its operating costs, specifically energy consumption, membrane replacement, and input chemicals (Keenan 1992). Reverse osmosis uses less energy than distillation techniques, such as multi-stage flash, which traditionally have been dominant.

Research and development of less costly desalting technology for the region should be a high priority. Saudi Arabia and Kuwait, which jointly hold about 40 percent of the global desalination capacity, can benefit from supporting regional research efforts in this field.

An Agenda for Regional Cooperation

Despite some national successes, the future challenges in meeting the growing demands for water are beyond the capabilities of individual countries. Water does not respect political constraints; rivers and groundwater aquifers cut across national boundaries. Regional rainfall, river water, lakes, and groundwater are all part of one system. The region faces drought every four years, and no national boundaries protect against floods or droughts. The expansion of economic activities inevitably leads to increasing demands for water. The answer cannot be found within one country's national boundary. Joint regional action on water would enhance peace rather than fuel conflict. Countries should focus on sustainable management of their water resources, recognizing the international nature of those resources in many cases, and addressing the serious environmental consequences beyond their borders of their water use. These consequences require immediate regional cooperation. This involves water planning based on hydrological units, such as an entire drainage basin. A concerted regional effort should focus in particular on the topics discussed below.

Exchange of information and data on the water sector

The countries in the region need to establish a regional information network and data base for hydro-meteorology and water use. An integrated regional information system is needed to regularly record and disseminate climatic data, including rainfall, and data from hydrological networks and river-gauging stations. The system should also include data on groundwater and land use planning. Experience elsewhere in the world suggests that the countries in the region would benefit from establishing an institutional framework for conventional remote-sensing data programs. This framework should set data collection standards, monitor and integrate water data from all sources, and generate regular and intermittent data compilations and studies. The regional systems should be responsible for the collection of data of national and international interest, and make information available equally to all concerned government agencies in the region. The recent advances in Geographic Information Systems (GIS) recommends the use of this technology to the region. A concerted effort is needed to strengthen the national hydrological institutions in the region and to assist them in establishing a regional information, training, and dissemination center.

Regional management

A regional data and information base is needed for better management of water resources. The vehicle toward this goal is integrated river basin planning, which recognizes the dynamics of the river and takes account of groundwater recharge implications. At headwaters, we need to plan to protect the hills and watersheds. In the middle sections, where land is flat and the flow is slow, we need to plan for efficiency and protection from evaporation and other losses, and in the lowlands, where groundwater is usually recharged and wetlands and lakes could form natural drainage and storage systems, we need to stabilize salt balances.

Effective water management of the Jordan River, for example, requires planning. Regional cooperation is needed to establish innovative legal, institutional, and organizational structures responsive to the region's water situation. They should address efficient water management, reduce pollutants in river basins, establish environmental standards for wastewater treatment and recycling, and monitor water on a seasonal basis to adjust allocations among competing sectors with limited disruption to the economic activities of the communities dependent on a river basin. Existing international law, however, provides no help in areas where political and military conflicts are dominant. A just and lasting peace in the region is essential for the realization of efficient management to optimize the use of water and to preserve its quality.

The technical issues are complex but manageable. Successful river basin planning is found, for example, in the Pacific Northwest of the United States, where a regional water management and conservation system along the Columbia River has been in effect for several decades. This approach is carried out according to a sophisticated supply management system using automated sensing and transmission of water resource data in real time from unattended remote sites to computerized base stations where it can be easily analyzed by water resource managers. Supply management is supported by conservation management through a variety of networks of automated agricultural weather stations.

This approach (also used in California) has been complemented by water bank programs, which have proven to be a viable technique for managing existing supplies, providing water for new users, and storing water for drought years. Under these arrangements quantities of water in reservoirs are pre-allocated to the different contracting entities, based on criteria developed by water experts. A given water distributor or large farm, for example, would "own" a portion of the water being held in the system, and be able to call upon it in case of need.

This concept is timely and relevant to the Middle East. The countries involved could become spaceholders in regional reservoirs with technical and legal contracts for water rights. With spaceholder contracts, reservoir storage is attributed to the specific spaceholders. In years of normal or surplus water supply, unneeded water simply remains in the reservoir under the control of the spaceholder. Surplus water can be marketed to other users provided there is an assurance that the existing contractors will not suffer shortages as a result of the new use. The commitments to the original users are always honored.

Converting existing reservoirs in the region into water banks requires an agreed upon legal framework, organizational structure, and management program accepted by the participating countries in the region. Water and legal experts, hydrologists, and political and economic advisors must be allowed to conduct necessary studies designed to assess the suitability of such innovative schemes as water banking and regional management as incentives toward a peaceful solution to the water problems in the region.

Modern water technology and engineering

A third area of cooperation among the countries in the region is the development and improvement of modern water technology. Desalination has been discussed above under non-conventional sources of water.

The countries of the region have given increasing attention to the possibility of intra-basin water transfers. Studies are underway to assess the possibilities of transferring water from Turkey to Jordan, Israel, and the West Bank. Egypt considered at some point building the peace canal across the Sinai Desert to carry water to Gaza. Israeli and Jordanian experts are studying the old dream of a Red Sea or Mediterranean Sea hydrological connection with the Dead Sea, which is the bold project of transferring seawater to the Jordan Valley raised in 1944 by W.C. Lowermilk's book, *Palestine: Land of Promise*. Regional studies on the scope, cost, and consequences of such enterprises would be very useful. We also need to study the dynamics of such transfers in terms of hydrology, engineering, economics, and environmental effects.

However, before considering large water transfer projects, the region needs to assess its present distribution network, its capacity for connections with other networks within and outside the region, and its methods of protecting water quality and quantity for future generations. Despite the severe water crisis in the region, no joint scientific studies have been undertaken or even attempted for the sustainable development of this precious resource. The answer to the water crisis should not be a quick fix through water transfer. The initial phase of an action plan should emphasize a joint institutional approach to sustainable management based on the exchange of information and experience and the design of viable options to benefit the countries sharing water resources.

Other areas of technological innovation are new instruments, engineering designs, and hydrological tools needed for recharging fresh groundwater aquifers, and for controlling pollution along river basins and associated groundwater systems. The region should focus on modern technology for artificial recharge, earth dams, and watershed management and for improving the micro-water catchments in the hills where uncontrolled water runoff and erosion can contribute to heavy water and soil losses. Associated with this technology is large-scale afforestation in marginal and semiarid zones, and fuller use and planning of the region's prime rainfed agricultural land.

Conclusions

Water in the Arab region is inherently scarce as a result of naturally arid climatic conditions. Population increase and economic growth have spurred higher demands for the limited water resources. The underlying historical perception by people in the Arab region and elsewhere in the world that water is a free natural resource supports the dominating influence of traditional, political, and social factors in the management and use of the resource. The increased use of the fixed resource in response to rising demands is not only reducing its availability, but also jeopardizing its quality. In view of the vital importance of water for sustaining life and promoting development, appropriate approaches and policies are needed to deal with the problems of water scarcity, and the challenges ahead.

Among the necessary actions that would enhance the management and use of the resource, the following are particularly important:

■ At the national level the water sector should be fully integrated with each country's overall planning for economic and social activities. Planning and management of water resources at the regional and river basin level should be begun.

■ Institutions involved in water development and delivery should strengthen the role of the users in water management and operate within a framework of clearly defined laws and policies.

■ Regional cooperation should be strengthened through establishing a network among national hydrological data banks and joint hydrological networks.

■ Water conservation is a necessity. Appropriate policies and actions are needed to promote management of water more as an economic commodity and less as a political asset or as a national privilege. Action is needed to:

◆ Reduce water losses from water supply systems and establish water banks to sell water to district blocks and community accounts, and to undertake the necessary maintenance works, and rehabilitate and manage the local water distribution sector.

◆ Promote cost recovery, encouraging strong users' participation, to increase revenues and encourage conservation and to meet equity considerations.

◆ Introduce water-saving devices in municipal water use, and modern methods in irrigation.

◆ Liberalize agricultural prices and remove constraints on private sector involvement in order to enhance the economic utilization of land and water resources. In this environment, cropping patterns would adjust in response to changing technology and market signals.

◆ Augment water resources through non-conventional sources and, in particular, water treatment and re-use. Other non-conventional sources, such as water harvesting and cloud-seeding, offer limited opportunities, although they should be examined further. Desalination of seawater, however, is becoming an alternative of considerable potential, compared with the costs of developing marginal sources of water.

The Arab world cannot rest on the subject of water, as long as pressures on this resource continue to rise and 40 percent of the population do not have adequate access to it. We certainly need continuing innovation and rationalization in our handling of water, but foremost and above all we need to develop and put into place, in our countries and our regions, a balanced system for the management of resources. We must work toward a framework for management functions that will integrate considerations of the present and the future, of technology and democracy, of economics and environmental preservation, of growth and security, into ongoing, real-life, informed, and prudent management of our land and water.

2

Transboundary Water and the Challenge of International Cooperation in the Middle East

John Waterbury

Use of the waters in the transboundary river basins of the Middle East presents a classic collective action challenge, one that has scarcely been met, much less resolved, anywhere in the region. In the last decade it has become widely recognized that there is a problem, but that fact in itself cannot propel any of the state and non-state actors towards international cooperation.

This essay seeks to explore the dynamics of bargaining for cooperation in the major international drainage basins of the Middle East. The collective action dilemma can be summarized thus: while it can be demonstrated that cooperation in the use of this scarce resource can make some actors better off without making any actors worse off, those who stand to gain little or nothing from cooperation have every incentive not

TABLE 1

Per Capita Surface Water Availability: *Select Middle East Countries*[1]

	Total Water/Year bcm	Population millions	Per Cap. Water m^3
Egypt	55.5	56	993
Ethiopia	82.5	52	1586
Iraq	91.2	17	5364
Israel	1.95	4.6	424
Jordan	.77	3.3	233
Syria	23.0	13	1769
Turkey	100.0	55	1818
West Bank/Gaza	.20	1.8	111

to cooperate and to force the costs of collective action onto those who stand to gain the most. While this set of studies is devoted to the Arab world, it must be remembered that the bulk of that region's transboundary water resources rises in non-Arab countries: Turkey, Iran, Ethiopia and the states of the Lake Victoria Basin.

The Nature of the Resource and of its Utilization

A false analogy is often drawn between the exploitation of the rivers of the Middle East and "the tragedy of the commons" in which an "open access" good, is exhausted as each individual user follows a narrowly rational strategy of using as much as possible since he or she cannot restrict the use of the resource by others.[2] By contrast, Middle Eastern rivers, both *de facto* and *de jure,* are common property resources. Access to them is not open and it can be restricted. First, rivers, and sometimes associated aquifers, are used consecutively not simultaneously, and at a given location in the basin the local riparian has the exclusive use of the resource. Second, within the territory of a given riparian country rivers are generally regarded as in the public domain, a kind of trust managed for society by a central political authority that can regulate use. Therefore the incentive for large numbers of individuals to use the resource before some other individual does is not as pronounced in river basins as in open access resource situations (e.g., fisheries, forests, the sea bed, etc.), and the possibility of regulated use among a very small number of sovereign national govern-ment actors in a basin is somewhat greater. However, whereas other forms of common property serving individuals and families, such as pasturage and wells, have ancient institutional arrangements governing their use, river basins seldom boast the same among the several states they traverse.

In the Middle East, as elsewhere, customary and religious law grants rights of appropriation only to the *use* of surface water, not to the water itself. Rights are based on use, and established use constitutes a recognized need. As we shall see recognized needs at both the local and the interna-tional levels lead directly to the legal concept of acquired rights. Thus usufruct rights come to the fore in bargaining over access to river water and its use.

Geopolitical Dynamics

The context for bargaining is set by two principal factors: topography and history. International law recognizes both implicitly or explicitly but does not assign priority to either. Relevant international law, which will be examined in greater detail below, is best conceived of as an instrument in a *rapport de forces* that is dynamic and changing but that is not itself legally determined.

Topography and history are closely related. It is generally the lower reaches of river basins that are first developed. Alluvial soil deposits and relatively flat terrain are conducive to settled agriculture. Population densities are higher, urban life more developed, and markets deeper than in the upper reaches of the basin. It is thus no surprise that the exploitation of the surface waters and aquifers of the basin is initially the most intense in the lower reaches.[3] This creates established patterns of land and water use that have subsequently been embodied in the legal principle of acquired rights.

In the 20th century the twin processes of rapid population growth and agricultural expansion and intensification have pushed cultivation and habitation into the upper reaches of river systems so that actual or potential conflicts of interest emerge. When upstream populations made relatively modest demands on basin water, it did not really matter to downstream populations by what principles of common, religious or, today, international, law those demands were justified. Since the 1920s (or even since 1898 if one recalls the Fashoda "incident") it does matter.

As upstream populations have increased their use of available water they have invoked legal principles to nullify or at least neutralize downstream claims to acquired rights. Their basic position is that the principle of first come, first served should prevail without regard to the needs of downstream populations. They may add that with modern storage and delivery techniques the upper reaches have greater economic potential than the lower. What we thus witness is the elaboration of two fundamental bargaining positions with important consequences for the presence or absence of cooperation. Downstream actors insist on acquired rights and the servicing of existing needs. They call for binding cooperative regimes that will constrain upstream states in their use of water and guarantee a supply of water to the downstream populations. The upstreamers counter with the sovereign and unrestricted right to first use on their territories, and the need to develop fully the potential of long-neglected portions of the basin.

For cooperation's sake, it is fortunate that few states can adopt consistent legal stances because their geographic positions are themselves varied. For example, Syria is mid-stream on the Euphrates and the Orontes but upstream on the Jordan. The Yarmuk forms part of its boundary with the Kingdom of Jordan. Syria may sympathize with Palestinian claims to full control of the surface waters of the West Bank, the bulk of which currently drain into the Israeli coastal aquifer, but Syria itself is the beneficiary of a cross-frontier aquifer (the Ras al-'Ain) that drains from Turkey into northern Syria. Similarly, Turkey controls the headwaters of the Euphrates and the Tigris but is the downstream state in the Orontes (a relatively minor consideration for Turkey given the Orontes' small annual discharge of some 400 mcm).[4] Israel, although it has captured the headwaters of the Jordan (but not of the Yarmuk), is faced with the possibility of giving up that control one day, and therefore cannot take a firm stand for sovereign upstream rights. Moreover, if that principle were extended to the aquifer that drains the West Bank, a future Palestinian state might challenge Israel's acquired rights.

There are three states, and a qualified fourth, in the international drainage basins of the Middle East that are major players with consistent positions. One is an upstream state, Ethiopia, aptly described as the water tower of eastern Africa. It receives no water from any other state but gives away water to all its neighbors. Turkey, with the minor exception of the Orontes, is similarly endowed and placed. Egypt receives all its water from outside its borders, and, with negligible rainfall, is vulnerable to any unilateral actions by its upstream neighbors. Finally, although Iraq has relatively abundant rainfall in the north, most of its surface water comes across its borders and none of its water goes to its neighbors.

The benefits of cooperation in river basin development are typically highly asymmetrical. Thus Egypt receives all its water from the Nile river while for Zaire its share of the Nile basin is of negligible importance to its overall water balance. The combination of asymmetry and several bargaining parties (for example, 10 sovereign nations, now including Eritrea, share the Nile basin) render *voluntary* cooperative solutions extremely difficult. When water is the sole focus of negotiations, gains and losses in a very real sense become too clear. Clarity can lead to paralysis and the non-resolution of the collective action dilemma.

When progress is achieved, asymmetries in benefits can be so great that one riparian may agree to bear all or most of the cost of a cooperative solution. Such was the case when Egypt financed the construction of the Jebel Aulia storage dam on the White Nile, just upstream of Khartoum in 1932, and of the Owen Falls dam in Uganda in the early 1950s. Much more recently, on September 3, 1987, Jordan and Syria signed an agreement for the construction of the Wihda dam on the Yarmuk, under which Jordan accepted the full cost of construction and the payment of compensation to those displaced by the project *in Syria,* and will also give Syria 75 percent of all power generated at the dam site. Only Jordan's desperate need for additional irrigation water can explain the price it was (and may still be) willing to pay to obtain Syrian cooperation. This accord, of course, has not been implemented, as Israel is not a party to it, and the price for Israel's adherence will be high; part in Yarmuk water, part in normalization of relations with Syria and Jordan.

One way out is through multi-good bargaining. The bargaining agenda is made more complicated so that asymmetries in the benefits of cooperation with respect to one good can be overcome or at least obscured by deals struck with respect to other goods. These may be quite straightforward, such as Iraqi oil for "Turkish" water, or less obvious: Palestinian recognition of Israel in partial exchange for recognized rights to West Bank water; Syrian control of Kurdish insurgent raids into Turkey in partial exchange for Euphrates water; Israeli technical know-how in water conservation for a part of the water saved (Kally and Tal, 1989); Egyptian good offices in raising external assistance for all states in the Nile basin in exchange for a binding understanding on Egypt's needs for water.

But what of *involuntary* solutions? Cooperation can be imposed. International relations theory refers to these as hegemonic solutions. A pattern of resource use is elaborated by the dominant power in the basin and imposed on the others. Colonial powers have been such hegemons, defining rights and obligations across the jurisdictional boundaries of their empires. The British devised an imposed regime for all of the Nile basin (although Ethiopia was never fully integrated into it) as did the Soviet Union in the basins of the Amu Darya and Syr Darya. In the region that concerns us here, I would argue that Israel has imposed a solution in the Jordan basin, that Turkey is in a position to do so in the Euphrates (and to a lesser extent in the Tigris), and that Egypt has acted as a quasi-hegemon in the Nile basin.

It will be noted that the three examples of hegemons cited include an upstream state (Turkey), a mid-stream state (pre-1967 Israel) and a downstream state (Egypt). Although geographical asymmetries are powerful, position in the basin does not therefore fully determine bargaining power. Unsurprisingly, other resources come into play (Naff and Matson, 1984; Frey 1992). If position, military power, and economic resources are all joined in one state, its writ will run in the basin, but military and economic power may be so concentrated in states in less favorable locations in the basin that they can impose their own solution (Israel) or that more favorably located states will think twice before tampering with a strong downstream state's "rights" (Ethiopia in relation to Egypt).

Somewhere between voluntary and involuntary solutions, and potentially associated with both, are *induced* solutions. This brings a third party into the bargaining process. It may be a superpower interested not so much in the resource issue *per se* but rather in solving it to further some other purpose. In 1972 the Soviet Union sought to reconcile Syria and Iraq, both countries in which the USSR had a large political and military stake, concerning joint use of the Euphrates. In this instance an induced settlement was not achieved. Similarly today, the U.S. has promoted multilateral talks on water issues among the belligerents in the Arab-Israeli theater on the assumption that progress toward a cooperative solution in this domain may create an atmosphere conducive to an overall settlement. Miriam Lowi (1992, 1993), however, sharply challenges this logic, claiming that the intensity of the larger dispute will overwhelm attempts to reach a cooperative solution to the water disputes.

To date the most effective agents of induced cooperation have been multilateral funding institutions, particularly the World Bank. The Bank has tried, to the extent possible, to condition its financing of large hydraulic projects in international river basins on the prior negotiation of cooperative agreements among the riparian states. This policy was successful in the case of the Nile in 1959 but so far unsuccessful with respect to Turkey's Southeast Anatolia Project (Turkish acronym: GAP), where development has gone ahead without an accord and without Bank funding. In the case of the Yarmuk, the World Bank (and the United States Agency for International Development) have been unable to induce an accord among Syria, Jordan and Israel.

Cooperative solutions, whether arrived at voluntarily, induced or imposed, are very difficult to achieve, and once achieved, difficult to maintain.[5] Only in the Nile basin have any significant successes been registered. But the recent literature that predicts acute conflict in the absence of cooperation mis-specifies the real dangers. It tends to invoke what I call the Fashoda syndrome, harking back to the famous incident in 1898 in which General Kitchener met Colonel Marchand on the brink of what seemed to be a possible colonial war for control of the headwaters of the Nile. The problem is that it is not at all easy for one riparian in a drainage basin to deprive another of significant amounts of water, and it is not at all easy to define the military goals to be pursued if there is a resort to force. Again, the nature of transboundary rivers creates special characteristics of military encounters that they may engender (cf. Biswas 1982; Gurr 1985).

First, to deprive another state of the use of an international river, a riparian must be able to cut off or divert the flow of the river. This costs money and may be as disruptive to portions of the population of the state undertaking the action as it is to the inhabitants of the targeted state. Second, the storage facility or diversionary works may be vulnerable to air, artillery or ground attack. Although the Geneva conventions and other principles of the conduct of war expressly forbid targeting such works, I have not found evidence that they consider cases in which the works themselves are the *casus belli* (International Law Commission (ILC) 1991). Finally, water cannot be stored indefinitely without endangering the storage facilities themselves.

For the most part, water disputes in the Middle East have remained below the military level with two exceptions. The first was in fact the Fashoda incident, which was resolved without bloodshed. The second was Israel's attack in 1964 on the works in Syria that were being undertaken to divert the headwaters of the Jordan. That project was abandoned. However, there have been a number of tense moments. In 1925, after the assassination of the Sirdar, Lee Stack, by Egyptian nationalists, Britain punished Egypt itself by allowing the Sudan to begin to design the Sennar dam on the Blue Nile and the irrigation grid that was to become the Gezira scheme (Waterbury 1979, p. 65). The next bellicose incident, aside from the Israeli attack of 1964, came in 1975 when Iraq mobilized for war as Syria began to fill the reservoir upstream of the Thawra (Tabqa) dam. The Arab League intervened to mediate the dispute. After the Camp David accords, there was talk of Nile water being delivered to Israel's Negev. This elicited a hostile declaration on the part of the Ethiopian government, to which President Sadat responded with a warning that any state that tampered with Egypt's water supply would risk a military response (Rogers 1991, p. 22). Finally, when Iraq occupied Kuwait in 1990, it is claimed that Lord Owen urged Turkey to restrict the flow of the Euphrates to Iraq. If such advice was given, Turkey did not heed it.

No matter how acute the crises that may emerge in the coming years over water supply in the Middle East, armed conflict is not likely to be an outcome. Unilateral diversions of the Euphrates have taken place in both Syria and Turkey without an effective military response from Iraq. Whatever arrangements emerge from the current negotiations over water in the Jordan basin, no riparian is likely to be able to challenge Israeli hegemony in the forseeable future. Only in the Nile basin, where Ethiopia, at the earliest a decade from now, might divert significant amounts of Blue Nile water for purposes of irrigation, could we see some resort to military force on the part of Egypt. But even here, without Sudanese cooperation in the use of airbases and airspace, it is hard to imagine Egypt carrying on a sustained campaign.

It is possible that cooperative arrangements will be put in place. In their absence, it is likely that various riparians will pursue costly unilateral solutions to their supply problems that will be indisputably suboptimal in economic terms. Resources will be misallocated within riparian states and across the basins as a whole. New supply-demand equilibria will not be achieved through warfare and seizure of water resources but rather through disruptive adjustments within riparian states. Agricultural production will be reconfigured or curtailed, a more rational regional division of agricultural production will not be explored, aquifers will be mined, and costly treatment plants installed to salvage or re-utilize a relatively fixed supply of water of deteriorating quality.

The Legal Framework for Bargaining

It is needless to go over in any detail the legal precedents typically invoked in bargaining situations, for that has been done voluminously elsewhere (inter alia, see Garretson 1967; Michael 1974; American Society of International Law 1977; Godana 1985; Utton and Teclaff 1987; Alheritière 1987; Mallat 1991). I stress once again that which one among various legal principles prevails in any given river system will be the result of the relative geo-political strengths of the riparians rather than of the abstract cogency of the principles themselves. At one end of a spectrum is the principle of unrestricted sovereign use of water "located" in one's territory, and at the other is that of a community of users any one of which can veto any unilateral innovation departing from accepted practice. Lying in the middle of the spectrum is the concept of limited territorial sovereignty, which appears to have earned consensual endorsement in the international law profession.

The two poles of this spectrum lay out principles that are relatively easy to interpret. On the one hand, we have the discredited but still invoked Harmon Doctrine laid down by the U.S. Attorney General Judson Harmon in 1895. He opined that there were no principles of international law that would oblige the U.S. to take into consideration Mexico's needs with respect to the Rio Grande. He continued that any concessions the U.S. might make to those needs would be based on "comity" rather than international law and thus would not constitute a legal precedent. On the other hand the concept of the community of basin interests, with an attendant right of any riparian to block unilateral actions by any other, is straightforward but so fraught with the prospect of paralysis that few nations have endorsed it.[6]

The middle ground, where most states feel most comfortable, is governed by principles that are the most difficult to define in operational terms. This ground is occupied by the concept of limited territorial sovereignty, itself rooted in the classical injunction *"sic utere tuo ut alienum non laedas":* "thus use what is yours so as not to cause harm to another." A corollary is that a riparian should make a good faith effort to obtain the acquiescence of other riparians in the basin to water projects it wishes to undertake, but that these riparians enjoy no legal right of veto. The Lake Lanoux arbitration of 1957 between France and Spain invoked "the rules of good faith" that enjoin any riparian to take into consideration the interests of other riparians *to the extent compatible with the pursuit of its own interests* (Michael 1974, p. 53). In turn the guiding principles of "good faith" are avoidance of causing appreciable harm to other riparians and the pursuit of equitable or reasonable use.

Herein lies the problem. There are no broadly accepted definitions of what is equitable, reasonable, and appreciably harmful. In the most important attempt to operationalize these concepts, the Helsinki Rules of 1966 (see International Law Association (ILA) 1967; ILC 1979) lay out 11 principles to guide water use. These range from the social and economic needs of populations, to established patterns of use, to geography. No single principle takes priority over another. Parnall and Utton (1976) have rightly dismissed these guidelines as more platitudinous than operational, but, like the offerings of others, their substitute criterion of "optimality" is equally opaque.

Let us look at some specific Middle Eastern examples. Turkey invokes equity in developing irrigation in the backward region of Southeast Anatolia while Iraq invokes appreciable harm to established patterns of use and the need to protect its sunk infrastructural and social costs. Who is to choose between the equity aspects of absorbing Soviet Jews into the Israeli economy and the harm that their demand for water may cause Palestinians? Is it "reasonable" for Egypt to cultivate sugar cane when the Sudan can cultivate it with much less surface water? How does one choose between long-term economic returns and short-term social welfare concerns, or between the welfare of future generations and that of the present? Peter Rogers (1991) has tried to operationalize Baumol's measure of superfairness with respect to international water resources, but his successful determination of a "core" is dependent on measuring only economic returns to the riparians involved.[7]

In various official reports and newspaper accounts I have extracted or inferred the following principles or interests advanced by Middle Eastern riparians to defend or lay claim to shares of transboundary waters:

> *economic potential, future acreage:* the Sudan, Turkey, Syria, Ethiopia
> *existing acreage:* Egypt, Iraq, Israel
> *existing population:* Egypt
> *per capita water availability:* Syria, Turkey
> *equity, national security:* Turkey, Lebanon, Egypt, Israel[8]
> *modified Harmon doctrine:* Turkey, Israel, Ethiopia

In the 1991 Report of the ILC the section on the law of the non-navigational uses of "international water courses" revisited and extended the Helsinki rules (ILC 1991, pp. 152–98). It is important to touch upon some of the issues treated in that report. First, it offered a definition of what an international water course is:

> *international water course* means a water course, parts of which are situated in different states;

> *watercourse* means a system of surface and underground waters constituting by virtue of their physical relationship a unitary whole and flowing into a common terminus;

> *watercourse state* means a State in whose territory part of an international water course is situated.

The ILC report rejected the proposition that a watercourse is international in a relative sense in that the use of water in some parts of the system may have no effect on other parts. Thus the draft resolutions put before the United Nations in 1991, based on the ILC work, stress the unitary nature of international watercourses, and, importantly for the Middle East, include aquifers within them. By extension it is not only the riparian interests in the river basin itself but all those in the larger drainage basin and catchment area who have a stake in its management.[9]

Beyond these points, however, the 1991 report does not break new ground. It condenses but does not modify the 11 guidelines of the ILA Helsinki Rules; it cites the obligation not to cause appreciable harm and the obligation to cooperate, largely if not exclusively, through the exchange of data. It sets forth rules of notification of unilateral initiatives, providing a six-month waiting period for replies and protests, and another six-month period of obligatory negotiations. Significantly the report makes no provision for failed negotiations; presumably in the event of failure the initiator is free to proceed, having made a good faith effort to obtain acquiescence.

The World Bank operates under a similar but more detailed set of guidelines (World Bank 1990). When the Bank is approached for credits for projects on international watercourses, it must first advise the applicant to notify other riparians of its intentions and to supply them with all relevant data. Six months after notification, the Bank will make a decision on whether the project may proceed. If there have been objections, Bank staff must assess their legitimacy. In some instances it may consult a committee of independent experts, but the World Bank must independently determine whether any appreciable harm is likely to be caused to the other riparians, whether or not any of them raise objections. If it is determined that the project will cause no appreciable harm to other riparians, the World Bank may advance financing despite objections that may have been raised.

Basin-By-Basin Bargaining

The Nile

So far as cooperative solutions are concerned, the Nile Basin with its 10 riparians (Egypt, the Sudan, Ethiopia, Eritrea, Uganda, Kenya, Tanzania, Rwanda, Burundi and Zaire) and gross contrasts in degrees of dependence upon the Nile among them, is a nightmare. By the same token the need for cooperation is acute for only two of the riparians, Egypt and the Sudan, and it is thus no coincidence that the only treaty binding the riparians of that basin is the 1959 agreement between Egypt and the Sudan.

Before the 1950s, Britain negotiated on behalf of its possessions in East Africa a number of treaties and understandings that in the final analysis served to protect Egypt's acquired rights. These instruments involved Ethiopia, under Italian control, and the Congo, under Belgian control. While no independent successor state has abrogated any of these instruments, all have said that they are not *necessarily* bound by them (see Odidi 1979; Waterbury 1982 and 1987; and Godana 1985).

Despite the chastisement of Egypt for the assassination of Lee Stack, the 1929 agreement between Egypt and the Sudan enshrined Egypt's acquired rights to 48 bcm of water, sufficient to irrigate 2.19 million hectares of land. At the same time, the Sudan was guaranteed 4 bcm in recognition of its *future* needs. Thus the 1929 Treaty embodied both of the fundamental claims mentioned above. It could do so because the system contained so much water — some 84 bcm on average — allowing the Sudan, as the upstream state, to build the Sennar dam in order to capture its share without threatening Egypt's rights. Because at that time all storage facilities were seasonal in nature, the real issue was not water *per se,* but rather "timely" water, i.e., the water Egypt needed in the summer months for its cotton, sugar cane, and eventually rice harvests. The 1929 Treaty guaranteed Egypt exclusive rights to the summer flow of the main Nile, and to enhance it Egypt undertook construction of the Jebel Aulia dam. The Sudan, by contrast, had to cultivate its cotton during the winter months.

This was a landmark agreement, not only for the Middle East but for the world. It fixed quantitative shares in an international river. Its significance may have been obscured by the fact that Egypt was a sovereign state in only a narrow legal sense, and Great Britain negotiated the treaty on behalf of the Sudan. Nonetheless, this treaty sought to operationalize the concept of fair and equitable use before the term was invented.

The 1959 agreement between Egypt and the Sudan superseded that of 1929, and modified the sharing ratio. In the intervening 30 years the needs of Egypt and the Sudan for irrigation water had grown dramatically. Seasonal storage facilities could not capture enough of the annual flood to meet the increased demand. After the military coup of 1952, Egypt advocated the Aswan High Dam as the solution to "over-year" storage, thus rejecting *for a time* the Century Water Scheme that had been elaborated by the British and would have placed storage facilities in Uganda and perhaps Ethiopia, and swamp-draining canals in the Sudan. The British plan, so the Egyptians believed, maximized their vulnerability in that the countries that stood to gain least from cooperation would control the infrastructure upon which Egypt's lifeblood would depend (see Waterbury 1979, pp. 63–115; Collins 1990, pp. 198–246). The High Dam scheme, almost entirely within Egypt's borders, was seen as the way to minimize Egypt's vulnerability.

The 1959 agreement was not the result of Egypt's need for Sudanese acquiescence in the construction of the High Dam, but rather of the Sudan's need for Egyptian acquiescence in the construction of the Roseires dam on the Blue Nile. The agreement apportioned the total average annual discharge of the main Nile, as measured at Aswan. Net of anticipated surface evaporation and seepage at the High Dam reservoir, this was calculated at 74 bcm, of which Egypt was allotted 55.5 bcm and the Sudan 18.5 bcm. The ratio of Egypt's share to the Sudan's had thus decreased from 12:1 to 3:1. A major benefit of the agreement and of the High Dam that is seldom recognized is that over-year storage ended Egypt's obsession with "timely" water and freed the Sudan to use its share of water according to whatever seasonal calendar it sees fit.

Equally significant was the stipulation that in the event there was, in any year, water in excess of the average, it would be shared 50-50 by the two countries. In 1920, H.T. Cory, an American civil engineer who had participated in the Nile Projects Commission Report of that year, had recommended that all water in excess of Egypt's acquired rights be apportioned equally between the two countries so that the Sudan's great agricultural potential could be realized. He cited both western and Islamic law to emphasize the public trust aspects of water management and to qualify riparian acquired rights (Collins 1990, p. 44). His 1920 opinion was finally embodied in the 1959 agreement.

That agreement led to the creation of the Permanent Joint Technical Commission (PJTC) on the Nile. It, too, is in several respects unique. Staffed equally by Egyptian and Sudanese engineers, it oversees implementation of the 1959 agreement, places monitors from the two countries at each other's major storage sites, and studies and contracts for all future projects. For over 30 years the PJTC has met quarterly whatever the state of Egypto-Sudanese relations. Had the low flood years of 1987 and 1988 been followed by yet another low water year, the PJTC would have had to devise a formula to apportion the *shortfall* below the average discharge. The agreement specifies no formula for that (Abu Zeid and Abdel-Dayem 1990).

For years Egypt has been fully utilizing all the water allotted to it under the 1959 agreement, although the Sudan, because of its economic disarray, has not. I suspect that Egypt has received, although perhaps not used, more than its allotment. Nonetheless, Egypt needs more water if it is to expand its irrigated surface. It has plans to bring some 1.2 million hectares of new lands into cultivation within a decade. After the Egypto-Sudanese accord of July 1975, Egypt hoped that part of those needs would be met through the construction of the Jonglei Canal, designed to channel Nile water and reduce its evaporative losses as it passes through the Sudd swamps of the southern Sudan. The additional water available to Egypt, with a similar amount realized for the Sudan, would be 2 bcm. Civil war in the southern Sudan, however, forced the suspension of the project.

On a broader plane, Egypt realized by the early 1980s that so long as its dealings with the other riparians in the Nile basin focused only on water, it would be extremely difficult to establish binding agreements with any of them. Then Minister of State for Foreign Affairs, Boutros Boutros Ghali, sought to create a multi-good bargaining situation by focusing on the overall economic development of the basin, involving improved communications, tourism, trade, and above all an integrated power grid linking Zaire, Ethiopia and other East African states to the Middle East and Europe through Egypt. With Egyptian and Sudanese coaxing, the Undugu Group of Nile states was formed in the early eighties as an informal forum to promote a basin-wide development organization. Egypt and the Sudan also promoted an important project as part of an attempt at "induced" cooperation with Upper Nile states. In 1967 the Hydrometeorological Survey of the Upper Nile and Lake Victoria Basin was launched. Its overt objective was to generate the data and train the personnel for a continuous detailed meteorological mapping of the Upper Nile watersheds. The less overt agenda, from the Egyptian and Sudanese points of view was to 1) bring all the basin states together in a common project framework, and 2) to help build the expertise in the Upper Nile states that would allow them to hold discussions with the Egyptians and the Sudanese on an even footing (Waterbury 1982).

Egypt's greatest fear over the years has been that Ethiopia might undertake development of its western, Blue Nile, watershed without obtaining Egyptian acquiescence. Geography would seem to make such development inevitable at some time. Ethiopia's former Water Commissioner, Zewdie Abate, wrote in 1992 (p. 7):

> Ethiopia has to feed its burgeoning population. Its overfarmed, over-populated highlands have exhausted their cultivable potential. It has to turn its efforts to the development of its western watersheds, where there is an irrigable potential land area of 900,000 hectares in the Blue Nile basin and 1.5 million hectares in the Baro-Akobo (Sobat) basin. Ethiopia, in its plan of development, must endeavor to investigate aspects of the local natural environment of its targeted irrigation sites, and their traditional exploitation.

Egypt also fears that Ethiopia and the Sudan might enter into agreements at the expense of Egypt. For example, storage facilities could be constructed in Ethiopia to supply irrigation water and electrical power to sites on both sides of the Ethiopian-Sudanese border. For the Sudan, for several reasons, development in the Blue Nile watershed is preferable to development in the White Nile basin, where Egypt has pinned its hopes on Jonglei and other projects (Waterbury 1979 and 1982; Guariso and Whittington 1987).

At the time of Sudanese independence in 1956, the Ethiopian government set forth its position on the Blue Nile in terms reflective of the

Harmon Doctrine (Collins 1990), and reiterated that position in 1977 at the Mar del Plata Water Conference (Waterbury 1982). On December 23, 1991, the new Ethiopian government drafted an accord of friendship and peace with the Sudan in which both sides affirmed their commitment to equitable shares in the Nile waters and the avoidance of appreciable harm to one another. Ethiopia declared its intention to become a full member in all organizations of the basin states with the objective of establishing a Nile Basin Organization. The two sides agreed to set up a joint technical committee for the exchange of data and to explore possible lines of cooperation. While this accord was not cause for alarm to Egypt, the statements of the Ethiopian Prime Minister of May 20, 1993, were. After a visit to Israel, he was quoted as saying that his government had advanced proposals to Egypt for a redistribution of Nile water. Ethiopia would thus find itself in support of a demand voiced over the years by many Sudanese (and provided for in the 1959 agreement) that the established allocation be re-negotiated. Any such renegotiation could only come at the expense of part of Egypt's existing share ('Auda 1993).

There is one further recent development in the utilization of the Upper Nile basin. Uganda has received approval of and financing for the raising of the Owen Falls dam to increase the production of hydropower there (World Bank 1991). Uganda will bear the full cost of the project. The raising will stabilize Lake Victoria at a high level. That additional storage capacity could enhance the efficiency with which the Jonglei canal will be operated, if it is ever completed. On the other hand, it may mean that Uganda will not entertain the use of Lake Albert as a storage site as favored by the Egyptians. In any event, what we have in this instance is the private provision of a public good to all the riparians in the White Nile system. Out of rational economic self-interest, Uganda is willing to pay all the costs of this public good, from whose benefits other riparians cannot be excluded, thus unilaterally solving a collective action problem.

Although Egypt is the downstream state in this river system, and utterly dependent upon it, it is also the most powerful economic and military power in the basin. It cannot project that power easily throughout the basin, but no other riparian, including Ethiopia, can afford to ignore it. It is for that reason that elsewhere I described Egypt as a quasi-hegemon. It cannot impose a solution, but it can coax and threaten its neighbors convincingly.

Despite that, no basin-wide solution has yet been achieved. Egypt has thus had to seek, on an interim basis, a national solution to its water needs. This means quite simply working within its current allocation of 55.5 bcm and stretching it further through more efficient on-field use, re-utilization of drainage water, careful conjunctive use of river and ground water, and reduced releases at the High Dam site for navigational purposes. For instance the 162,000 hectare project in the al-Arish area of the Sinai is to rely on water-saving pressurized delivery systems, rather than open irrigation canals. In any future round of negotiations Egypt's success in conservation could be turned against it; other riparians may argue that further conservation and, indeed, shifts in the kinds of crops cultivated, could reduce Egypt's established needs for water yet more.

Realistically Egypt can plan on, at the very most, the completion of the Jonglei canal but not on its twin, Jonglei II. I suspect that all the other upper Nile projects in and around the Sudd swamps are dead. By contrast, increased use of irrigation water in Ethiopia may more than off-set the net gain of Jonglei I. Sometime in the next decade Egypt is likely to see a modest reduction in the amount of Nile water available to it (Whittington and McClelland 1992). It could be, then, that the cooperative solution that Egypt has for so long sought is not designed to make it better off without making any other riparian worse off, but rather that it seeks to gain recognition of rights that will minimize what looks like an inevitable reduction in supply.

The Euphrates

If a large number of actors militates against cooperation (Hardin 1982, p. 13), as in the Nile Basin, so too does the presence of only three actors, Turkey, Syria and Iraq, in the Euphrates. It is a "terrible threesome" in which any one actor will see itself as holding the swing vote. It will hold that vote for the highest possible concessions from the other two. To the extent that all three actors follow this logic, no solution will be achieved.

Complicating the possibility of a cooperative solution further, but simplifying the *rapport de forces,* is the fact that Turkey is, today, a nearly pure hegemon in the basin. With the destruction during the Gulf War of Iraq's military base, Turkey has become by far the dominant economic and military power in the basin. It can project military force throughout the basin, and, more positively, can offer economic incentives to obtain acceptance of its solution.

That solution hinges on the GAP (Kolars 1991). This project in southeast Anatolia will involve 22 dams, of which three have been completed and a fourth is underway, for power generation and the irrigation of some 1.6 million hectares. Two thirds of the irrigated surface will be in the Euphrates basin and the other third in the Tigris. Turkey has not obtained formal agreement from Iraq and Syria to the construction of any of these dams, and both the latter have frequently protested this fact. For its part Turkey has offered a unilateral solution: the "guaranteed" delivery of a flow of 500 cubic meters per second (cumecs) on average at its border with Syria. The guarantee applies only to the Euphrates; there is absolutely no understanding on the Tigris. Moreover the guarantee, made in 1987, was only to Syria. Iraq has been left to its own devices to assure its supply. On an annualized basis 500 cumecs works out to nearly 16 bcm or about half the average annual pre- construction discharge of the Euphrates at the border with Syria. The average annual discharge in Iraq is somewhat higher, about 33 bcm at Hit, because the Khabur and Balikh tributaries join the Euphrates in Syria. These also rise in Turkish territory, and there is no understanding among the riparians pertaining to them.

Syria and Iraq do have some bargaining chips. Turkey, before 1991, had large markets in Iraq, and, in a general sense, would like to develop markets throughout the Arab world. Antagonizing two prominent Arab states will not further this goal. Second, until 1991, Iraq could offer Turkey oil delivered through a pipeline running to a Mediterranean terminus in Turkey. Turkey closed down the pipeline during the Gulf War (demonstrating thereby the danger of cooperative solutions that involve integrated infrastructure). The prospect of Turkey helping to develop the pipeline transport of Central Asian gas and oil to Europe may mean that this Iraqi bargaining chip has lost some value. For its part Syria partially controls Turkey's most serious internal security problem, that of the Kurdish populations of southeast Anatolia. Syria has supported the Kurdish Workers Party (PKK), which has engaged in armed action against Turkish security forces and against Kurds accused of "collaboration" with the Turkish authorities. Demonstrating that Syrian support to the PKK is a good in this multi-good game, Turkey's Foreign Minister Hikmet Çetin, on August 3, 1992, warned that Turkey would honor the 1987 agreement (guaranteeing the 500 cumecs) only if Syria honored the security accord that had been negotiated in April 1992.

The World Bank has been unable to induce a solution to the collective action challenge in the Tigris-Euphrates theater. It participated in the financing of the Karakaya dam in Turkey because this hydropower project did not appreciably affect the flow of the Euphrates, but it has refused to participate in the other dams because appreciable harm to the downstream riparians is clearly at issue. The Bank's refusal, however, has had no effect; Turkey is committing over $20 billion of its own resources to the GAP, and the smell of large export contracts has elicited supplier credits from several western nations. The most recent project on the Euphrates, the Birecik dam, is to be a buy-operate-and-transfer project, financed by a consortium of Turkish, Belgian, German, and French interests, with Chase Investment Bank, UK, as financial advisor. For induced solutions to come about, it helps if the inducees are financially poor.

Let us step back from the current situation and look briefly at its historical roots. When Great Britain held sway in Iraq and France in Syria between the World Wars, no accord was struck with the independent Republic of Turkey to define reciprocal obligations in the Tigris and Euphrates basins. There was no pressing need for such an accord since only Iraq had acquired rights (typical of the downstream, early developer), and neither Syria nor Turkey had plans for the development of their portions of the system. A treaty of Friendship and Good Neighborliness was negotiated between Syria and Turkey in 1926 (Iraq was not included) and in March 1946 a similar bilateral treaty was signed between Turkey and Iraq (Syria was not included). The latter was significant in a number of respects. First, according to al-Khairu (1975, p. 557), Turkey, by this treaty, recognized Iraq's acquired rights, at that time some 13 bcm, to irrigate about one million hectares in the lower Euphrates. Second, the treaty recognized Iraq's right to undertake works to secure its water supply and to control floods. The treaty's first protocol noted that the most appropriate location for such works would be *in Turkey,* and that their cost should be absorbed by Iraq (al-Khairu 1975, p. 556-57). Had this suggestion been implemented, Iraq would have found its major water control infrastructure outside its borders, parallel to Egypt's situation if the Century Water Scheme had been carried out in the Nile basin. Turkey today could claim, although I have no evidence that it has, that it has undertaken at its own expense precisely the kinds of works provided for in the 1946 treaty.

In the mid-1960s both Turkey and Syria began to draw up plans for exploiting their portions of the Euphrates. A first round of tri-partite talks took place in 1965. They did not lead to an accord, but each country, for bargaining purposes, put forth maximum demands: Iraq, 14 bcm; Syria, 13 bcm; and Turkey, 18 bcm. These demands exceeded the annual yield of the Euphrates by 15 bcm (Bari 1977, p. 238; Waterbury 1990). At the same time, as Turkey moved toward construction of the Keban dam, a World Bank mission drafted a report on the water resources in the Euphrates system that became something of a baseline for all future talks. Iraq-Syrian talks in 1966 and 1967 were marked by Syria's rejection of the principle of acquired rights, although Syria accepted that Iraq was entitled to 59 percent of the Euphrates flow in normal years. Subsequently the two countries discussed a formula that would guarantee Iraq two thirds of the flow in normal years, but invert those proportions for any excess flow. No accord was ever reached. Another futile attempt was made in 1972, with Iraq invoking its cultivated surface and Syria pointing to crop water use intensities. Syria ultimately would not sign a draft agreement, and Iraq sought Soviet mediation. The Soviet report used the 1965 World Bank data and recommended that the Tabqa (al-Thawra) dam in Syria, which it was helping to construct, and the Keban dam in Turkey be operated in such a way as to protect Iraq's acquired rights. Syria objected to the report's findings.[10]

As both the Keban and Tabqa reservoirs filled, Iraq threatened military action against Syria. The Arab League tried to mediate the dispute, but once the two reservoirs had filled, Iraq's water crisis eased. In 1980 all three riparians agreed to establish a joint technical commission for the exchange of information. It has met only sporadically since, but it represents the closest thing to a tri- partite cooperative accord in the basin. When Turkey began to fill the Atatürk dam in January 1990, leading to a month-long stoppage in the flow of the Euphrates, Syria and Iraq negotiated an understanding that Syria would guarantee Iraq 290 cumecs at their border, or a 42/58 split of the 500 cumecs Syria is to receive. One may legitimately question whether or not Syria intends to honor this commitment, as discussed below.

Since 1946, Turkey's public position has hardened and is distinctly Harmonesque. The 1976 International Law Commission yearbook reported that Turkey asserted full sovereignty over watercourses within its territory and argued that the physical nature of water could not affect its legal regime (ILC 1979, p. 23). Turkey favors the definition of international *river* laid down in the Vienna Conference of 1815 because it would tend to exclude from consideration tributaries to the river. Indeed, Turkey does not accept that the Euphrates is an international river, preferring the term "transboundary" (Beschorner 1992, p. 56).[11]

In 1992, when the Atatürk dam was inaugurated, Prime Minister Süleyman Demirel responded in a press conference to questions about the interests of Iraq and Syria :

> "This is a matter of sovereignty. We have every right to anything we want... Water resources are Turkey's, and oil is theirs. Since we don't tell them 'Look, we have a right to half of your oil,' they cannot lay claim to what's ours... These crossborder rivers are ours to the very point they cross the border."
>
> (*The Turkish Times,* August 15, 1992, p. 5)

Turkey cites the relative abundance of water in Iraq and attributes any water deficits in either Syria or Iraq to poor management and agricultural practices (Beschorner 1992, p. 61). Thus, as if under Attorney General Harmon's dictum on the Rio Grande, Turkey concedes 500 cumecs to the downstream states out of comity, not because it is legally obliged to do so but out of good neighborliness.

Despite this unyielding legal stance, Turkey would like acquiescence to the solution it is imposing through engineering works on its own territory. In July 1987 Prime Minister Özal of Turkey and Prime Minister Abdul Raouf el-Kassem of Syria negotiated a Protocol on Matters Pertaining to Economic Cooperation between the Republic of Turkey and the Syrian Arab Republic. It was in this protocol that the figure of 500 cumecs was set down, but the protocol is equally significant for its attempt to create a multi-good situation involving cooperation in petroleum, gas, power, banking and trade, as well as water. Article 7, moreover, stated that "The two sides shall work together with the Iraqi side to allocate the waters of the rivers Euphrates and Tigris in the shortest time possible." Following on Prime Minister Demirel's visit to Damascus, January 19–20, 1993, the two countries reaffirmed their commitment to the 1987 Protocol and announced their intention to make 1993 the decisive year in drawing up the allocation of the Euphrates' waters. 1993 has come and gone.

Within 15 years it is likely that Iraq will see its supply of Euphrates water cut by half. Turkey is projecting that GAP projects in both the Euphrates and Tigris basins will utilize 11 bcm. That may underestimate evaporation at reservoir sites, transit losses, and real on-field use. On the other hand, some irrigation water, in degraded form, will return to the system through drainage.

For its part, Syria clings to the goal of food security, seeking to achieve self-sufficiency in grain production no matter how faulty the economic logic of that course. It aims to raise the irrigated surface in the Tabqa project to about 600,000 hectares and to add another 300,000 hectares in the Khabur and Tigris basins. It anticipates that its demand for water will rise from 12 bcm today to 20 bcm by 2010 (Mikhail 1992; Hannoyer 1985). Two thirds of all new Syrian irrigation projects will depend on water from the Euphrates and Tigris basins. It is because of these plans that one must be skeptical regarding Syria's pledge of 290 cumecs to Iraq.

Iraq might thus be facing a Euphrates discharge at Hit of roughly 11 bcm or 2 bcm less than its asserted acquired rights. It would have to compensate for this loss, as well as cover any additional needs of its own, by transferring water from the Tigris to the Euphrates basin within its borders.[12] This, I believe is the probable outcome in the long run, and it would represent Iraq's abandonment of hopes for a basin-wide cooperative solution to its water needs.

In this instance of imposed cooperation, Turkey will be better off and Iraq worse off, with Syria's position not altogether clear. The only cooperative solution that would make some better off and none worse off would involve run-of-the-river power stations and reservoirs in Turkey, a regulated flow of the Euphrates enhancing the *utilizable* discharge of the river in Iraq, and scaled-back irrigation in the GAP scheme.

The Jordan Basin

The bargaining context of this basin has been more intensely studied than either the Nile or the Euphrates and is the prime focus of the current multilateral talks on Middle Eastern water. Here I will concentrate on only the most salient features of the contemporary situation. Everywhere else in the Middle East domestic or non-agricultural use of water is a small proportion of total use. However, in Israel over 20 percent of total consumption is domestic and industrial while in Jordan the corresponding proportion is 35 percent. In other words, while both countries *must* make painful adjustments in the way water is used in the agricultural sector, such adjustments will not avert a crisis. Indeed, as just about all observers have noted, the crisis is already here as aquifers are exploited at rates above that of their recharge.

Israeli policy-makers have been torn between seeking a long-term, basin-wide cooperative solution for Israel's needs and, to the extent possible, "going it alone." The first solution would be inconceivable in the absence of formal peace with all its neighbors. The second might well preclude peace in that it would entail continued control of the headwaters of the Jordan through occupation of the Golan and, perhaps, direct or indirect control of the West Bank watershed. The latter supplies as much as 25 percent of Israel's total water needs through its coastal and northeast aquifers.

Those who anticipate relinquishing the Golan and the West Bank/Gaza in exchange for peace must necessarily also think in terms of a basin-wide accord involving the Hasbani, the Dan, the Banyas, the Yarmuk and the West Bank watershed. A few voices have been raised to suggest that as much as 100 mcm of the Litani could be lifted and then dropped into the Jordan headwaters. It has also been suggested that Israel could trade water conservation technology to Egypt in exchange for a portion of the savings in Nile water (Waterbury 1992). These proposals are based on an acceptance of Israel's vulnerability as a mid-stream state, although it is obvious that Israel will continue to enjoy the military might to neutralize that vulnerability, although at a cost.

The counterview, espoused especially by the Likud (which frequently exaggerated Israel's dependence on West Bank water), is to retain direct or indirect control of the territories and their water resources and to prevent Jordan and Syria from pursuing joint development of the Yarmuk. In this view Israel would continue its imposed solution, first adumbrated with the construction of the national water carrier[13] and continued with the occupation of the Golan and West Bank in 1967. The imposed solution could be sustainable in the long term by increased efficiency of water use in agriculture, extensive reuse of municipal and industrial waste water, and an increase in desalination of sea water. The logic of this position has been baldly put by Ploss and Rubenstein (1992, p. 20):

> Some political scientists have proposed exchanging territorial con-
> cessions by Israel for the right to import water from Syria or even
> Turkey. But such ideas are conducive to war, not peace. ...Water
> cannot be a consequence of peace; it is a condition for making peace.

Israel, and to a lesser extent Jordan (see Lowi 1992; and Abu Taleb et. al. 1992), have already begun to deal with their water problems in a manner that may be forced upon the rest of the region in the near future. The drought years of the late 1980s forced Israel to rethink its agricultural sector. The rethinking involves increased water charges to improve efficiency, shifting from water-intensive to less water-demanding crops, and taking acreage out of production.[14] Increased reliance on desalination may meet the growth in domestic water demand.

Jordan cannot effectively challenge Israel's imposed solution, nor can water conservation alone solve its water problems. At the very least it must cooperate with Syria in the development of the Yarmuk, and, as we have already observed with respect to the 1987 Syro-Jordanian accord on the Wihda dam, Syria exacts a high price for cooperation. By the same token, Syria is not desperate for a cooperative understanding with Jordan, because it is unilaterally developing its portion of the Yarmuk watershed without one. Nor is Syria desperate for an arrangement with Israel that does not include the return of all or part of the Golan Heights. Syria's need for cooperation is focused squarely on Turkey, not Israel, and with its "guaranteed" 500 cumecs, it may feel it has all the cooperation its water sector needs.

It is unlikely that Israel's imposed solution can survive a Middle East peace intact. Like Egypt, Israel must anticipate some reduction in supply on the Yarmuk and from the West Bank aquifer. A cooperative agreement, as in the Nile basin, would serve to lock in a basic recognized supply, not actually increase it. Jordan stands to benefit from any agreement that allows construction of the Wihda dam, and under any possible scenario Syria will not be worse off. If cooperation was extended outside the basin such that Egypt or Turkey supplied water to Israel, Palestine and Jordan, then those three riparians would be better off and Syria no worse off. Only those in Turkey or Egypt who might have used the transferred water might be worse off, but they could be financially compensated.

Conclusion

We should be surprised when formal cooperative arrangements involving more than two riparians emerge anywhere in the Middle East. To date there is only one allocative accord, that of 1959 between Egypt and the Sudan, involving only two of ten riparians in the Nile basin. Turkey has promised Syria an allocation of water pending a final accord among all three riparians in the Euphrates basin. Syria has made a similar promise of about 290 cumecs to Iraq. Israel, in the absence of any understanding, has adhered selectively to formulae on the allocation of the Jordan's waters set forth in the Johnston negotiations of the mid-1950s.

Where there is no solution to the collective action problem, each riparian will pursue a maximizing strategy of investing in storage and delivery infrastructure in its own territory. Unless significant steps toward greater efficiency in water use are undertaken, this strategy will be unsustainable if all other riparians are also following it. Where there is an imposed solution, those who must acquiesce to it will have to change dramatically the ways in which they use water. This is most obvious in the Euphrates basin where the discharge of the river may be reduced by half in the next 15 years. In short, all strategies, including those of the hegemons, will involve national rather than basin-wide solutions.

Economic optimality, both national and basin-wide, will probably be thrown to the wind. For instance, in pursuit of food security both Syria and Egypt may use capital inefficiently to subsidize agriculture (by not charging for water and by high support prices), at the expense of sectors of the economy where returns to capital are higher. Within international basins water will not go to those regions in which each unit of water used brings the highest economic return. Infrastructure will be duplicated rather than integrated, and the scarce resource, water, will not be used for the maximum benefit of the maximum number of the basin's inhabitants. This, more than water wars, is the probable outcome of non-cooperation. The notion of an opportunity cost for water, as well as for capital, has not been widely adopted in the Middle East, and, as a result, neither factor is used efficiently.

There is no *a priori* set of criteria by which to assess the gains to cooperation. The costs of achieving cooperation, in threats and bluffs, laborious negotiations, ineffective monitoring, etc. may outweigh the gains. Net gains and losses are an empirical question. On the other hand, the returns to cooperation should be measured by multiple criteria: the recognition of rights and thus enhanced predictability in use; infrastructural integration that may inhibit the resort to force among riparians over water or any other issues; the encouragement of greater regional specialization and trade in agricultural products; and, to the extent that multi-good bargains are struck, cooperation in more than one domain. The absolute quantities of water that cooperation provides may be small to negligible, but the losses of water to particular riparians in the event of non-cooperation, and the economic inefficiencies associated with autarchic supply solutions, could range from large to enormous.

Notes

1   These figures are drawn from the period 1990/91 from various sources in the bibliography. They do not correspond exactly to figures presented in Gleick (1992, p.17). Discrepancies may arise out of measures that take into account only the discharge of rivers, but not of aquifers or rain run-off. With respect to the West Bank and Gaza, total water availability is on the average 700 million cubic meters (mcm), but the Arab population has effective access to only 200 mcm of it.

2   Bromley and Cernea (1989) point out that the "tragedy of the commons" is a misnomer in that commons were usually carefully regulated through traditional institutions. It is when these institutions break down or are destroyed by the state or markets that the tragedy occurs.

3   It is important to note, however, that the lower reaches are more prone to flooding, and disease vectors, such as malaria, are more widespread. It is also harder to provide for the defense of local populations against marauders and invaders. In the absence of a strong centralized political power, populations may abandon the lower reaches of a river basin for more easily defended niches in the upper reaches. The lower reaches of the Tigris-Euphrates have historically been characterized by periods of de-population and re-population.

4   The discharge of the Orontes and the Yarmuk are about the same, but are little more than 1 percent of the annual discharge of the Euphrates. Thus, although Syria uses about 90 percent of the flow of the Orontes, Turkey is not particularly

perturbed by the absence of an accord protecting its rights. It can then cite its own tolerance as a downstream state on the Orontes as a model of lower riparian behavior for Syria and Iraq on the more important Euphrates river.

5 Hegemons are seldom the unitary actors the term implies. Robert Collins (1990) shows how divided British colonial officialdom was in the early 20th century with respect to the Nile. Officials in Uganda, the Sudan or Egypt often came to promote "their country's" interests. The hegemonic solution thus tended to be the result of bargaining among colonial dependencies rather than a blueprint produced in London.

6 To return to the incident of 1979 in which Ethiopia objected to Egypt's alleged proposal to transfer Nile water to the Negev, the Mengistu regime was essentially claiming a veto right consonant with the extreme formulation of the community of interests. This is tactically understandable but strategically curious in that Ethiopia has consistently invoked principles much closer to the Harmon Doctrine. Trying to have it both ways is also, of course, a well-understood practice of international relations.

7 Superfairness obtains when no party to a cooperative solution would prefer to trade her benefits for those of another party. In Rogers' application, returns to the use of water, i.e., a single-good bargain, determine the net benefits. It might complicate formalization, but simplify reaching a solution, if the core were determined in a multi-good bargain.

8 This rubric is not self-evident for Turkey and Lebanon. For the first, the GAP project is both an equity and a security issue in that all the "neglected" populations of southeast Anatolia are to benefit, but especially the Kurdish populations, parts of which currently constitute a threat to Turkey's internal security. Likewise, Lebanon needs to use the Litani to develop the Shi'ite south of the country, upon which the future stability of the "third" Lebanese republic will depend (see Kubursi and Amery 1992).

9 In this sense, Iran has a stake in the catchment of the Tigris, and Zaire and Eritrea in the Nile, although the two rivers do not traverse their territories.

10 Disputes over data nicely demonstrate what is typically on the bargaining table in such negotiations. Iraq claimed 1.2 million hectares under cultivation, while Syria asserted that Iraq had only 835,000, claiming the difference had been lost to production due to salinity. Intensities of water use per hectare also lead to disputes; is it legitimate for Iraq to calculate water use on the basis of irrigation needs plus the flushing water required to cleanse acreage of salts?

11 That I use transboundary in the title of this essay should not be construed as endorsement of the Turkish position. I do not fully understand that position except that it probably reflects Turkey's preference for minimal constraints on riparian sovereignty. Those preferences are embodied in the legal term "successive rivers" and are reflected in the concept of relativeness in the definition of international watercourse.

12 One of the unknowns facing Iraq is the amount of water Turkey will use when it turns to development of its portion of the Tigris basin. Syria, also, has plans

to irrigate some 150,000 hectares using Tigris water from its short run of the river forming part of its border with Iraq downstream of Cizre in Turkey. However, while 98 percent of the discharge of the Euphrates originates in Turkey, 61 percent on average of the discharge of the Tigris originates in tributaries within Iraq.

13 As far as I know the national water carrier represents the only extra-basin transfer of water in the Middle East. It was carried out without any consent from the other riparians. Because Lake Tiberias, from which the national water carrier originates, was under full Israeli suzerainty, this unilateral abstraction of water was tantamount to an endorsement of the Harmon Doctrine.

14 A hectare of cotton in the Middle East may use 10,000m3 of water in its growing period. That is equivalent to the yearly domestic water consumption of 100 city dwellers.

3

Overall Perspectives on Countries and Regions

J. A. Allan

Introduction: Water in the Arab Middle East and North Africa

The availability of water has always shaped life and livelihoods in the Middle East and North Africa. Throughout history this region's peoples have been short of water in the summer except for those living on the banks of the region's major rivers. Huge tracts are waterless throughout the year. The river valleys, together with the favored lands that received winter rains along the Mediterranean coast and on the plains of northern Syria and Iraq, have always been the most popular places to reside and to develop crop production, as well as to build the economic and social systems that in turn led to the creation of a sequence of civilizations, usually hydraulically based. The rest of the region has been sparsely populated, and the economies of the dry tracts made few demands on scarce water resources until the past half century, and especially the past two decades, when new technologies transformed the capacity to lift and move water.

The peoples of the Middle East are remarkable not so much in the scale of the water shortages that they have faced as much as in the numerous ways in which they have continuously adjusted to deficiencies in water supply since the days of hunter gatherers. Concern has been expressed since the Bronze Age about the reliability of water supplies in the Tigris-Euphrates region. In Egypt of the Hyksos period (c. 1600 BC) and in Mameluke Egypt (15th century AD) expeditions were sent to the upper Nile for this reason, as well as in the Ottoman period. The British were particularly attentive to Egyptian concerns in the late 19th century and in the 1929 Nile Waters Agreement they were able to impose what proved to be unsustainable constraints on their upper Nile colonies in East Africa not to use Nile tributary waters. It is only in the past forty years, however, that ministers, officials and journalists have identified water deficiencies of economically strategic significance, and have raised the level of regional hydro-paranoia so that it has become a potentially significant de-stabilizing element in the affairs of Arab countries.

By concentrating public attention on the growing water gap and de-emphasizing the demographic explosion that is its cause, and at the same time being less than frank about the remarkable economic adjustments that national economies have achieved in the recent past, the political leaderships and the region's selectively critical media have slowed the pace at which any understanding of the real status of resources and of the real economic options has been assimilated by leaders and peoples. One of the adjustive measures adopted to date, the unsustainable overuse of water resources, is economically and ecologically damaging, and will in due course be modified in favor of strategies of water demand management. The importation of food on the other hand is only an economic challenge.

The present juncture in the region's water management history is particularly important because such a thoroughgoing suite of adjustments is required to re-shape water allocation policies to accommodate the economic and political challenges posed by water shortages. To date national leaderships and water managing institutions of the individual countries of the region have been unable to set new directions at the pace demanded by the growing water resource gaps. Implementing new policies, there is no doubt, will be unpopular and counter to the interests of key players in the politics and the economies of the region.

Nevertheless, throughout history adjustments to the low availability of water have been the rule in the region. The first adjustments consisted of the location of populations and economic activity in favored sites. As demand for increasing volumes and new patterns of water availability emerged in this century, adjustments took the form of the development of new water sources. However, since the early 1970s, many countries have been faced with such water shortages that they have had to seek solutions outside the narrow confines of their national water budgets, drawing upon their respective larger national economies and ultimately the international economy.

The approach taken in the following discussion will be first to draw attention to the limited availability and the growing demand for water.

Secondly, the discussion will show how adjustments are made:

■ by increasing the availability of water as well as moving it to preferred locations through engineering and technological interventions. This heretofore dominant approach will play a decreasing role in future.

■ by improving the productivity of water 1) by moving and delivering it less wastefully, 2) by deploying into the growing fields more efficient methods of water utilization by crops, and 3) by reducing water losses in systems distributing water to domestic and industrial users. This conservation aspect has only been addressed to a limited extent in the Middle East and North Africa and will be a major area of focused innovation and investment in the future.

Thirdly, it will be shown that a suite of new approaches will enable the peoples of the region to utilize water according to different and more realistic assumptions concerning its value and its place in their economies. The new approaches will also permit a re-evaluation of water's strategic importance and its place among their interests internationally. These new approaches will include:

■ the adoption of principles of allocative efficiency that lead to the use of water first in the sectors that bring the best returns to water; that is, in industry and services rather than in agriculture. Within each sector, they lead to using water for productive activities that generate sound economic returns — for example, crops that command a high price on world markets rather than those like sugar, wheat and rice where competing external producers have access to free or nearly free water. More controversially, such principles would urge using water in those parts of river basins where it is most effective, rather than where intense evaporation and transpiration lead to waste. Such allocative principles are unlikely to gain currency as long as conventional attitudes to resource ownership prevail and in the almost complete absence of water markets for the water flowing in international river systems.

■ a number of technological and institutional instruments that together will make principles of demand management usable. The technologies include metering and analyzing water use, and water treatment and re-use. Improved institutions will be even more important, needed to ensure the effective deployment of both the engineering and economic and financial instruments (markets for water), and regulatory and legal frameworks. Water use in municipalities, industries and agriculture will be reduced so that there will be sufficient safe water to sustain the health as well as diverse and adequate livelihoods for the peoples of the region. It will also be possible for the environment to be secure and sustainably managed (Lutz and Munasinghe 1991; Pearce et al. 1989; Pearce and Turner 1990).

Table 1 is an interdisciplinary framework that shows the major factors to be considered in the approach that scientists and political figures necessarily must adopt if they are to provide systems of water use that will be economically sound, socially and politically acceptable, safe in terms of ensuring health, and environmentally and ecologically sustainable.

The political leaderships in the region have not yet adopted the principles implicit in such an approach in all respects, but they have been obliged to adopt a number of them despite claiming in their public statements on food security that they have not (Allan 1983). This contradiction has been possible because water is only part of the natural and other resource endowments that make up a national economy, and some economies have been sufficiently strong, through oil revenues or through external support in the example of Egypt, to overcome the potential constraints of water deficits.

Economies are characterized by their capacity to substitute for scarce resources with resources mobilized in other sectors. Indeed, its capacity to substitute for a scarce resource is a crude indicator of the effectiveness of management of a national economy. The non-intuitive notion of substituting something else for water in a national economy is a difficult one for politicians, officials and especially ordinary water users. In practice such substitution has been achieved in a variety of ways, albeit unrecognized, in the Middle East and North Africa. Some oil-rich countries, for example, use their energy to draw usable water from saline or brackish water by desalination, although out of the approximately 150 bcm of water used annually by Arab countries, desalination produces less than 1 percent.

The system that has indirectly mobilized massive volumes of "substitute" water, perhaps fifty times as much as desalination, is the familiar device of world trade in food. The author estimates that in the early 1990s, 50 bcm of water per year were imported "in food" into the region — equivalent to the annual flow of the Nile in Egypt and to about 30 percent of the region's total annually available surface water. Such figures show that international economic systems in which the region's public and private enterprises play their part are extremely significant in enabling the peoples of the region to enjoy their "entitlement to food." (Sen 1981)

The governments of the region have been particularly successful in achieving the economic and political stability which depends to some extent on their capacity to deliver this entitlement, which is no longer

TABLE 1

Goals and Principles of Water Allocation and Management:
A Framework for the Analysis and Development of Policy and Practice

Goals of Water-Using Activity	Guiding Principles	Policies	Institutional Instruments	Engineering
Facilitation of political circumstances to enable optimum resource use	Minimization of conflict; promotion of co-operation in the areas of water use at all levels.	Conflict resolution; identification of reciprocal arrangements to promote economically and socially beneficial water use and the installation of such arrangements.	Water sharing arrangements: • traditional and new; recognition of water rights and of the ownership of water; consultation between legislators, officials (local, national and international) • under "democratic" institutions; introduction of new economic and legal instruments to shift access to water to the most beneficial users and uses.	Earth observation (remote sensing); *in situ* monitoring and information systems.
Productivity (*"Development"*) Allocative efficiency Productive efficiency	Returns to water, sustainability of water supplies.	Investment in sectors, activities and crops that bring optimum returns. Demand management.	Water pricing, agricultural subsidies, crop pricing and other intervention. Advanced pricing systems imply water metering. Agreements both local & international.	Large and small civil works for water abstraction, treatment, delivery and distribution, recycling, water metering. Water efficiency studies and water management.
Equitable use	Social benefits.	Identification of the social benefits and disbenefits of water use and the promotion of beneficial uses.	Land reform, water regulation, new legislation, reduction of illegal water use, changes to traditional rights.	Water control systems, irrigation management.
Safe use	Provision of adequate volumes & quality of water.	Identification of appropriate systems—traditional and new—promoting the safe provision of water use, re-use and disposal.	Monitoring, legislation, regulating institutions (traditional and new).	Planning for future demands, water control systems, water treatment, maintenance for reliability.
Environmentally sound use (*"Conservation"*; *"Cultivate the world as if we live forever"*).	Sustainable use of landscape and amenity including intangibles.	Identification of appropriate systems—traditional and new—for sustainable water use.	Monitoring, legislation, regulating institutions (traditional and new).	Quality monitoring, water treatment, wastewater treatment, waste disposal.

possible on the basis of indigenous water. The collective achievement of Middle Eastern and North African governments in this regard contrasts markedly with the capacity of the governments of other regions, sometimes better endowed environmentally, to provide food entitlements for their peoples.

Water Supply in the Arab World

The water endowment of the countries of the region

It would be extremely difficult to set out a water budget of the Arab world and its countries that would withstand scientific scrutiny, much less attract political consensus. Scientists are aware that many of the data have not been adequately researched, and more important, that heroic assumptions are involved in the estimates of groundwater.

Also especially difficult is evaluating rainfall, a very important part of the water budget for some countries. A proportion of such precipitation infiltrates the soil profile and is retained there. Such water, occurring naturally in soil profiles, is among the most precious of all renewable natural resources as it provides the essential starting point in the food chains of both natural and agro-ecologies. Naturally occurring water coming as a free good does not fall equally in volume or reliability on the populations residing on the Earth's surface, and the countries and communities of the Middle East and North Africa are particularly ill-provided. In addition, a high proportion of all rainfall falls on tracts that cannot be cultivated because of the steepness or roughness of the terrain. As a consequence the fraction of rainfall that becomes usable soil moisture is only a tiny proportion of the water naturally precipitated. Of the tens and sometimes hundreds of cubic kilometers of water that fall naturally on the countries of the region, only between one and ten percent ends up in the tissue of vegetation and crops of economic significance. This cannot be estimated to change significantly in future, since climate change is too difficult to predict (Conway 1993). In any event, such change would be small in scale compared with other elements on both the supply and the demand side of water budgets in the region.

The element of national water budgets that is most reliably monitored in Arab countries is gross surface run-off. Surface run-off data can, however, provide misleading impressions of the water security of a country. This is because the capability of a water resource to support economic activity and to undergird further economic multiplication depends on a large number of associated geographical and economic variables. It is becoming increasingly clear that successful and economically effective development of such surface resources depends on social and political circumstances as much as natural endowment. For example, traditional perceptions and political institutions, both of which evolved in periods of water surpluses, are proving in combination to be a very dangerous inspiration for those allocating and managing water in the contemporary Arab world where the problem is how to respond radically and constructively to accelerating water deficits.

Geography also plays a role, but by no means a determining one. The shape of the terrain and the location and seasonality of rainfall and surface run-off determine whether water can be stored, and together with technology, whether its distribution can be economical. For example Egypt and the Sudan have few sites to engineer water storage; Ethiopia has many. Some geographies make the use of surface water difficult within the countries in which the water reaches the surface. For example the springs at the foot of Jebel Sheikh — the Hasbani, the Dan and the Banias — were of little use to Syria before 1967 because of their location and elevation. Some of these waters were for a period in the mid-1960s destined for use by Jordan, until the project to transfer the water to the Yarmuk River was interrupted by Israel. Many countries in the region have water in relatively low-lying surface water systems while their main populations prefer to live in high and otherwise more habitable tracts. Jordan's surface water in the Yarmuk and the Jordan rivers represent the majority of national water, but it flows at elevations 1000 meters below Jordan's large concentrations of urban population; the costs of lifting and moving the water to the cities would be over $1.0 per m³. Libya offers another example of inconveniently located water; its future water lies 1000 kilometers south of its major population zone on the shore of the Mediterranean (Allan 1989b).

Table 2 provides estimates of the water resources of the countries of the region. The reliability of the data is limited, but they provide a perspective on a number of types of water economy. The water economies can be categorized according to whether they are in surplus or in immediate or long-term deficit.

The information summarized in Table 2 and the linkage to economic circumstances in Table 3, make it evident that water availability is a serious and urgent issue for Egypt, Jordan, Yemen, Syria and the Maghreb countries. It is important for all of them with respect to water for agriculture, and all but Egypt have difficulties in supplying domestic water users in the cities in the summer. Libya, among the oil rich countries, also has problems with urban water supply, as does Iraq in its current economic

T A B L E   2

Water Resource Estimates in Middle Eastern and North African Countries

	Surface Water				Groundwater		Rainfall		Desalination capacity 1990 bcm/annum	Approximate volume of accessible fossil groundwater reserves 1990 bcm	Proportion of total available water allocated to agriculture 1988–90 %	Annual requirements for full self-sufficiency including agriculture 1990 bcm/annum
	Average surface inflow to country bcm/annum	Average internally derived surface flow bcm/annum	Average surface flow to another country or sea bcm/annum	Average available surface water bcm/annum	Average groundwater abstraction rate–1990 (including falaj) bcm/annum	Average total water available bcm/annum	Average national rainfall volume bcm/annum	Available for agriculture, etc. bcm/annum				
Middle East												
Bahrein	0.0	0.0	0.0	0.0	0.2	0.2	0.1	0.0	?	0.0	20	1.0
Iraq–1980s and '90s	16+25+10	3.0	9.0	43.0	2.0	45.0	60.0	4.0	?	?	85	18.0
Occupied Territories	0.0	0.0	0.0	0.0	0.3	0.3	2.0	0.1	0.0	0.0	50	2.0
Jordan	0.2	0.1	0.03	0.5	0.4	0.1	2.0	0.3	?	?	90	4.0
Kuwait	0.0	0.0	0.0	0.0	0.1	0.1	0.0	0.0	?	5000	2	0.7
Lebanon	0.0	3.8	0.5	3.3	0.5	3.8	12.0	2.0	0.0	?	?	4.0
Oman	0.0	0.0	0.0	0.0	0.2	0.2	0.2	0.1	0.04	?	65	0.4
Saudi Arabia	0.0	0.0	0.0	0.0	3.5	3.5	4.5	1.0	1.0	?	80	8.0
The Sudan	103.0	10.0	55.5	18.5	0.5	0.5	20.0			?	95	22.0
	Sudan loses half the White Nile flow–19 bcm annually–and most of the internally derived surface flow by evaporation in the Sudd. Also about 10 bcm lost annually at Lake Nasser/Nubia.											
Syria	17.0	0.5	12.0	5.5	1.5	7.0	7.0	6.0		?	75	10.0
	The natural flow of the Euphrates would bring Syria c29bcm annually. By 1990 the average flow was c17bcm by the hydraulic works of Turkey.											
UAE	0.0	0.2	0.0	0.4	0.4	0.8	0.5	0.0	0.3	?	50	1.0
Yemen	0.0	6.5	1.0	5.5	2.5	8.0	48.5	4.0	0.0	?	90	13.0
North and Northeast Africa												
Egypt	55.5	0.0	12.0	43.5	3.0	55.5	2.0	0.4	0.2	30.0	88	70.0
Algeria	0.0	18.0	6.0	8.0	2.0	10.0	35.0	5.0	?	60000	72	30.0
Libya	0.0	0.1	0.1	0.0	0.0	1.8	8.0	0.5	?	2500	90	7.0
Morocco	0.0	25.0	5.0	20.0	5.0	25.0	126.0	15.0	?	?	91	30.0
Tunisia	0.0	2.5	1.0	1.5	1.8	3.3	12.0	0.8	?	10	90	9.0

Table compiled by the author from various sources.

TABLE 3

Water Economies Categorized by National Economic Strength

Economies in water surplus	Economies in immediate or long-term water deficit
Oil economy	*Oil economies*
Iraq	Saudi Arabia
	Kuwait
	UAE
	Qatar
	Oman
	Libya
	Partial oil economies
	Egypt
	Yemen
	Syria
	Tunisia
	Algeria
Non-oil economies	*Non-oil economies*
Lebanon	Jordan
The Sudan	Morocco

Compiled by author.

and infrastructural predicament after the 1991 Gulf War. All the oil rich Arab countries except Iraq have water deficits, but such deficits can be accommodated by their capacity to import water, or to use expensive water such as that pumped from deep aquifers or desalinated from saline or brackish supplies.

Augmenting water supplies in the future

At least in public statements, the governments of Arab countries, both political leaderships and officials, are still very much committed to augmenting national water budgets with new water. The most popular means to achieve this end in the past, especially in this century, was to control surface flows by creating multi-purpose reservoirs. Although it is still very appropriate for upstream neighbors in the Nile and Tigris-Euphrates catchments —Turkey is well advanced in creating such structures (GAP 1990) and Ethiopia would very much like to embark on a similar program (Abate 1992 and 1993)—this approach is no longer an option for the Arab countries. By the end of the 20th century there are few unbuilt major dam storage sites for electricity or agricultural purposes in the arid and semi-arid Arab countries. Those that do exist, such as the Meroe (Fourth Cataract) site in the northern Sudan, are economically and internationally very controversial. Storing water at sites where evaporative losses will be three meters depth of water annually is no longer regarded as economically feasible in river basins where water has become scarce.

With the exhaustion of conventional engineering practices to store and distribute flowing surface water, governments have had to seek new water in other parts of the hydrological system. A favored approach by those who would benefit has been the drainage of wetlands. The best known project is the Jonglei Scheme, which was designed to yield annually four cubic kilometers of water to be shared equally by the joint investors in the project, Egypt and the Sudan. Construction stopped in the early 1980s as a result of violent opposition by the local communities who did not want their livelihoods and ways of life changed by the draining of the swamps of the Sudd and especially not by the proposed second stage of the project (Howell et al. 1988; Collins 1990).

In contrast, the plan to drain the swamps in the lower Euphrates in Iraq is going ahead rapidly at the time of writing. Here the politics and the purposes are different from those in the south of the Sudan; mobilizing new water is only a minor factor. In southern Iraq traditional interests are being ignored by a central authority, the Iraqi government, which can coerce local populations to accept the environmental impacts of drainage (North 1993). Motivated by perceived national security problems, the Iraqi leadership intends to prevent cross-border infiltration using the marsh waterways that in some cases cross the border with Iran, and wishes to keep elements of Iraq's Shi'a opposition from using the marshes as a refuge.

Elsewhere, the option to drain marshes is being severely restricted by the international community, which has during the past decade become increasingly concerned about ecological issues. Wetlands are no longer regarded only as water resources ripe for development. They are seen as needed for the survival of local wildlife and often essential for the intercontinental and inter-regional migration of species, especially birds. The economic impact of such changes of opinion is difficult to quantify, as are the benefits of wildlife and species diversity, but they have clearly moved the decisions of international agencies in the direction of non-intervention, especially after the 1992 UNCED conference in Rio de Janeiro (UNCED 1992).

The Arab region has massive groundwater reserves, both shallow and deep, renewable and fossil. They became accessible on a large scale several decades ago with the emergence of technologies to detect and extract them and of economies able to provide the necessary capital to develop them. For many years now, groundwater has been a major segment of the region's water budget. Unfortunately it is expensive to appraise both the quantity and the quality of large-scale aquifers and there are no comprehensive and reliable data on the extent and usability of the region's groundwater (Edmunds and Wright 1979; Lloyd 1990; United Nations 1973 and 1982). The groundwater resources are nevertheless very extensive indeed, especially in the vast Intercalaire formations that underlie parts of Algeria and Tunisia (UNESCO 1972). The groundwaters of Libya are relatively well researched and the Libyan authorities have experimented with *in situ* development as well as with the construction of major water carriers to move water from remote desert locations to centers of population at the coast (Allan 1981; 1983; 1988; 1989a and 1989b).

New water: Long-distance transport — The Libyan pipeline experience

One Arab government has gained experience in financing and engineering a water transport system at a national scale. Libya's Great Man-Made River is a bold attempt to address the water deficits of its coastal tract by moving water from its aquifers in the deserts of the south, where *in situ* development has proved to be socially and institutionally difficult as well as non-economic. The giant pipelines have been designed to convey between 2 and 3 cubic kilometers of water per year, which far exceeds the present approximately 1.5 cubic kilometers of national usage (Allan 1989a).

Libya's experience of the 1980s in addressing its water crisis is a parable that other leaderships in the region should observe closely. Libya's revolutionary government seized power in 1969. The costs of resource mismanagement in its early years were only background noise amid the rise, sometimes very rapid as in 1973 and 1979, of income derived from oil. For an increasingly prosperous Libya, it was possible to contemplate high cost solutions to its water problems. The Great Man-Made River project was embarked upon with a first phase costing about 15 percent of the year's oil revenues earned in 1980 ($23 billion), and further phases and related investments to cost at least four times this sum. Devoting about 12 percent of GDP over a period of a decade to securing the nation's strategic water supplies was not an unreasonable prospect. Unfortunately the parable took a very different direction from that anticipated during the decision-making in 1979–1980, in that during the 1980s Libya ceased suddenly to be a rich nation. Revenues fell to one-quarter of their 1980 levels and the strategic investment in water security began to dominate the national economy even though the construction schedule was extended to last two decades rather than one.

The most disturbing feature of the parable is not so much the dramatic change in economic fortunes as the unwillingness of the Libyan leadership to shift its perception of the value as well as of the cost of developing its water resources with respect to the changed economics of the 1980s. The intent of the Libyan government at the time of writing is still to devote a high proportion of the costly water, delivered at the coast at an estimated cost of $1.0 per m^3, to irrigated farming despite the implied economic folly of using over $10,000 of water per hectare (not to speak of other inputs) to produce crops yielding only $1,000 to $2,500 per hectare.

The Libyan example is an extreme case of the political interests of leaders and those of minority constituencies over-riding medium- and long-term national interests. It is also an example of the inability of leaders and opinion formers to disaggregate water supply according to cost and to distinguish among water demands according to the capacity of various uses to bear the real costs of water inputs. The special significance of the Libyan case is that the extreme contradiction between the current politically driven policy to mis-allocate water and the underlying economic costs of developing and transporting deep and distant fossil water is to be exposed during a period of economic weakness rather than camouflaged in an era of continuing oil-derived prosperity. The economic stress of operating a system delivering water to a non-viable irrigation sector will be unsustainable and the consequences of mis-allocation will not be submerged in the surge of funds available in the 1970s. Libya's rich country approach to water resource development no longer fits its economic circumstances.

The Libyan experience tells us a great deal about the relevance of long distance water carriers to solving the water resource problems of the region. They can deliver significant volumes of water, approaching 2 bcm per year in the east of Libya and about 1 bcm in the west. These volumes would be significant in all Arab countries except those served by the Nile and Tigris-Euphrates systems. But most important is the high cost of the water. Only domestic and industrial users will be able to bear the costs of such water, and then only those in strong economies.

Two other national pipeline projects are being initiated, both to deliver expensive water. The water for Aleppo, about 80 mcm per year, is already conveyed by pipeline from Lake Assad. In Jordan the proposed project to move water from the Disi aquifer on Jordan's Saudi border is at an advanced stage in what is always a protracted suite of authorization and financial preliminaries. The destination of the water will be the domestic and industrial users of the Amman urban area who suffer annual water restrictions, severe ones in drought years.

International water transfers by pipeline have significant potential but to date have not become operational. This is surprising in that relatively small volumes of water can be of great significance to countries trying to manage on water budgets of under 1 bcm annually. Jordan and most of the states of the Arabian Peninsula make do with less and Palestine would be another entity with very modest water resources. But international water carriers are perceived as strategically vulnerable. The fears have been given substance by Israel's military intervention, which halted the Arab initiative in the early 1960s to prepare for the transfer of water from the Banias spring, one of the major springs feeding the Jordan, across Golan and into the Yarmuk to increase the volume of water available to Jordan's irrigation projects in the Jordan Valley. Most Arab leaders feel that similar fates would befall other international projects. The Turkish Peace Pipeline proposed in the early 1990s could convey significant volumes of water — about 2 bcm — from sources in southern Turkey of no immediate value to Turkey (Gruen 1993).

These cases highlight the difficulties facing those making policy on water in the region. The sources of new water are never unencumbered. They are either politically insecure, limited in quantity, or ill-located and therefore difficult to develop. Sometimes they are subject to a combination of these disadvantages.

New water: Long-distance transport: Sea movement

Interest in the transport of water from water surplus regions will be stimulated in the region as a whole by the studies and innovative engineering experiments in Turkey. Turkey has regions enjoying water surpluses but also wants to support areas, on the mainland and in neighboring Cyprus, that suffer seasonal and increasing water deficits. As a result technologies are being tested within the national economy of Turkey that will demonstrate the relevance of sea transport of water over distances significant also to users in Arab countries on the southern shore of the Mediterranean and beyond. The water could be modestly priced, on the order of $US 0.20 per cubic meter (including the costs of development, operation and the provision of capital), according to the designers of the Medusa Bag project. This is an unattractive cost for agriculture but a very favorable cost to supply coastal urban settlements and coastal industrial complexes such as those in the countries of the Maghreb, Libya and even in the more distant Gulf (Savage 1993). The volumes of water being discussed by Turkey are very significant to many national entities in the Arab world. The transport of 250 mcm per year over 650 kilometers has been considered in reaching the cost of $0.15–$0.20 per cubic meter, with additional water handled in the same system costing only $0.10 per cubic meter (Savage 1993). Amounts of only 100 mcm of water annually are very significant to Jordan and the putative Palestine as well as to the cities of the Maghreb and Libya.

New water: Desalination

Desalination is another source of potential new water for a limited range of uses. There are Arab countries such as Abu Dhabi where desalinated water is used as if it were not costing over $1.0 per cubic meter in sectors with negative economic returns. And Abu Dhabi is the extreme example of the imaginative use of expensive desalinated water to provide green amenity in its urban areas and alongside its major roads. El Ain in Abu Dhabi is a remarkable garden city in the middle of the desert served by a mixture of groundwater and desalinated water. A feature of the Abu Dhabi water economy is that reused municipal water is an element in the water budget and processed waste water is a significant contributor to the supply for the gardens and green areas in the cities of the Emirate.

Desalination is a technological option for all Arab countries but only an economic one if there are oil or other revenues to subsidize building and operating the desalination plants. Operating costs vary according to the cost of the energy used and it is being suggested that plants could be built to produce desalinated water for less than $1.0 per cubic meter. Oil exporting countries have cheap energy and the option to devote their energy resources to desalinating water. Desalination has the advantage of providing a measure of security in water supply, but as shown in the 1991 Gulf War the desalinating plants are extremely vulnerable to attack from the air or by missile and they can be seriously threatened by off-shore pollution.

An important part of the desalination scene is the improvement of brackish water, rather than the very saline seawater off-shore of Arab countries. Brackish waters range from those with 1500 parts per million (ppm) of total dissolved solids, which are usable on many crops but not conventionally for human consumption, to waters with 5000 ppm or more. These waters are very much cheaper to improve for municipal and industrial use than seawater with 35,000 or more ppm of total dissolved solids. Countries with serious water deficits and limited economic options, such as Jordan, are looking closely at the purification of brackish water as a viable possibility.

New water: Treatment and re-use of wastewater

The re-use of water occurs naturally, especially in irrigated farming, and the use of treated industrial and municipal wastewater is becoming common at least in pilot schemes. The hydrological cycle moves water naturally from irrigation channels to the soil and then into groundwater flows and storage. The Egyptian irrigation systems have been engineered under particularly favorable geomorphological circumstances in naturally well-drained alluvial deposits that allow the re-use of water between two and four times during the flow of Nile waters from Aswan to the Mediterranean Sea. These water re-using practices in Egypt's irrigated farming enable an overall water efficiency of over 70 percent, which compares very favorably with any country, even where advanced systems of water distribution and delivery of water have been deployed.

Another way of looking at water re-use is to estimate how much water it "adds" to a system in which there is significant water re-use. In a country such as Egypt it could be estimated that the water applied annually to its irrigated fields is much greater than the nominal 38 bcm shown in national statistics, out of the country's total use of 55.5 bcm. Depending on the assumptions concerning consumptive use by the plant and the returns to groundwater and the drainage channels, the volume spread, sprayed and dripped onto Egypt's crops would not be less than 60 bcm and could be as high as 80 bcm. These figures, when used to calculate the water applied per hectare, produce an estimate of 20,700 to 27,600 cubic meters per hectare, which is close to the numbers calculated to be necessary for effective irrigated crop production by consulting engineers in the international agencies (Aboukhaled et al. 1975, pp. 41–51).

Egypt is exceptional in the region in its agricultural water re-using practices, however, partly because water in such quantities as in the Nile system exists elsewhere in the Middle East only in the Euphrates and Tigris systems in Syria and Iraq. Here, however, the hydraulic circumstances are different in that effective drainage is much more difficult to engineer and the soils are saline and expensive to reclaim. In other Arab countries surface flows are minor in scale and nowhere are the hydraulic and soil circumstances as favorable as in Egypt for agricultural water re-use. The effect on water quality of re-using water in irrigation systems is well understood: there is a progressive deterioration in water quality with the number of soil profiles infiltrated by the water. Water that enters the system at Aswan at 300 ppm of total dissolved solids emerges at various points in the system at progressively higher levels until it reaches the drains at the northern end of the Delta at toxic levels containing over 2000 ppm.

Most governments in the region are looking at the treatment and re-use of urban and industrial wastewater. Only the Gulf countries have had the capacity to incorporate these important recovery systems in their water management practices, partly because they have the resources to implement the expensive engineering but also because the volumes of water to be treated are modest. In Egypt, Cairo has recently attended to handling its wastewater with a new major sewage system and close consideration is being given to the construction of sewage treatment that would yield 2 to 3 bcm of water annually of a quality adequate for crop production. That people throughout the world are uneasy about consuming products produced with treated urban wastewater is a socio-political challenge for governments and national and international agencies attempting to encourage the utilization of such treated water. That people adamantly oppose the use of domestic water treated in proposed plants does not prevent them from simultaneously feeling an irrational sense of security when consuming water treated in older plants that for decades have improved many-times-used Nile water, very far from pure by the time it reaches Cairo.

The volumes of water handled in urban and industrial wastewater systems are in absolute terms small. Urban and industrial water use is only between 10 and 20 percent of the total water in the Arab world. In Egypt, even if all urban wastewater in the country were to be treated, such water would only amount to between 3.6 and 5.4 percent of the total water in the system. As indicated in the preceding paragraph the estimates of the amount of water utilized in Egypt varies according to whether one considers only the water that enters the system (55.5 bcm per year) or the total water that is utilized through "natural" re-use and engineered agricultural re-use (60–80 bcm per year).

The notable feature concerning Egypt in Table 4 is that the potential gain from the various endeavors to mobilize expensive new water will be a small proportion of the water in the system. The calculations underestimate the contribution of any water that might be jointly developed by Egypt and the Sudan in that any water that enters Lake Nasser/Nubia is worth at least 50 and as much as 100 percent more than its original input volume as a result of re-use in the Egypt's irrigation system. On the other hand the losses in evaporation during storage in Lake Nasser/Nubia could also be considered and these would reduce the volume contributed by between 5 and 15 percent. Lest it be assumed that the Egyptian case can be extrapolated it is relevant to compare the position in immediately upstream Sudan, which contrasts markedly because of environmental

T A B L E 4

Existing and New Water Potential in Egypt and Jordan: The Great Importance of Nile Re-use

(in bcm)(L = low estimate; H = high estimate)

	Existing Water				New Water			Total of all water, including reused water
	Water available	In Agriculture		Total	From improved systems and irrigation efficiency	From Jonglei (unlikely to be completed)	From urban wastewater treatment	
		Single use	With re-use					
EGYPT Usual estimates	55.5	38	—	55.5	L: 5 H: 10	2 2	2 4	64.5 71.5
Accounting for agricultural re-use	55.5	38	L: 22 H: 42	77.5 97.5	L: 5 H: 10	2 2	2 4	86.5 113.5
JORDAN Usual estimates	0.8	0.6	—	0.8	0.1	Fossil sources L: 0.1 H: 0.2	L: 0.2 H: 0.4	1.2 1.5

Sources: Water available data from the Ministry of Public Works and Water Resources of Egypt and the Jordanian Water Authority. Other figures are author's estimates.

circumstances there. There is effectively no natural water re-use in the Sudan's major irrigation project, the Gezira Scheme, because of the nature of the black cotton soils, which prevent deep infiltration. Water stays on or near the surface and there is a tendency for excessive applications of water to be wasted through evaporation.

The Jordanian case is also very different. New water from municipal re-use, though small in absolute volume, would be very significant in terms of the national water budget. The treatment of urban wastewater would bring at least as much new water as the water saved through improvements in the systems of distribution. The important figures to note, however, are the proportional contributions likely to be made to the total national water budget by such new sources of water. Relatively small volumes of water of 100 mcm per year are significant to national entities such as Jordan and the putative Palestine, as well as to the Arab Gulf states.

New water: from fractured rocks zones

The widespread search for oil since the 1940s has revealed that there are vast reservoirs of groundwater throughout northern Africa and Arabia (Wright et al. 1971; Pallas 1980), unfortunately often in aquifers that are difficult to develop. In the past two decades it has been argued that there are sub-surface zones or conduits traversing vast distances within the surface formations of the sedimentary mantle of northeastern Africa (Ahmad 1975). The issue of whether some of the remote groundwater is still being recharged, as argued by Ahmad for the Kufrah aquifers in the same work is unresolved because techniques to detect subsurface conduits have not yet been developed.

Subsurface flows of water have been given attention in the recent past as the result of the successful detection of zones of geological stress in western California in which the fractured material forms significant subsurface channels for groundwater flow sufficient to augment the water supplies of the severely water-stressed communities of coastal California. The same techniques, involving the processing of satellite imagery and structural geological surveys, are being deployed in water-scarce Jordan, where it is hoped that the zones of stress associated with the major rifts of the Jordan Valley and Dead Sea will contain subsurface water (Anderson 1993). Unfortunately, the stress zones being examined are for the most part far below sea level, and in addition ill located vis-à-vis the centers of population and water demand. Moving the Disi groundwater from the southern border to Amman is already a deterringly expensive prospect; to develop water from still deeper levels in the fractured zones is even less likely to be economically feasible.

An alternative to new water: Water provision for more numerous future users by re-allocation

Later sections will deal with the possibilities, much more promising than all those discussed above, of gaining water for the more numerous Arab water users of the future by managing the demand for water. This means essentially re-allocating water from economically water-inefficient applications such as irrigated agriculture to meet the inescapable duty of future governments and water authorities to provide domestic water for the larger populations of the future, especially urban populations in the decades ahead. Fortunately, in addition to the gains for the cities of water re-allocated to them from inefficient farming uses, there are a number of available but expensive technologies for supplying water that could be more widely deployed than they are now, since municipal users can generally pay much more per unit of water than can cultivators.

Taking water from existing systems to supply additional users in other sectors where higher economic and social returns exist will be an increasingly important strategy, but it has not yet entered the policies of Arab governments or the water managing institutions of Arab countries. For the reasons already discussed in the preceding paragraphs as well as in the following analysis of the traditional place of water in the economies and cultures of the region, such policies are difficult to adopt and deploy. For those who consider that new water is the only solution and that the political problems of re-allocation are insurmountable, re-allocation is not yet a relevant option. Those, on the other hand, who consider that serving the interests of as many effective water users as possible is the major issue, see the re-allocation of water as a major feature of their future water policies.

Neither the delivered cost of water nor the capacity to pay for it is uniform

One of the problems facing those who attempt to discuss water in the Middle East at a macro-scale is that there is an inevitable tendency to consider national water budgets and the augmentation of supply as if all water uses could be considered in a single category. However, water can be categorized — or disaggregated — according to a number of criteria associated with its supply, as well as with another suite of criteria having to do with demand. On the supply side water availability can be categorized in terms of volume, reliability, quality, location and cost of delivery, as well as whether it comes as rainfall or from indigenous surface flows, cross-border surface flows, groundwater storage, cross-border subsurface flows, re-used water, desalinated water, imported water as water and finally imported water incorporated in other imported products.

With respect to demand, water can be categorized according to very similar criteria to those relevant to supply. These have to do with volume, reliability (both environmental and political), quality and cost for various uses, locations and sectors. The economic context in which water supplies and water demands are expressed is also relevant.

Intuitively it would be appropriate to match the characteristics of demand to those of the supplies currently available or to those supplies that could be delivered through technological and/or economic interventions. The analyses in Tables 5 and 6 illustrate the different types of water that could be available to states on the supply side as well as the characteristics of the major differentiated water demands in national economies.

Where there are unlimited supplies of water of high quality there is no problem in matching supply and demand. Such circumstances only existed in the past in the Middle East. By 1970 all Arab countries were facing difficulties, in some cases already serious ones such as those in Egypt and Jordan, in matching supply and demand, despite the heroic efforts to augment and control the reliability of supplies through works such as the Aswan High Dam. In practice cheap water is always limited, and new water is almost always going to be expensive or very expensive in arid countries, unless there is some breakthrough in the application of solar energy to water desalination.

The problem lies in the tendency of those responsible for addressing the national water deficits and devising ways to augment national water supplies to approach the water issue as if all water was the same and all users equally entitled to free water. If, however, one approaches the problem of water deficits with the analyses provided by Tables 5 and 6 as a basis for matching disaggregated supplies and disaggregated demands it is clear that no Arab country has a water shortage with respect to its municipal and industrial needs. There is sufficient water, some of it expensive, for non-agricultural uses. In each case in which new expensive water is or will be needed to meet the increasing demand for municipal and industrial water, either it will be able to be re-allocated from agriculture, or the high costs of developing it can be afforded.

The problem lies in the politics. No political leadership nor any traditional high volume users of water are prepared to disaggregate either the supply or the demand for water.

TABLE   5

Water Availability with Respect to Cost of Delivery and Reliability, and Relevant Uses in Arab Countries

		Reliability			Cost of	Water
	Volume	Environmental & Technical	Political	Quality	Delivery	Type
Free or low cost water suitable for all types of use including irrigated farming						
Rainfall	*	*	***	****	*****	
Indigenous surface flows	*	**	*****	****	****	
Cross-border surface flows						
• no control	* to ****	*	*	****	****	
• engineered	* to ****	*****	*	****	****	A1
Cross-border sub-surface flows	* to **	***	****	*****	*****	A2
Groundwater: renewable • from less than 100 m depth	* to ***	(short term) **** (long term) *	****	** to ****	****	A3
Groundwater: non-renewable • from less than 100 m depth	* to *** (long term) *	(short term) ****	****	** to ****	****	A4
Expensive water suitable for water uses able to bear high water charges						
Groundwater: renewable • from over 100 m depth	* to ***	(short term) **** (long term) * *(no long-term role)*	****	** to ****	*	B1
Groundwater: non-renewable • from over 100 m depth	* to ***	(short term) **** (long term) * *(no long-term role)*	****	** to ****	*	B2
Re-used water	*	*****	*****	****	*	B3
Desalinated water	*	*****	*****	****	*	B4
Imported water as water	*	*****	**	****	*	B5
Imported water in other imported products	****	*****	***	*****	*	B6

* very unfavorable   ** unfavorable   *** appropriate   **** favorable   ***** very favorable

T A B L E   6

Types of Water Demand in Arab Countries and the Nature of the Water Demanded

		Reliability				
	Volume	Environmental & Technical	Political	Quality	Cost of Delivery	Water Type
Free or low cost water required ($0.0 to $0.01 per cubic meter)						A1, A2,
Irrigation	*****	****	****	***	*	A3, A4
Medium to high cost water can be used ($0.02 to $1.00 per cubic meter)						
Industry	*	*****	*****	**	****	B1 – B5
High Cost Water can be used ($1.0 or more per cubic meter)						
Municipal use For green amenity						
• in weak economies	*	****	****	**	(****)	B1
• in strong economies	*	****	****	**	*****	B1–B5
For domestic consumption						
• in weak economies	*	****	****	**	(****)	B1–B5
• in strong economies	*	****	****	**	*****	B1–B5

*	Very low	****	High	
**	Low	*****	Very High	
***	Medium	(*****)	Very High but too expensive to deploy	

Water Demand in the Middle East and Northern Africa: Emerging Patterns

Evolving patterns of demand

Water demand and supply are in practice closely linked, because upward or downward adjustments in the demand for water can have a very dramatic impact on the levels of supply required in an economy. The pattern of water demand in the Arab region has changed throughout history, and until the 19th century the available supplies of water were adequate for the needs of the economies and peoples of Arab countries. From the beginning of the 20th century until now, the challenge has been seen as meeting water demands by engineering new water.

For the past two decades the region's leaderships and peoples have been in a phase of continuing to assume that past water demand management practices were appropriate to meet the challenges of the late 20th century and beyond. They have further assumed that new water will be found to meet rising demand, despite sharp evidence to the contrary, such as the region-wide and rapid rise in food imports. Because of this misreading of the position, and possibly more importantly because of the effectiveness of the international food trade and because of the ability of Arab countries to pay for food imports, the issue of water demand management has been neglected. An examination of water demand management options is particularly important because the political economy of water in Arab countries will be increasingly dominated by the development and application of such policies, despite current resistance by political leaderships and vested interests throughout the region.

Water demand in Arab countries is driven by four main factors. The first is the level of population and its rate of increase. The second is the standard of living of the populations of Arab countries and their expectations. It is the third factor that is by far the most significant, however. This is the way that water is allocated 1) between economic sectors: agriculture, industry and domestic, and 2) at lower levels of an economy, at the level of the farm for example, between the different crops. Fourthly, water demand is affected by the efficiency of the systems used to distribute it. A measure of efficiency is the level of unaccountable losses in municipal systems. Losses of under 20 percent are regarded internationally as a reasonable level of management efficiency. In the Middle East and northern Africa unaccounted-for losses of over 50 percent are commonplace.

Measures to modify and improve patterns of water demand

The issue of water demand must be addressed because if levels of per capita use are not modified in Arab countries then the doubling of the levels of water needed in 20 to 25 years to feed, maintain the health and satisfy the amenity expectations of more numerous people will destabilize economies and unnecessarily increase the dependence of economies on imported food.

The consequences of the severe pressure on water resources have been avoided for the past two decades by importing food, and this practice will continue to be the central element in the water policies of the region for the foreseeable future. Figure 1 indicates the proportional dependence of some of the economies of the region on water imported in food. Both the oil rich such as Saudi Arabia and economies with small or no oil revenues such as Egypt and Jordan are dependent on "imported" water in food.

Changing water allocating behavior at the state level, as well as the practices of water use by individual families, will be a major challenge for political leaderships and for concerned officials. The subject is a major one and can only be briefly outlined here. The two major issues to be addressed are the allocation of water at all levels of the political economy of water and secondly the improvement of the systems of delivering and using water.

FIGURE 1

Water Uses of Some Arab Countries Indicating the Dependence on Water Imported in Imported Products, Especially in Food

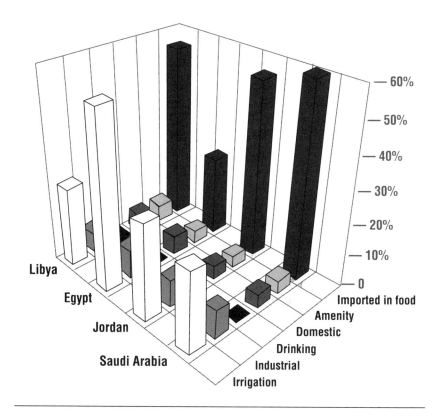

Application of principles of allocative efficiency

Economic principles are impossible to advocate and pursue when the real costs of inputs and their real value to an economy, as well as to individuals within it, cannot be identified. Since water is with few exceptions treated as a free good in the region, and this is especially true in the sector that uses the most water, agriculture, the vested interests in continuing the practice of free water provision are very powerful indeed. As a result it is impossible to have the economic value of water immediately recognized as significant, let alone to establish it as a basis of future water policy. Nevertheless it has to be emphasized that the principle of "returns to water" is a powerful underlying economic principle, which if ignored when a resource is scarce or finite, will return to haunt users in all parts of an economy.

Allocation is a political act and it is for this reason that users and politicians prefer to ignore the issue of the cost of delivering water and especially the consequences of pricing water at its real cost of delivery; beyond doubt, major changes in livelihoods would be involved. Nevertheless international experience urges the adoption of policies based on such principles.

System improvements to reduce water demand
The subject of the efficiency of water distribution in agricultural, industrial and municipal systems is a complex one as evidenced by the discussion above of irrigation water re-use in Egypt. Water infiltrated from irrigated fields can either enter groundwater and drainage systems which provide effective re-use and further re-use, or it may enter a saline aquifer and be permanently degraded and lost to an economy. Water in municipal systems is conveyed through pipes that leak, is used wastefully by consumers, and in systems which are metered it is common for a proportion of the water to be tapped from the system without payment. These unaccountable losses can be over 50 percent; while the data are everywhere very imprecise, it is generally recognized that a 50 percent loss estimate in this region is a reasonable starting point for discussion.

System improvement can be achieved first by investing in more effective distribution systems through replacement and repair of piping, as well as by institutional remedies such as the implementation of water charges based on metered use. This last is a major engineering challenge in itself. The costs of system improvements can deter implementation and the introduction of water charges will always be resisted by users especially where they have been accustomed to receiving free or nearly free water.

A Brief Overview of the Region's History of Water Management
Since antiquity farmers and engineers have been moderating the natural flows of surface water and lifting water from the ground in order first to provide domestic water supplies, secondly to provide water for livestock and thirdly, and in volume terms most importantly, to enable crop production. It is the latter activity that has demanded the most ingenuity and effort because crop production requires large volumes of water, normally over 90 percent of the water utilized by a community enduring an arid climate.

A version of the history of water use in the Middle East and northern Africa is shown in Table 7. It illustrates a number of important develop-

T A B L E   7

Past Water Resource Allocation and Management Practice in the Arab World and Future Options

	Natural Water	Natural Water	Natural Water	Natural Water	New Water	New Water
	Surface storage, and withdrawals from groundwater	Distribution and drainage: Agriculture	Distribution: domestic and industrial	Quality assurance	Waste disposal and water reuse	Desalination
Limited management of water supplies						
Pre-19th Century	No control, some natural surface storage. Traditional groundwater use	Minor canalization. Natural re-use of Nile in agriculture. Traditional systems: qanat and falaj	None	None	Limited: Some careful management at some periods in Cairo	None
Limited supply management — significant advances in distribution and drainage						
19th Century	Minor control works on surface water	Major canalization in late 19th century	Limited piped supplies in part of century	None	Limited	None
Increasing supply management of surface water — beginnings of major groundwater use						
1900-1950	Major dams on Nile constructed. Groundwater development	Major canals and drainage	Major cities provided	Limited	Limited	None
Control of surface water supplies and heavy use of groundwater						
1950-2000	Aswan Dam, and other major dams on Tigris-Euphrates. Widespread construction of minor storage works. Major groundwater use.	Water carriers in Libya (also in Turkey)	Urban provision – including system improvement; reduction of system waste. Beginning of water charging. Efficient systems in agriculture.	Improving	Initiated in most cities, e.g., Egypt. Some rural provision.	In strong economies (Gulf)
Demand management to adjust demand to scarce surface water and scarce and expensive groundwater						
2000-2050	Major dams in upper Niles. Total control.	Additional water carriers	Urban and rural piped supplies	Widespread	Widespread re-use of urban water for agriculture and other sectors	In strong economies

ments, especially during the present century. The evolution of management responses suggests that by the late 20th century the option of augmenting water supplies with new water from the countries of the Arab world was no longer available. New water within national boundaries, either does not exist or could be only a minor source of water for most Arab countries. Meanwhile the surface water supplies in some of the region's most important rivers are likely to be diminished by the use of Nile water by Ethiopia and of Euphrates and Tigris waters by Turkey during the next half century.

The major shifts in the approaches to water development and use in the Arab world during the past 10,000 years came very slowly until a century ago. Within the last one hundred years, however, there have been important technological developments and a progressive rise in population. Moreover, in the past half century the economic fortunes of the different Arab countries have both changed and diverged greatly. Some countries, but not others, can now deploy advanced technologies to lift water from great depths and to move it in major water carriers. With the very important and strategic shift in the Arab economies from being self-sufficient or better in food, to being food importers, the position of water in these economies changed. For a dry country, importing food is, in effect, importing water. And at the time of writing, possibly the most important factor of all is beginning to affect current and planned water allocation and management: an evolving awareness of the real economic role of water and of the need to derive sound and efficient economic returns from scarce water used in national economies.

The evolution of Arab water strategies, their changing contexts, and impending transitions

The major shifts in water management policy and practice during the past century can be summarized in Table 8, which indicates the major issues given priority by those devising water policy as the development context changed. The changes in context include the technological context, the pressure on natural water resource endowments from population size, the ability or lack of ability to substitute for water, as well as the awareness of the economics of water. There are some additional factors, which are as yet poorly understood and even more difficult to operationalize in any comprehensive analysis, not to speak of operationalizing them in the real world of actual water allocation and management. These include the allocation of safe and adequate supplies, maintaining equity in access to benefits from water, and the use of water in an ecologically sustainable way.

T A B L E   8

Major Shifts of Approach in Managing and Allocating Water in the Arab World.

From the management of supply	To the management of demand
Old solutions Increasing the volume and timeliness of water availability: • dams, water storage, canals, pipelines • enhancing natural re-use in agriculture	*New solutions* Reducing waste: • reducing evaporation from surface storages • reducing water leakage in systems • increasing water treatment and the re-use of water
From ignoring the economics of water use in water allocation and management	**To the incorporation of economic principles in water allocation and management policies**
Old and current practices Assumption that water is a free good	*New practices* Attempts to charge for water in all sectors. Recognition of principles of environmental economics.
From traditional inequitable practices	**To the recognition of principles of equity**
Old and current practices Some traditional practices were devised according to principles of equity especially at the local level of allocating and managing irrigation water. The international division of water has generally been inequitable.	*New practices* The adoption of concepts of entitlement to secure supplies of domestic water. The recognition of the value of water and the need to allocate water to sound economic uses. At the international level the adoption of Helsinki and ILC Rules.
From ignoring ecological impacts and the sustainability of water-using practices	**To the recognition of principles of ecological sustainability**
Old and current practices Little recognition of the consequences of water and soil mismanagement	*New practices* The adoption of Environmental Impact Assessment methods

The most important context for future water demand is the growth of population and the predicted demographic transition. Demographers argue, on the basis of the economic and social experience of industrialized economies, and of the more limited experience of developing economies such as those of the Middle East, that falling death rates and declining birth rates will at some date in the 21st century bring the region's population to a low, or even a zero rate of growth. Until this "transition" is complete at some unknown future time, the economies of the region will have to arrange for the supply of adequate water for the domestic and industrial needs of a continuously growing population, albeit one growing more slowly as time passes. It will also be necessary to supply sufficient food to increasing numbers of people, partly from the region's own agriculture, but mainly by food imports paid for by revenues earned in the industrial and service sectors.

The shape of these future demands is little understood. Meanwhile the responses to the initial economic, social and political challenges of adjusting to the water deficit reflected in food imports have been predictably cautious and attended by anxiety on the part of officials and politicians and a great deal of wild prediction with respect to the capacity of indigenous water to meet current and future demands.

The scale of the problem can be illustrated, if not precisely defined, by considering the likely increases of population and related food demands of the region in relation to past trends in water use. Figure 2 indicates the shape of past and future water use and puts into perspective the capacity of the region's water resources to meet the real water needs of the last decade of the 20th century and those of the 21st century.

Those responsible for water resource allocation and management will have to participate in a number of transitions. The first, and by far the most important one, is the demographic transition. The increase in the demand for food signifies an implied demand for the water required to produce that food, water which cannot be supplied within the region. The food demands can only be met by mobilizing economic activity that will generate revenues with which to purchase food on the world market. The

FIGURE 2

Estimated Levels of Water Demand and Use in the Arab Region 1900–2080.

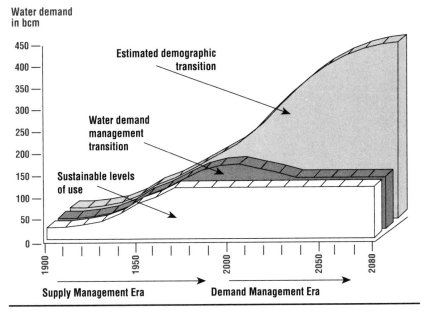

region will increasingly have to substitute for water by this economic device, assuming, of course, that there is no technological breakthrough to make water available cheaply. It is likely that by the later part of the 21st century at least twice as much water will be imported in food as will be withdrawn from the region's surface and groundwater resources.

The second transition is the one away from supply management, which has been the dominant approach since the beginning of the 20th century. The policies and practices of supply management cannot be economically expanded in the future in any Arab country, although there are a number that are pursuing unsound strategies in developing new water supply for applications that are made economically unviable by the use of such water. Arab countries are experiencing the end of the supply management phase and are entering what can be termed the transition to demand management.

Demand management has begun to preoccupy those advising on the water sector and lending money to it. Demand management addresses the need to improve the efficiency of water use in all sectors by introducing instruments and regulations that draw attention to the value of water. The approach also steers users to allocate water to sectors and activities that achieve the best returns to scarce water. Political resistance to demand management procedures has meant that they have not been adopted comprehensively by any Arab government, although a number have begun to introduce some instruments such as charges for water in, for example, Jordan, Tunisia and Morocco. The transition to demand management will require very substantial investments, since the technologies for municipal water re-use, industrial recycling and water treatment are initially expensive, as is the rehabilitation of defective municipal water distribution infrastructures.

The most important feature of Figure 2 is the relative significance of water management policies within the regional resource management economy. The successful development and implementation of demand management will greatly occupy governments and international agencies, as well as those responsible for installing the required new physical and institutional infrastructures. Nonetheless, its contribution to the real water needs of the region will always be subordinate to the measures taken within the regional and the international political economies to ensure the importation of water, mainly in food produced in other areas where water is available and cheap.

Demand management in the national water economies of the Arab world

The overall perspective on past and future water demand and supply generalizes a diversity of individual country circumstances. The only Arab country with large surface water resources, Egypt, has, as already indicated, a remarkably effective "natural" system of irrigation water re-use and is making a start at urban re-use. However, as urban and industrial water use is only a small proportion of total water use (10–15 percent) "new" water from recycled urban and industrial water will never be a major proportion of the Egyptian water budget. The recycled water would, however, be as important as other new water that Egypt aspires to mobilize, such as that which would come from a completed Jonglei Scheme. In practice the recycled water is certain to become available while the Jonglei water is very unlikely to do so, but it will only represent a very small proportion of the current Egyptian water gap and an even smaller proportion of its future much larger water gap. Egypt conforms to the regional scenario suggested by Figure 2 in its increasing dependence on food imports to fill its water gap.

The most extreme water allocation and management problems arise in the countries of the Jordan catchment where the overuse of water in the agricultural sector has led to drawdowns in groundwater and the deterioration of groundwater quality. In these circumstances all the current users as well as those of the putative Palestine are adopting and will increasingly adopt demand management practices. They will also re-allocate water between sectors, first to reduce water consumption and secondly to improve the levels of returns to water. Jordan and neighboring Occupied Territories/Israel have substantially less than half of the 1,000 cubic meters per head per year needed to be self-sufficient in food, and will have much less in future. Existing demographic trends as well as politically stimulated immigration would exacerbate the situation. The re-use of urban water for irrigation will be an important element in future water policies whatever the political boundaries. It has been calculated that the future water requirements for the 10 million people who could be living in the areas of Israel and the Occupied Territories would require only one bcm of water per year for domestic purposes if their use was to be demand-managed to the level of 100 cubic meters per person per year. If the one bcm of such water use was to be effectively recycled and made available to irrigated farms and projects, up to 700 million cubic meters of water per year could be delivered to the agricultural sector, which is over half that which is being used currently. While neither the existing volume of water nor that which could be added by recycling would be sufficient to meet the food needs of the 10 million future inhabitants of Israel and Palestine —10 bcm of water would be needed to achieve such self-sufficiency— the proposed scenario involving substantial, or even total, recycling of water used initially for domestic purposes would be a rational approach to maximizing the utility of the scarce water of these economies.

While such proposals have much to recommend them in the longer run, the immediate challenge to those wishing to promote agreement over water use in the Jordan catchment in the context of the Peace Talks is to address the accumulated resentment, especially of the Palestinian community, as a result of the opportunist hydropolitical interventions of Israel during the past five decades. Compensation is anticipated as well as future equitable distribution.

At the time of writing, 1994, there was little evidence of demand management practices in the Arab countries of the Jordan catchment. In Jordan opposition to the concept of charging for water in the irrigation sector was loud and politically persuasive. Charges for water were being levied in the tracts served by the Jordan Valley Authority but these were only nominal (Salameh et al. 1993), and could not be regarded as a significant instrument to steer farmers to water conservation practices or to concentrate their attention on the need to allocate water to crops which could bear the real costs of such water. On the other hand the policy of urban water management in Jordan was, in one aspect at least, beginning to be shaped by sound principles. Progressive charges were being levied for domestic use in the cities and towns of Jordan such that higher levels of water consumption were charged at a higher tariff than the basic supply. These tariffs were being reviewed and charges for water were increasing (Salameh et al. 1993). In the area of urban water conservation the position is unclear and to date water charging has not had a significant impact on the level of unaccounted-for losses in the urban distribution systems of Jordan. Such data are extremely difficult to assemble and the figure in popular currency of 50 percent of unaccounted-for losses in the urban systems of Jordan is a very soft number. Water charges alone will not bring about improvements to the level of the commonly adopted target of 20 percent losses and it will be necessary to invest substantially in the rehabilitation of the distribution infrastructures as well as to change users' behavior. Where the necessary investment funds should come from is a controversial issue. Notions of structural readjustment that drive the politics of many international bodies would suggest the consumers should pay for both the capital and the operational costs of services such as water provision. Such a policy is not politically feasible in the short term and it will be necessary to fund infrastructural improvements through government-generated loans. In economies lacking the capacity of the oil rich, the role of the international agencies in promoting sound water demand management practices will be very important indeed.

The other Jordan catchment state, Syria, has not shown any evidence of being influenced by principles of demand management. The 1967 War denied Syrian access to the Banias source of the Jordan, a flow which Syria had planned to divert to the Kingdom of Jordan. During the years since the 1967 conflict Syria has made more and more use of the waters of the upper Yarmuk catchment for irrigation, based on the impoundment of minor tributary valleys. The impact on the flow of the Yarmuk has been substantial and has changed the hydrological options of downstream Jordan and Israel. As yet, Syria has no incentive to adopt principles of demand management of its Yarmuk water as it has not exhausted the supply management opportunities on the Yarmuk, which to date have been judged to be economically viable. Meanwhile, since the negative economic impacts of such management are felt only by downstream neighbors, the need to consider alternative policies has no urgency in Syria.

Other Arab countries with major surface water resources, those of the Tigris-Euphrates basin, have encountered major difficulties in developing the waters for irrigated crop production. The reasons are environmental, having to do with the landforms, soil formation history, and the related tendency to water-logging and salinity. Environmental circumstances are very different from those in the Nile valley. In addition there are institutional issues that have also proved to be intractable. Both Syria and Iraq have policies that reflect their agricultural expectations based on the further supply management of water. Evidence of the past three decades suggests that the utilization of additional water in irrigated farming will advance slowly, especially in the enduring constrained economic circumstances of Syria, as well as in Iraq, which has moved from the position of not needing to develop an economically viable irrigated sector when oil revenues were plentiful up until 1990, to the current position where it too has limited options with respect to development spending. Urban water supply should not be a problem in either country, but in Syria urban water had become a problem in the second half of the 1980s, and by the mid-1990s the management problem had not been successfully addressed. In Syria there is an urgent need for every type of institutional and engineering instrument to be deployed to encourage conservation by consumers and system improvements by responsible authorities. The adoption of demand management principles is not a high priority in Syria or Iraq as they still enjoy the availability of supply management options.

The oil rich Arab countries have been able to substitute energy for water to cover the modest demands of domestic users by mobilizing cheap energy to convert seawater and brackish water. The demands of municipal and industrial users have also been met by desalination technology. Such provision is at the high cost end of the supply curve (see Rogers, in this volume) and is an option only available to economies with a strong flow of revenues such as those from oil. All of the oil rich governments have searched for new groundwater, and in the cases of Saudi Arabia and Libya, ambitious projects have been installed to utilize and transport such water. The principles of demand management are nowhere evident in these countries as the expensive water is being used, or is destined, for the irrigation of crops that command a lower price than the delivered cost of the water applied.

The Maghreb countries are among the most progressive in the adoption of demand management practices. Charges are levied for water in the cities of Tunisia and Morocco. Irrigation water is increasingly charged in Morocco and charges of $0.10 per cubic meter were being accepted by farmers to gain access to new groundwater developed by government departments (Wild 1994). The Maghreb countries are leading the way in experimenting with new engineering and institutional infrastructures and their experience will point out directions for the less favored water scarce Arab countries of the Middle East.

Any discussion of demand management in the Middle East will tend to become outdated quickly. Perceptions are changing rapidly and while political inertia is real and has determined a slow pace of adoption of new approaches, policies and practices, there is sufficient evidence that beginnings are being made to adapt economic policies to economic realities even if at the moment the public statements of political leaders and officials belie this trend. By the end of the century major engineering investments will be underway to implement demand management policies and the management of urban water leading to wastewater re-use will be increasingly common. Re-allocation of water from agriculture will continue and there will be modest compensating volumes of water being pumped to the agricultural sectors from urban re-use. Meanwhile opinion is by no means monolithic on the issue of water. Those with vested interests in continuing to have access to free, or to very low-cost water advocate predictable arguments, but there are increasing signs that there are influential individuals ready to develop newer policies and to implement them, including the adoption of principles of demand management when the political circumstances are ripe.

Political and Institutional Issues Affecting Water Allocation and Management

By way of conclusion, let us briefly consider a critical social factor that provides a persuasive explanation of why current policy and practice on water are as they are in the Middle East and North Africa. The analysis is especially useful in explaining why leaders and officials in Arab countries address partial agendas, and avoid issues of crucial importance if strategies are to be adopted that are economically and ecologically sustainable.

The analysis requires that the major problems of allocating water be identified and then differentiated in terms of the sectoral and institutional contexts they affect, and especially in terms of the interests of users and beneficiaries of water. Trying to understand why there are preferred as well as unpopular policies on water, we are led to consider the levels and kinds of divisiveness engendered by the policy options. Table 9 is a first attempt to gain insights into why there are filters that prevent the consideration and development of comprehensive and sound policies.

Table 9 groups the problems that should be addressed according to whether they have or have not been taken into account by the leaderships and political elites of Arab countries. The first four problems have been addressed, sometimes for many thousands of years, and generally speaking have generated little controversy and little political heat. Where the promotion of these policies requires that investment budgets be allocated to one project or another, there could be conflict at whatever level in the political process the competition for resources was expressed. But the overall inspiration and direction of policy in addressing the first four problems is clear and generally perceived to be for the general good.

The last two problems are, however, very differently viewed by policy makers because they require steps that would generate resistance and hostility among users, directed at policy makers and the implementers of policy. A very important factor in the political economy of water allocation and management in Arab countries is that there are only very rudimentary, and by no means comprehensive, market mechanisms to help users, not to speak of leaders, legislators, officials and engineers, to understand the value of water. As a result any changes in policy have to be based solely on political arguments and political processes without the reinforcement of preferences expressed by well-informed users in market places or through democratic institutions. The leaders and governments of Arab countries

TABLE   9

Activities Allocating Scarce Water, with Political and Institutional Consequences

Problem	Arena and institutions affected by the problem	Politically divisive or not – in current information circumstances
Issues addressed by political elites, government institutions and international agencies		
1 International "share"	International relations	Non-divisive nationally (**conflictual abroad**)
2 Finding new water	National: public bodies involved in search; also international agencies	Non-divisive
3 Developing new water	Public and private bodies & international agencies	Possibly **divisive** because of competition for benefits
4 Using existing and new water as effic- iently as possible (Productive efficiency)	Public and private bodies & international agencies	Non-divisive
Issues not addressed by political elites and government institutions although international agencies may show concern		
5 Efficient intersectoral water allocation (Allocative efficiency)	Private interests, the political elite & government institutions; also international agencies	**Extremely divisive** because there would be losers and these losers are powerful
6 Introduction of engineering and institutional (economic) instruments to improve the efficiency of water use according to sound principles of demand management	Private interests, the political elite & government institutions; also international agencies	**Extremely divisive** because major changes in user's behavior would be required

have been able to meet food (and water) gaps by importing food, and as a result have been able to avoid addressing the two most important approaches needed to deal effectively with the urgent and increasingly important problems of achieving sound water allocation and management. If one were to re-order the issues listed in Table 9 in terms of their relative importance for the future economic and ecological security of Arab countries, problems five and six would be the first, rather than the last, to be addressed.

In the introduction to this review it was suggested that remedies for the economic distortions and destabilization of economies and ecologies of Arab countries caused by the mis-allocation and mis-use of water would be addressed during the coming decade. The period will be a transitional one in which water allocation and management policies will be re-oriented. It has been the purpose of this discussion to promote and accelerate this essential process.

4

The Central Region: Problems and Perspectives

Yahia Abdel Mageed

Introduction

The Arab region is one of the most dry and water-scarce zones in the world. Its average water availability is 1,700 m³ per capita, which may be compared to the world average of 13,000 m³ per capita. Although it spans a wide diversity of climates, extending from equatorial humid tropics in its southern tip through Mediterranean weather to the temperate zone in its north, 40 percent of the land area is desert.

The water scarcity condition that characterizes most of the Arab region continues to pose a serious constraint on development and the environment. At the dawn of history the region witnessed spectacular ancient fluvial civilizations in Egypt in the Nile delta, and in Mesopotamia on the banks of the Euphrates and the Tigris, but it has also experienced the destruction of land through overuse and failed management of irrigation. Clive Ponting in *A Green History of the World* describes the rise and fall in the latter basin of the Sumerian empire, the first civilization to develop writing. Huge canals siphoned off considerable amounts of water to irrigate vast expanses of land. Continuous forest clearing for the expansion of irrigated agriculture induced soil erosion, local climatic changes, reduction of rainfall and desertification. In addition, the irrigated lands suffered from impeded drainage and rising water tables, eventually leading to the collapse of environmental stability. If one visits the Mesopotamian desert now it is hard to imagine that these very soils supported the Sumerian civilization 4,500 years ago.

Since the beginning of this century, the Arab region, due to its strategic location at a crossroads of the world, became a stage of complex, large-scale political and economic interventions, particularly after the construction of the Suez Canal and the discovery of huge fossil fuel reserves. Water was an important tool and played a pivotal role in these interventions. Today, the water sector poses a constraint to development and carries very large potential risks for the future, not only from the difficulty of satisfying rising and increasingly diversified demands for water, but from the vulnerability of the environment and the water resource base to damage from the misuse of water.

In recent history, the Arab region has witnessed major projects to harness its water resources to meet the increasing needs of its populations for food and for industrial and domestic water supplies. These projects include a number of impoundments in the major basins of the region, spectacular among them the High Aswan Dam, the Great River project in Libya, which extracts fossil ground water reserves and transfers them to the coastal areas, and the Jonglie Canal project — uncompleted due to circumstances in southern Sudan — aimed at the long distance transfer of water from the Sudd region to the water-scarce zone in the north of Sudan and in Egypt.

The Arab region's population growth ranks among the highest in the world. Present population, estimated at about 200 million, is expected to reach 400 million by the year 2010 and 757 million by the year 2030 using the present natural rate of growth of 2.5 percent per year (Arab League 1982). The region is now witnessing huge demographic changes. Urban and industrial centers along the river banks and the coastal areas are growing at enormous rates. It is estimated that almost two-thirds of today's population live in urban and industrial conglomerates, exerting heavy pressure on the environment and straining water supplies and sanitation facilities. Irrigated and rain-fed agriculture have expanded into marginal lands, and due to the absence of adequate infrastructure and the increasing population pressure, pastoral systems have broken down in many areas in the region. All these developments are leading to serious environmental consequences, likely to disturb the resource base and the sustainability of production. The annual food gap in the Arab region increased from $2.5 billion in the mid-1970s to $13 billion in the mid-1980s (AOAD 1986). Food imports in 1990 were reaching $23 billion per year, taking into consideration increased consumption and inflation, according to the Tripoli Declaration of The Ministerial Council of the Arab Organization for Agricultural Development (AOAD).

Following the 1973 confrontation in the Middle East and the sharp rise in the price of petroleum products, food self-sufficiency and food security became catchwords for most of the region. The central region, particularly

Sudan, came to be seen as a focal area where untapped land and water resources offered vast opportunities for the investment of Arab funds to realize goals of food security and self-sufficiency for the Arab world, but so far this has not been achieved.

The Central Region

The Arab central region includes Egypt, Sudan, Somalia and Djibouti. Unlike the other regions, the central region is composed of two distinct sub-regions, separated by the Ethiopian plateau. The Ethiopian plateau is also the main source of water to both sub-regions through the Blue Nile, Atbara and Sobat rivers of the Nile system, and the Wadi Shebelli and Juba systems. In geographic, climatic and hydrologic features the Somali/Djibouti sub-region relates more to the southern part of the Arab Peninsula than it does to the Sudan/Egypt sub-region. For the Somali/Djibouti sub-region, population interactions across the Gulf of Aden are dominant, while the Ethiopian highlands act as a barrier against such interactions with the Sudan/Egypt sub-region. For these reasons, this article concentrates on Sudan and Egypt. The water supplies of both sub-regions, apart from direct rains, are transboundary resources shared among riparians.

Central region land and water resources

The central region constitute 30 percent of the land area of the Arab region and has a population of 85 million people, 38 percent of the total population of the Arab region, as shown in Table 1.

Table 2 summarizes the water resources of the central region in relation to Arab region totals. It is, however, cautioned that these global figures be taken as indicators rather than as quantities of water reliably available for

---

TABLE   1

Land Resources (in million hectares)

	Land Surface	Agricultural Land	Forest	Pasture
Arab Region	1401.4	198.2	130.3	174.2
Central Region	416.8	72.51	99.8	55.0
Egypt	100.2	4.5	—	—
Sudan	250.5	59.0	91.0	24.0
Somalia	63.5	9.0	8.8	29.0
Djibouti	2.3	0.01	—	2.0
Central Region as a Rounded % of Arab Region	30.0	36.0	77.0	31.0

Source: AOAD 1986.

T A B L E   2
Water Resources of the Central Region (in bcm)

	Rainfall	Surface Water	Ground Storage	Renewable Ground Water
Arab Region	2211	295.60	7730	42
Central Region	1304	91.85	6039	8.8
Egypt	15	*60.50	6000	4.5
Sudan	1094	*23.00	39	1.0
Somalia	191	8.15	—	3.3
Djibouti	4	0.20	—	—

Source: Khouri et al. 1986, pp. 575-634. modified.

use. Agriculture accounts for 80 percent of the use of this water, and agricultural use, whether rainfed or irrigated, cannot be based on average annual yield, but only at best on an 80 percent reliable year, meaning the amount of water that can be relied upon to be available in four years out of five, which is very much less than the average. The quantities of water above are not always available in the places and times when they are to be used.

The asterisked figures in Table 2 represent the surface water that reaches Aswan; they include the 10 bcm that are lost by evaporation in the High Aswan Lake. When that subtraction is allocated equally between the two countries, 55.5 and 18.5 bcm are available for use by Egypt and Sudan, respectively, according to the 1959 Nile Waters Agreement. It is planned, however, to increase the yield of the Nile by about 25 percent through conservation projects in the Egypt/Sudan sub-basin region to meet future demands.

In many studies and documents it is estimated that the average annual water resource of the Arab region is 338 bcm, of which 296 bcm are surface water and 42 bcm are renewable groundwater. The present water use is estimated at 172 bcm, of which 140 bcm are from surface water. The demand in the year 2030 could reach 435 bcm with a gap of about 100 bcm (Fadda 1989).

If we take an 80 percent reliable year, the gap will be much bigger than this figure, although this does not apply to the Nile since over-year storage in the High Aswan Lake guarantees the average yield of the Nile of 84 bcm at Aswan.

While the central region remains the richest in terms of land and water resources within the Arab region, it is the poorest in terms of GNP per capita, ranging between $600 in Egypt to $120 in Somalia, compared to an average of $1720 in the Middle East and North Africa as a whole (World Bank 1992).

Egypt and Sudan are the principal downstream riparians in the Nile basin (shown schematically in Figure 1); the Nile River itself is the main source of surface water supply to both of them. Sudan is the meeting place of all the Niles originating in the Ethiopian highlands and the equatorial lakes beyond the borders of Sudan and Egypt. The river is unified in Sudan to become the main Nile before crossing Egypt to the Mediterranean Sea. Over the years, despite natural barriers such as the six cataracts in Sudan's north and the Sudd swamp region in the south, the waters of the basin have brought about significant political, cultural and economic relationships between Sudan and Egypt on the one hand, and between these two countries and the higher riparians on the other hand. Water use patterns and trends, and indeed economic development, in the basin have to a great extent been influenced by the hydrological characteristics of the river system, and by political and economic events and circumstances that prevailed in the basin at different times in history.

The basin's natural characteristics

The Nile basin has a total area of 290 million hectares; its monthly and annual natural discharges vary considerably. The annual discharge that reached Aswan in normal years (1912–1955) amounted to 84 bcm, based on the following contributions:

Blue Nile	59%
Sobat	14%
Atbara	13%
Bahr El-Jebel	14%

The contribution of the Ethiopian plateau amounts to 85 percent of the flow of the Nile at Aswan. Out of the annual flow of other waters totaling 46.6 bcm entering Sudan (equatorial lakes: 34.6 bcm, Bahr El-Ghazal system flow : 12 bcm), only 15 bcm reach the White Nile. A total of about 34–40 bcm form the vast swamps of the Sudd region of Bahr El-Jebel and Bahr El-Ghazal, which are largely lost by evaporation and evapotranspiration.

The annual discharge at Aswan in the present century varied from 104 bcm (1946) to 42 bcm (1913). This sharp variation is mainly due to variations in the river systems originating on the Ethiopian plateau. The seasonal variations of the Blue Nile for the period 1912–1982 range from 6200 cubic meters per second (cumecs) during flood to 125 cumecs in the dry season. The White Nile system's much lesser seasonal variations range from 525 cumecs to 121 cumecs, and this is mainly due to routing effects

FIGURE 1

Schematic Representation of the River Nile Basin

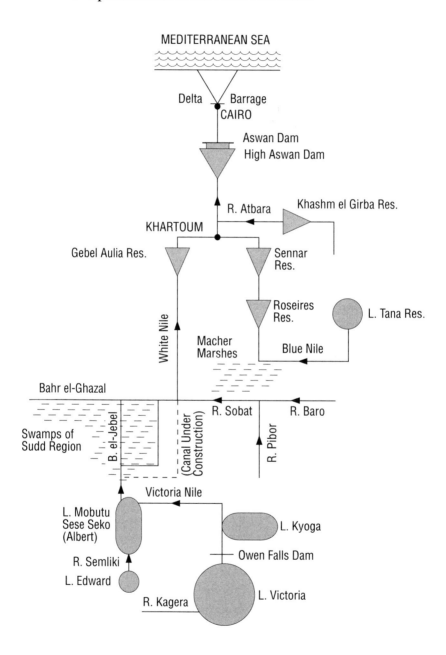

at the equatorial lakes and the Sudd region. Out of the total annual flow at Aswan almost 65 percent is contributed during the flood time, August to October.

Annual rainfall in the central region ranges from 1500 mm in the southern tip to nil in the north, except for the narrow coastal area in the northern tip of Egypt and the Red Sea coastal area that receive some winter rainfall. The duration of the rainy season ranges from 7 months in the southern Sudan to 3 months in the central Sudan. Direct rainfall is an important source for forests, pasture and rainfed crop production in the Sudan.

Over the years these natural characteristics have had a marked influence on the patterns of water management and development in the region, which have aimed to meet the increasing and diversified needs of the population. Egypt, being the most downstream state and having no other sources of water supply, suffered the most from the annual and seasonal fluctuations of the river. Starting millennia ago, basin irrigation was the earliest form of irrigated agriculture, practiced by diverting flood waters to the low-lying fertile soils of the delta through a system of canals and dikes. This practice was later transferred to north of the Sudan. However, it did not modify the devastating floods of high water years, nor augment the low flood years that result in shrinkage of the irrigated basins and a consequent shortage of food, two dangers that have always threatened Egypt. These were the problems that tempted the Egyptians to penetrate into the upper reaches of the basin in search of flood control and storage sites to meet their increasing needs for water. Over the years annual storage reservoirs were built to store water from the tail of the flood to augment the low dry season flows. Multi-year storage came into being with construction of the Aswan High Dam to guarantee the average annual flow of the river. Since the beginning of this century plans were also in hand for conservation works in the Sudd region of the Sudan to reduce the evaporative losses there and thereby increase the yield of the river.

Development and hydropolitics

Following the fall of the Turkish rule in the Sudan in 1885, the basin became a theater for European interventions and domination that continued up to the middle of this century. There was a division into zones of influence. Most of the basin — Egypt, Sudan and the East African countries — fell under the influence of Great Britain. At the turn of the century, Great Britain concluded three agreements that applied only to upper riparians of the basin. They prohibited any construction without

British consent on the tributaries of the Nile above the Sudan and affecting the flow of that river. These included the 1891 agreement with Italy, the 1902 agreement with Ethiopia and the 1906 agreement with the Congo Free State (Teclaf 1967). Sir William Willcocks, the famous Nile architect, declared in 1903 that with the aid of concrete, steam and electric power, of huge dredges and dynamiting equipment, more could be constructed in two decades of the modern era than during an entire dynasty of rulers in antiquity employing corvée labor and armies of prisoners. Already in 1890, Willcocks had planned the Aswan Dam on the Nile for irrigation and navigation, one of the first multi-purpose projects in the world. The Aswan Dam, completed in 1902, marked the beginning of expansion of modern irrigation in Egypt. It is an annual storage dam arresting part of the flood in order to increase river yield during the dry season. The main goal of the British policy was to expand cotton production for the textile industry in Britain.

Following the reconquest of the Sudan in 1899, further plans were made for the expansion of irrigation, mainly for cotton production, both in Sudan (the Gezira) and Egypt. These plans included heightening the Aswan Dam, providing storage facilities for Egypt in the Sudan at Jebel Awlia on the White Nile, and building Sennar Dam on the Blue Nile for irrigation of the Gezira projects. The plans were interrupted by World War I, but immediately after the war were put into motion. The set of projects necessitated new arrangements for apportioning the Nile waters between Sudan and Egypt, for examining plans of basin-wide significance that emerged for over-year storage in the equatorial lakes and Lake Tana, and for providing for flood control in the Sudan, as well as for assuring timely water for Egyptian needs. To these ends, the 1929 Nile Water Agreement was concluded between Egypt and Great Britain, the latter acting on behalf of Sudan and the East African countries.

The working arrangements based on the agreement provided Egypt with full rights over the whole natural flow of the river during the dry season from January to July, with only very limited withdrawal by the Sudan permitted during this period. The agreement also stipulated that the East African countries would not construct any works in the Equatorial Nile without consulting Egypt and the Sudan. The British government used its good offices to facilitate the establishment of over-year storage in the equatorial lakes linked with the conservation projects in the Sudd region to increase the Nile yield. It gave the Egyptians facilities in the Sudan and in the East African countries to gauge the river and oversee its control.

This agreement limited the development of irrigation in the Sudan to the Gezira project, where cotton is the main crop over an area of 400,000 hectares at a cropping intensity of 50 percent. During the Korean War's sharp increase in cotton prices, further irrigation areas were developed totaling 160,000 hectares in the Blue and White Niles, drawing only on flood waters from July to the end of December. This situation created a potential conflict between Sudan and Egypt in the apportioning of Nile water. In the period of the independence of the Sudan, this question became over-politicized, particularly because there were political movements in Sudan calling for unity with Egypt and others favoring an independent Sudan in coalition with Britain. The Nile water question became an important issue in creating rivalries between the two countries, particularly after the 1952 revolution in Egypt.

This was a deliberate British policy. The 1929 agreement should not have stood in the way of Sudan's irrigation expansion. It is known that the irrigation development potential of the Sudan is in the vicinity of the Blue Nile and Atbara rivers, and no appreciable development can be made without provision of storage dams to store part of the flood waters for use during the dry seasons. Therefore, it was not the 1929 agreement that restricted such development, but a deliberate delaying policy. Up until the time Egypt and the Sudan re-resolved the Nile question in 1959, the Sudan had established rights to only 4 bcm, compared to Egypt's 48 bcm, while 32 bcm of the flood water found its way to the sea.

After Egypt's 1952 revolution, the concept of over-year storage in Aswan emerged, seeking to control the full flow of the river and to gain 32 bcm of water otherwise lost to the sea (actually a net benefit of 22 bcm, since 10 bcm is lost to evaporation from Aswan Lake). The Aswan High Dam project, despite many unjustified criticisms, has resolved the conflict that the 1929 agreement progressively created between the two countries by lifting the restriction on abstraction during the dry season and guaranteeing an average flow of the river of 84 bcm. The 1959 Nile water agreement between the two countries divided the net benefit of 22 bcm, 14.5 bcm to Sudan and 7.5 bcm to Egypt, making their shares 18.5 and 55.5 bcm, respectively. Under the agreement there was formed the first joint commission between the two countries ever to be established in the basin, laying down the foundation for basin-wide cooperation. The agreement also recognized the riparian rights of the other basin states.

1959, during the Cold War era, was part of a period extremely politicized by issues having nothing to do with water, although water issues were used to deepen the rivalries between the basin countries as dictated by the world politics of that time. The 1959 Nile waters agreement resolved the most critical conflict in the basin between the two downstream countries, Sudan and Egypt. Since its conclusion 34 years ago, no harm has been inflicted on the upstream countries of the basin. Present withdrawals in the territory of upper riparians are not felt downstream. There are plans, however, in the future for irrigation of 130,000 hectares in Uganda, 57,000 in Kenya and 200,000 in Tanzania in the catchment of the White Nile system (UNDP 1989). In Ethiopia, according to a 1984 FAO study, the potential for irrigation identified in the Blue Nile basin includes 100,000 hectares of perennial irrigation and 165,000 hectares of small-scale seasonal irrigation. The irrigation water requirements in these humid zones are very much less than those in the arid and semi-arid zones in Egypt and the Sudan. Therefore, this should not form an obstacle to basin-wide cooperation.

Discussing the problems of water in this sub-region in the context of water in the Arab region should not obscure the merits of considering the water in this sub-region in its basin-wide context, although the geography of the southern part of the basin leads away from the Arab world. The basin is a hydrologic unit and an economic unit. It has vast potential for socioeconomic development and for increasing the well-being of the basin societies, particularly in the agricultural sector, which is the primary sector for economic growth in most of the basin countries. The overall Nile basin has a potential to contribute towards the Arab region's food requirements and to offer other agricultural raw materials. The basin has vast untapped water and water energy potential in its upper and lower reaches awaiting investment to mobilize the agricultural resources for food self-sufficiency within the basin, and to create food surpluses to meet market opportunities readily available in the Arab region.

Egypt and Sudan have almost utilized their shares allocated by the 1959 Nile waters agreement between them. Egypt's present water use and availability in bcm is shown in Table 3.

T A B L E   3

Egypt's Uses of Water, Compared with Availability (in bcm)

Water Use		Availability	
Irrigation	49.7	Nile	55.5
Municipal	3.9	Groundwater	2.6
Industrial	4.6	Agricultural Drainage	4.7
Navigation/Regulation	1.8	Deep Fossil Waters	.5
		Treated Municipal	0.2
TOTAL	59.2	TOTAL	63.5

Source: Abu-Zeid and Rady 1991.

By the year 2000, Egypt's water requirements will be about 69.4 bcm. In the Sudan, the total committed use from the Nile waters up to 1990 is 16.17 bcm from its share in the Nile waters amounting to 20.5 bcm (18.5 bcm at Aswan). Most of the projects that were planned during the late seventies and eighties, including heightening of the Roseires Dam on the Blue Nile, the Setit projects, and associated expansion in irrigated areas including Rahad II and Setit irrigation projects (totaling 400,000 hectares) were postponed due to economic difficulties.

According to these plans in Sudan and similar plans in Egypt, both countries went ahead to implement the Jonglie Canal project in the mid-1970s, the first conservation project in the Sudd region, which aimed to increase the river yield by 4.0 bcm at Aswan. However, with almost two-thirds of the canal completed, construction was suspended in 1983 due to disturbances and political conflict in the southern Sudan.

The Sudd region in the Sudan is the largest fresh water swamp in the world, where almost 50 percent of the waters originating in the equatorial lakes (totaling 14 bcm), are lost in the Bahr El-Jebel swamps. Almost all the rainfall in Bahr El-Ghazal, amounting to about 15 bcm, is lost in the Bahr El-Ghazal swamps and never reaches the Nile. Over 12 bcm are lost in the Machar marshes from the Sobat system and the eastern torrents originating in the Ethiopian highlands. Since the beginning of the century, there have been plans for an Equatorial Nile Project such as the Jonglie Canal providing conservation works to reduce losses of the water that is needed to meet the increasing needs in the arid and semi-arid north. Egypt and Sudan have plans to ultimately increase the Nile yield by 18 bcm to be divided equally between them, according to the 1959 agreement.

However, these conservation projects have complex political and environmental implications within the Sudan and for the upper riparians. Within the Sudan, development in the south very much depends on sound environmental management of these swamps. The water conservation projects within the swamps to increase the Nile yield will ultimately create an environment conducive to development in the south. As for storage works required in the equatorial lakes on the territory of upper riparians, the merits of such projects for them need to be justified within a system of very strong Nile basin cooperation. Despite all these complexities, the Sudd region's waters remain a promising source to meet the increasing needs of the basin and particularly those in Egypt and Sudan.

Development and the environmental challenges

Since the conclusion of the 1959 Nile waters agreement, Egypt and Sudan have embarked on major programs to control the river and to expand irrigated agriculture and hydropower generation.

In Egypt, with the completion of the Aswan High Dam, the land reclamation program that started in 1953 gained momentum. This was coupled with a deliberate policy of industrialization utilizing the 2100 MW of hydropower that became available from the high dam. The land reclamation program included expansion of irrigation by about 1 million hectares in areas outside the old congested valley lands. These extensions are planned all over the country, in the east Delta and Sinai, middle Delta, middle Egypt, upper Egypt and the New Valley. Water needs for these new areas will be met by waters from the high dam, by the reuse of drainage waters, and in limited amounts from groundwater in the New Valley and the coastal areas. These sources are in addition to waters that will become available from the conservation projects in the Sudd regions in southern Sudan.

Since the completion of the high dam in the mid-1960s, this land reclamation program has met with many constraints and difficulties. They were caused mainly by the war with Israel and the economic crisis associated with it, which has effects remaining to our present time. The total area reclaimed to date is about 400,000 hectares, mainly in the western Delta. The same period witnessed the growth of enormous urban and industrial conglomerates in the old lands and associated demographic changes. Agriculture's share of the GDP fell from 34.3 percent in 1955 to 20 percent in 1990, while its share of employment fell from 56 percent of the population to 10 percent over the same period (Abu-Zeid and Rady 1991). Land loss to top soil skimming and urbanization averaged 12,000 hectares per year. There is still a big gap between the areas reclaimed and those cultivated due to constraints on settling farmers, and a major portion of the lands have not yet reached target production. Due to intensive cropping, Egypt's old lands are suffering from water logging and salination, and marked decline of productivity.

It could be concluded that the High Aswan Dam has certainly provided flood protection for Egypt and a safe water supply in the summer, making a double harvest possible throughout the country. It has also provided inter-annual storage to boost lean years. However, hopes for the irrigation of additional land have been only partly fulfilled (Meybeck, Chapman and Helmer 1989, pp. 243-252).

Depriving the old valley lands of the silt that now accumulates in the lake behind the dam has necessitated a marked shift towards artificial fertilization. Increasing cropping intensities have been practiced in the old lands after the construction of the high dam, and since water continues to be a free commodity in Egypt, farmers tend to overwater their fields. These factors have led to more water logging as drainage programs have progressed too slowly. However, the long-term problem will be salt. Salination is increasing rapidly and agricultural yield and consequently revenues are decreasing as a result of soil degradation.

The major challenge facing Egypt now is the absolute need to better develop and manage very limited natural resources of water, land and energy to meet the needs of the population, which is increasing at 2.5 percent per year. The population of Egypt was 36 million in 1960, 56 million in 1990 and is expected to be 70 million by the year 2000. Egypt has to import two-thirds of its wheat and vegetable oil. The nutrition gap between domestically produced food and national consumption in Egypt, which must be filled by imported food, is estimated at 40 percent.

The availability of good quality water in Egypt is also strained by the increasing discharges of untreated industrial, domestic and agricultural wastes into the fresh and marine water systems. Lake Manzala is a good example of this. One study estimates that the sixty-six agricultural drains into the lake that are monitored carry an annual discharge of 3.2 bcm. This includes raw sewage from 5000 rural centers and semi-treated or untreated wastewater from Cairo and other urban centers (World Bank 1990). Use of different types of fertilizers increased sharply between 1960 and the present: nitrogen from 192,000 to 791,000 tons, phosphates from 48 tons to 190, and potash from 2,000 tons in 1960 to 7,600 tons in 1986.

In the face of these challenges, the Egyptian government has undertaken a number of programs and adopted certain policy frameworks. These included the Irrigation Management System (IMS), the Structural Replacement Program, sponsored by USAID, as well as the Canal Maintenance Program, financed by the World Bank. These programs mainly aim to improve the efficiency of irrigation and drainage networks.

The most important policies adopted include:

■ The Irrigation Development Policy: The National Irrigation Improvement Program (NIP)

■ The re-use of drainage water policy

■ The groundwater use policy in the Delta and desert areas

■ The canal maintenance and prevention policy

■ The aquatic weed control policy.

(Source: Abu-Zeid and Rady 1991)

For protection of water quality, Decree No. 48 was issued in 1982, but it continues to face many difficulties in bringing about treatment of the sewage and industrial wastes that are the main pollutants of the Nile system in Egypt. This problem is a major and growing constraint on the availability of water of good quality, and is raising a health threat.

Since the early seventies, a number of institutional reforms and policy frameworks for managing the water resources in Egypt were established. An important institutional development is the creation of the Water Resources Center (WRC), which has 11 institutes doing research on the management of water resources. Through various technical assistance initiatives, WRC is carrying out a number of research programs and is developing scientific and technical capabilities to face the challenges in the water sector.

In the Sudan, the management and development of water resources is more complex than in Egypt, since in Egypt, the main source, the Nile, is fully controlled at Aswan. Almost all the control infrastructure is complete, moreover, and irrigated agriculture, the main user of the water, is the dominant production system in Egypt.

All the Niles originating in Ethiopia, in the equatorial lakes, and in the Sudan, have different hydrological regimes and socioeconomic systems associated with them from the Sudd region to the desert north. In the Sudan as a whole, rainfall, the main source of water for the pasture and forestry systems and for rainfed agriculture, supports the majority of the population. The main socioeconomic systems, including agro-forestry, sedentary agriculture systems and nomadic and pastured herding systems that prevail in the different zones, are dependent primarily on the rainfall regimes. These vary in magnitude and duration from the annual 1000 mm in the south, to the desert north, which receives no rainfall and where the only sources of water are subsurface and deep groundwater aquifers in the Nubian sandstone.

Agriculture is the Sudan's primary sector for economic growth. About 80 percent of the population is engaged in agricultural activities, ranging from irrigated cultivation to rainfed crop and animal production. These production systems have vast land and water potential to meet Sudanese needs, and beyond that to contribute significantly toward the Arab region's food needs. But over the years, due to many complex factors, Sudan's role of food supplier could not be realized. With the persistent drought spells that have stricken the African continent since the mid-1970s, many parts of the Sudanese water-scarce zones suffered food shortage and many pockets were hit by famine in 1984.

Since the beginning of the century, coinciding with the reconquest of the Sudan in 1899, Nile waters became the focus of action for the development of irrigation in the Sudan. The Sennar Dam, which was completed in 1924, provided storage water for about 372,000 hectares in the Gezira, the most modern irrigation system in Africa. As dictated by the British policy, the main crop in the Gezira was cotton on a four-course rotation with cropping intensity not exceeding 50 percent, a system which influenced Sudanese irrigation development for many years thereafter. The production system was based on a partnership among the investing syndicate, the government and the farmers. The share of the government constituted the water rate.

Up to the independence of the Sudan in 1955, irrigation development in the country followed the 1929 Nile Water Agreement as explained earlier. During this period the traditional rainfed agriculture was the main source of food supply (sorghum). Wheat needs at that time were limited to a small sector of the population.

With the conclusion of the 1959 Nile Waters Agreement and the construction of the Aswan High Dam, the Sudan's share of Nile water increased from 4.0 bcm to 18.5 bcm at Aswan (20.5 bcm in the Sudan). Restrictions imposed by the 1929 agreement on withdrawals from the natural flow between January and July were lifted. Between 1959 and the present the Sudan's irrigation system expanded from .6 million hectares to 1.82 million hectares.

The cropping patterns and crop-water relationships between the government and the farmers in this newly productive land generally followed the Gezira pattern. Storage of water to realize this expansion was provided by the Roseires Dam on the Blue Nile and Khashin El Girba Dam on the Atbara River.

This horizontal expansion of irrigated land in the last two decades was backed by a vertical expansion of intensification and diversification programs to meet the increasing needs for wheat and edible oil. Plans were done to raise the crop intensities of the Gezira project, and the Managil extension, totaling 810,000 hectares, from a 50 percent cropping intensity level to 100 percent. All new projects were designed to 100 percent and even 150 percent cropping intensities in the northern parts of the country to achieve food self-sufficiency. This policy has increased the need for more storage facilities in the Blue Nile and Atbara rivers. Plans were prepared to heighten the Roseires Dam and for future storage dams on the upper Atbara and Setit. It became evident that to meet the water requirements it would be necessary to provide more water by increasing the Nile yield through conservation projects in the Sudd region. The mid-1970s witnessed the start of the implementation of the first phase of the Jonglie Canal. The same period witnessed the emergence of mechanized rainfed farming in the east of the Sudan, in the Gadaref area and the southern parts of the Blue Nile province.

Following the 1973 war with Israel, and the sharp rise in fuel prices, food self-sufficiency and security emerged as a preoccupation of the Arab region. Sudan, which at the time was experiencing a period of political stability following the 1972 Addis Ababa accord that halted for a time the seventeen-year war in the country's south, received much attention as a major potential contributor to achieving such food security. The Arab Fund for Economic and Social Development in 1975 made an important initiative in this direction, aimed at a comprehensive 10-year program to develop Sudanese agriculture to meet Sudan's own needs and to create a surplus for the Arab region market. This ultimately led to the establishment of the Arab Authority for Agricultural Investment and Development (AAAID) in the Sudan in 1978. The authority started modestly, but very soon was caught off guard by the political complexities within the region, and by economic and political instability within the Sudan itself. The program of the Authority was finally interrupted after the Gulf War. However, AAAID is continuing its management of the projects implemented in Sudan and is doing feasibility studies according to its mandates in other countries of the region.

In 1973, the Arab League established the Arab Organization for Agricultural Development with headquarters in the Sudan. This organization has assessed the natural resources of the Arab region and developed a strategy for Arab food security and a comprehensive program to achieve its goals. Nonetheless, due to economic and political constraints, the food gap in the Arab region as a whole continued to increase sharply.

During the 1980s, the Sudan's natural resource base witnessed major challenges. A host of economic, social and political factors, together with the persisting droughts that prevailed in the last two decades, have led to critical environmental problems. Population and animal pressures are increasingly straining the resource base, as over-grazing, excessive tree felling and expansion of agriculture into marginal lands are leading to soil degradation and desertification. The rainfed areas are witnessing production declines due to lack of proper water management in both the high rainfall areas and in water-scarce zones. The sustainability of the irrigation systems is threatened by lack of funds for operation and maintenance, by poor irrigation efficiencies, and by increasing intensification of cropping.

Due to the drought spells and the war in the south of the country, enormous demographic changes are taking place in Sudan. Rural-urban immigration is increasing sharply, exerting increasing pressures on domestic water supply facilities and creating health hazards.

Many efforts are being made in terms of institutional reforms and the adoption of many institutional reports, but still management of water resources in the Sudan remains fragmented, over-focused on Nile surface water, and lacking in a comprehensive approach. All efforts will remain futile without a broader vision of water and water-related issues, and without sufficient funding and capacity-building to tackle these environmental challenges.

The Perspective

The problem of water in the central region is critical. This critical situation cannot be judged by the gap in quantity of water, as is reflected in much literature. That is an over-simplification of the problem, which in reality is much more complex. It goes beyond just a gap in water quantity, and needs to be seen in the context of emerging environmental problems:

- The major river basin of the central region is a transboundary one, and its water originates outside the region. Therefore, water needs to be considered in the basin context and requires cooperation among the basin states, most of which are outside the Arab region. It seems at present that despite the many efforts in this respect, the accumulated over-politicization of basin relationships is obstructing the benefits of such cooperation and may lead to water conflicts that will endanger the maintenance and sustainability of water resources for the benefit of the basin societies.

- Water scarcity is resulting from the vulnerability of the arid and semi-arid climate in the face of population growth. There are increasing pressures on a finite water availability, and an increasing point demand for water for urban centers.

- In the last two decades, many pockets in the central region (Sudan, Somalia) have been hit by persisting drought leading to very grave environmental consequences, including dislocation of population, particularly in the pastoral systems.

■ Water pollution is emerging as another threat to water resources systems in Egypt. It is threatening water availability, and affecting marine life in the coastal areas. The main sources of pollution are untreated domestic, industrial and agricultural wastes.

■ Water-related soil degradation and the spread of desertification is another threat facing many parts of the region. This includes waterlogging and salination resulting from poor irrigation management and poor drainage systems. Rainfed agriculture, which constitutes a major food production system, is suffering from many ills connected with soil-water management, variability of rainfall and the shocks of drought.

■ Extensive tree felling and over-grazing due to poor water supply facilities are accelerating the spread of desertification and demographic changes such as urbanization..

■ The region is exposed to global environmental problems associated with global warming and sea level rise, the pollution of international waters and the depletion of bio-diversity. There are scientific speculations that part of the region could be flooded within the next century, displacing tens of millions of people (MacNeil 1991, pp. 56-57).

Actions Needed

Confronted by these environmental challenges and threats, the countries of the central region need to act at both the national and regional levels to develop water policies and strategies that find a balance between short-term acute needs and the long-term perspective. Short-term strategies are necessary immediately where the sustainability of life-support systems is at risk. Long-term strategies have to be developed and implemented taking a multi-sectoral approach and integrating water-related issues with the broader national economies. The ultimate goal is a strategy for integrated and sustainable development of water resources.

It is necessary to review and set priority rankings of water and environmental problems (of a physical and technical nature) that affect the region's development. Then comes the analysis of the nature of these problems and the derivation of objectives for sustainable solutions.

Based on these analyses, capacity building programs can be formulated. Existing water institutions need to be restructured to move from narrow professional biases to multi-disciplinary functions. They should enhance their capabilities to undertake environmental assessment and management.

Environmental assessment of water resources needs to be done in the context of the water body (river, lake or aquifer). It is a continuous systematic evaluation of the main factors influencing water management and water-related environmental quality. The tasks that need to be performed will include:

■ Continuous assessment of the institutional capabilities at national and basin-wide levels, including manpower, research, etc.

■ Development of a unified basin-wide system for environmental assessment of water resources and water-related issues, including water quantity and quality, and consumption and demand trends in all aspects of water use.

■ Basin-wide environmental impact assessment (negative and positive).

■ Continuous assessment of pollution sources and the pollution of water resources.

■ Socioeconomic surveys and environmental assessment in the basin to assess quality of life, and the conditions of the river system and its coastal and marine life.

■ Development of data collecting and sharing programs, and development of channels to disseminate information to all sectors and levels of the public, with encouragement of a participatory process.

Environmental management is the master key to the sustainable development of water resources. Such management should take into consideration both the assimilative capacity of the environment, and realistic and feasible economic development goals in implementing a water development program.

National laws and regulations pertaining to the protection and development of water resources need to be elaborated and enforced. Such laws need to relate to deforestation, soil and water conservation, rural and urban health, and industrial and agricultural activities to ensure protection of the waters and their living resources from pollution.

Supporting measures need to be undertaken, including promotion of public awareness and participation, education and training, dissemination of information, and the restructuring of existing water institutions to enable them to undertake the new responsibilities for sustainable management and development of water resources.

5

The Arab Mashrek: Hydrologic History, Problems and Perspectives

Yahia Bakour and John Kolars

Introduction

The environment shared by the countries of the Arab Mashrek — Syria, Lebanon, Jordan and Iraq — marks a transition. To the north are better watered lands, primarily eastern Turkey, which is the major source of the Mashrek's international rivers, and to the south are the truly arid lands of the Arabian Peninsula and northeast Africa.

The Mashrek countries experience greatly varying rainfall from one year to the next, which reflects the changing paths of the jet stream and its attendant rain-bearing westerly winds. By the same token, north–south shifts in the sub-tropical high pressure zone responsible for the extreme aridity centering upon 30° N latitude, bring unpredictable series of drought years to the Mashrek, which can be followed by years of bountiful rains. Such was the case when heavy rains and snows in the winter of 1991–1992 ended a disastrous multi-year drought. This region therefore presents a particular challenge in the management of water resources and food production.

Variation of precipitation through time, and the limited availability of groundwater, discussed below, are important but incomplete measures of the hydrologic challenge confronting the Mashrek. The concept of the water balance[1] also provides an important indicator of the problems that face the region's water managers. Figure 1 shows the water balances for selected stations throughout the Middle East. Note the small potential evapotranspiration (PE) and very small deficit (i.e., drought) associated with stations in Turkey, the enormous PE and consequent large deficit for stations within the peninsula, and the intermediate values shown for Beirut and other locations within the Mashrek.

F I G U R E   1

Water Balances for Selected Stations
(See also Appendix to this chapter)

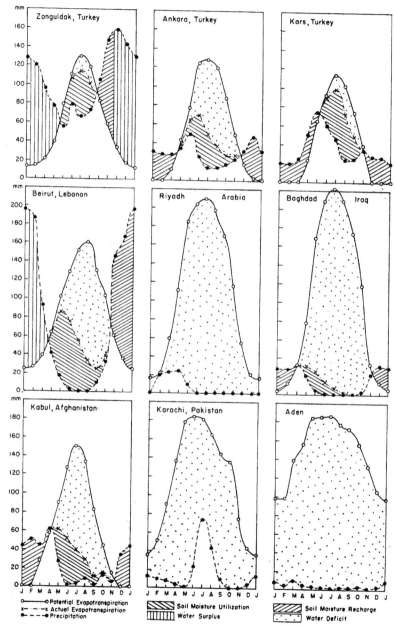

Source: Thornthwaite et al. 1958.

The latter intermediate conditions, coupled with temporal variance in which some years may resemble peninsular drought and other years a more northerly situation, summarize the Mashrek's water management challenge. The annual variability, notably, concerns the very stuff of human survival. Human beings are the most adaptable of creatures. They survive in the extremes of Arctic cold, tropical jungles, tiny archipelagos far at sea, and in remote desert oases. The one characteristic that all such extreme environments have in common is predictability.

Figure 2 illustrates this point. This figure shows the diminishing of annual precipitation from north to south on a transect leading from Zonguldak, Turkey, to Aden. A second curve indicates the increasing variance in rainfall as desert conditions are encountered. That is, the drier the climate the more unpredictable precipitation becomes.

At first inspection, the viewer may think that deserts are the least predictable of all environments. A moment's thought, however, reveals that as the predictability of rainfall decreases, the predictability of drought becomes more and more certain. If one lives in the Sahara or the Arabian Peninsula, one knows what to expect, and prepares for it. But, just as the twenty-five or fifty year drought wreaks havoc on farmers accustomed to rainfed agriculture, so does the unforeseen cloud burst threaten desert inhabitants with disaster. A heavy and unexpected rainstorm at the Jalo Oasis in Libya collapsed houses, covered fields with sheets of water-borne gravel, and damaged the crops that ordinarily thrived on water from deep wells (Abdusalaam 1978).

Further examination of Figure 2 shows that the zone of greatest unpredictability is at the intersection of the precipitation and variance curves. To the north is Turkey with its rainfed agriculture and age-old and numerous villages. To the south are the traditional widely scattered oases and now the modern sprinkler-fed farms of the Arabian Peninsula. Between lies the region that is the concern of this paper, the semi-arid Arab Mashrek, where "seven years of plenty are followed by seven years of want." This is a region where water has shaped human history, and which today presents some of the gravest challenges to modern water management.

The Role of Water in the History of the Region

Religion and ancient traditions make clear water's importance in the Arab Mashrek. It is written in the Holy Koran:

"We made from water every living thing."
(XXI, verse 30)

FIGURE   2

Rainfall and Its Variability

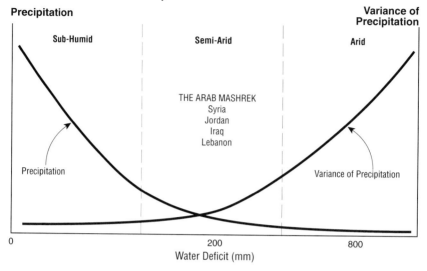

Precipitation

Variance of
Precipitation

Sub-Humid   Semi-Arid   Arid

THE ARAB MASHREK
Syria
Jordan
Iraq
Lebanon

Precipitation   Variance of Precipitation

0   200   800
Water Deficit (mm)

General Characteristics		
Enough water Usually enough land	Fluctuations — variability runs of good and bad years	Enough land Never enough water
Seasonality (warm/cold)		Extreme aridity
Drought	Extreme seasonality	Floods
Low yields per unit area		Low yields per person
Dispersed populations		Concentrated populations
Control of land (traditionally by cavalry)	Raiding	Control of water (fortified towns)

Traditional Development (Key words: kismet, tradition)		
Rainfed agriculture	Dry farming/nomadism	Hydraulic works
Transhumance (movement up and down mountains)	Village/tent symbiosis	Qanats, wind towers, etc.

Modern Development (Key words: productivity, balance of payments)		
Mechanize!	Settle the nomads/ get rid of goats	Make the desert bloom
Tractors, new seed, monoculture	Enclosure, policing, reforestation	Deep drilling/big dams, wholesale irrigation, desalination, air conditioning

In the Old Testament it is recorded how Moses struck the rock with his rod and brought forth water. The story of Noah and the great flood is a part of that same legacy. That scarcity of water and food has long been a problem is expressed in Genesis (13:6):

> "And Lot also, who went with Abram, had flocks and herds and tents. And the land was not able to bear them, that they might dwell together, for their substance was great, so that they could not dwell together."

It is addressed in the traditions of the Holy Koran:

> "To the man who refuses his surplus water, Allah will Say: 'Today, I refuse thee my favor, just as thou refused the surplus of something that thou hadst not made thyself.'"
>
> (Caponera 1954; quoted from Muhammad Ibn Isma'il,
> al-Bukhari, *Les Traditions Islamiques, Vol. II,* p. 108)

The first intensive human manipulation of water supplies for agriculture and domestic purposes is linked to the domestication of crop plants, particularly grains, in the Mashrek. It was there, somewhere in the so-called Fertile Crescent, that these most important of all inventions occurred. It can be deduced from the presence of querns and mortars that wild grains were being gathered and processed as early as 10,870 BP at Shanidar Cave in Iraq.[2] Wild einkorn and wild barley were being roasted at Mureybet, a site on the Euphrates River in Syria by 10,000 BP. Those plants did not occur naturally at such low elevations, and their presence at Mureybet indicates that they were imported from some distant source. Domesticated grains were being raised with the aid of irrigation at Jericho in the Jordan region sometime during the 1000 years before 9300 BP (Beaumont 1989, pp. 64–66), and Braidwood has identified the production of domesticated emmer wheat at Jarmo in the foothills of Iraq sometime between 8450 and 7950 BP (Braidwood and Howe 1962).

It is clear that simple irrigation techniques were in practice at Jericho, but by 6000 BP the production of grains had moved from higher elevations to Mesopotamia, that is, the flood plains of the Euphrates and Tigris Rivers. Progress at managing these rivers was at first slow, but by 4000 to 3000 BP city states were able to muster the skills and labor to build major irrigation and flood control works (Adams 1965, Chapter 4)[3]. A fascinating story is recounted by Herodotus regarding the Persian king, Cyrus. When one of his sacred horses was drowned in the Gyndes (Diyala) River, the king is said to have punished the river by diverting its flow into 360 separate channels, making it so weak that even a woman could get over it without wetting her knees (Adams 1965, p. 60). Although no dense ancient irrigation network has been found, a plausible explanation of such an act by Cyrus is that it refers to creating irrigation for agriculture rather than to ritualistic retaliation.

Thereafter, the empires that rose and fell were all ultimately based upon domesticated agriculture. From that time to this, it is possible to distinguish two approaches to irrigation in this region: massive works on major rivers, which demanded the coordination of large numbers of people (Wittfogel 1965; Adams 1965), and smaller ventures utilizing more modest streams and springs. While diversions from large streams constituted the major source of water for the large-scale projects, different sorts of aqueducts, wells, and tunnels served both large and small systems. Employing many styles of agriculture, the Mashrek's Tigris and Euphrates basin became, and was to remain, the granary of the Middle East.

An account of the development of agriculture on the margin of Mesopotamia is found in Robert Adams' *Land Behind Baghdad — A History of Settlement on the Diyala Plains* (Adams 1965). He traces the beginnings of large-scale water management to the Neo-Babylonian and Achaemenian periods after 626 BCE.[2] These early settlements that occurred along smaller water courses on the Diyala Plains gave scant warning of "the explosive developments to come" (Adams 1965, p. 61). In the Seleucid and Parthian periods, which were ushered in by Alexander the Great's conquests, the settled area of the Diyala plain expanded greatly. All the waters of the Diyala were used, excepting only those that high dams would make available in the 20th century, and Adams estimates that during Sassanian times (225–637 CE)[2] agriculture reached its areal limits, largely through the use of irrigation.

Nevertheless, Sassanian administration began a slow decline which ended with the Muslim conquest of 637 CE. Though subsequent Arab administrations brought some renewal of irrigated agriculture, the centuries that followed were marked by increasing deterioration of the canal network. Following the Mongol invasion in the mid-thirteenth century, the Diyala region fell into near ruin, although the destruction wreaked by the Mongols may have been more a *coup de grace* than a critical, crushing blow. The Abbasid and Ilkhanid regimes could do little to reverse this state. Nevertheless, urbanization assumed new significance, and after recovering from the slaughter of its inhabitants by the army of Hulagu the Hun in 1258, Baghdad never lost its importance. By the time of the Ottoman Empire, the cities of the Mashrek were deemed prizes worthy of conquest, although the agricultural potential of the region was further reduced through Ottoman tax farming and neglect.

Unused waters continued to flow unchecked to the Gulf until the advent of modern technology, introduced in part by British and French interests following World War I and the end of Ottoman rule. With independence following World War II, the nations of the Mashrek saw further dramatic increases in water use. Thus, the contemporary water crisis in the Arab Mashrek stems from the introduction of modern technologies — pumping from deep wells, sprinkler systems, and vast irrigation development schemes, which in turn, are driven by dramatic increases in the population of the region, and by the "revolution of rising expectations" with its never-satisfied demand for higher standards of living and consumption. The fragmentation of the area into contending nation-states has exacerbated the entire problem.

The Geography and Hydrology of the Contemporary Arab Mashrek

Before discussing the developmental and political implications of water use in modern times it is necessary to lay out the parameters of water sources, water availability and water demand in the Mashrek today.

The topography of the largest part of the Mashrek region in broad outline consists of a central alluvial plain formed by deposition from the Euphrates and Tigris rivers. This lowland stretches from the headwaters of the Arabian Gulf northward into Syria and to northern Iraq. The twin rivers now join near Qurna in Iraq to form the Shatt al-Arab, which flows 109 kilometers before entering the Gulf with a combined natural volume of 82 billion cubic meters (bcm) per year (Kolars 1992, Tables 4 and 5), nearly that of the Nile at Aswan (84 bcm) (Beaumont, et al., p. 98). The plain, which rises to a gently rolling plateau into which the valleys of the Euphrates and Tigris Rivers are incised, has been known traditionally as Mesopotamia, the Land between the Rivers; the plateau is called the Syrian and Iraqi Jezirah. Beyond the lowland, a semi-circle of foothills merges with higher mountains, the entire system forming a giant U-shape open to the south. The Zagros Mountains, which run from southeast to northwest within Iran but close to the border of Iraq, are the source of many streams. Farthest southeast, the Karun River flows entirely within Iran but joins the Shatt al-Arab 72 kilometers from the Gulf. Preceding upstream along the Tigris from Qurna, the Diyala, Adhaim, Lesser Zab, and Greater Zab Rivers all enter from the left bank. While the headwaters of these streams are found essentially within Iraq, their existence is due to extraterritorial, highland catchment areas to the east. The Tigris thereafter flows from Turkey, which provides about 38 percent (18 bcm) of the total, annual mainstream flow. The tributaries mentioned above provide 31 bcm annually (62 percent). (These data and those that follow are shown in Table 1 through Table 5. They are taken from Kolars 1992.)

Turning to the Euphrates system, the mountain barrier described above swings westward through southeastern Turkey from Iran. The mountains, known in Europe and America as the Anti-Taurus, are made up of a series of peaks and lesser chains. They are the source of left bank tributaries that enter the Euphrates River in Syria. No streams add water to the Euphrates in Iraq with the exception of irregular transfers of flood water through the Tharthar Canal and Reservoir, and runoff from infrequent rains.

The most important of the Syrian tributaries on the Euphrates is the Khabur River, which joins the mainstream near Deir es-Zor, bringing a

T A B L E   1

Sources and Uses of the Tigris River (mcm per year)

	Pre-Anatolian Development Project	Post 2000 AD	Natural Flow
Flow From Turkey	18,500	18,500	18,500
Removed in Turkey	0	6,700	
Entering Iraq	18,500	11,800	
Inflows to Mosul	2,000	2,000	2,000
Greater Zab	13,100	13,100	13,100
Lesser Zab	7,200	7,200	7,200
Other	2,200	2,200	2,200
Sub-Total	43,000	36,300	43,000
Reservoir evaporation	0	(4,000)	
Irrigation (to Fatha)	(4,200)	(4,200)	
Return Flow	1,100	1,100	
Adhaim	800	800	800
Irrigation (to Baghdad)	(14,000)	(14,000)	
Return Flow	3,600	3,600	
Domestic Use	(1,200)	(1,900)	
Diyala River	5,400	5,400	5,400
Irrigation	(5,100)	(5,100)	
Return Flow	1,300	1,600	
Sub-Total	30,700	19,600	49,200
Reservoir evaporation	0	900	
Irrigation to Tokuf	(8,600)	(8,600)	
Return Flow	2,200	2,200	
		(to outfall drain)	
Total Shatt Al-Arab	24,300	14,100	49,200

T A B L E   2

Sources and Uses of the Euphrates River (mcm per year)

Natural Flow	Observed at Hit, Iraq	29,800
	Removed in Turkey (pre-GAP)	820
	Removed in Syria (pre-Tabqa)	2,100
	Natural flow at Hit	32,720
Pre-Keban Dam	Flow in Turkey	30,670
(before 1974)	Removed in Turkey	(820)
	Entering Syria	29,850
	Added in Syria	2,050
	Removed in Syria	(2,100)
	Entering Iraq	29,800
	Added in Iraq	0
	Iraqi Irrigation	(17,000)
	Iraqi return flow (est.)	4,000
	To Shatt al-Arab	16,800
Full Use Scenario	Flow in Turkey	30,670
(circa 2040)	Removed in Turkey	(21,600)
	Entering Syria	9,070
	Removed in Syria	(11,995)
	Return flow and Tributaries (Turkey, Syria)	9,484
	Entering Iraq	6,559
	Removed in Iraq	(17,000)
	Return flow in Iraq	4,000
	Deficit to Shatt Al-Arab	(6,441)

T A B L E   3

Sources and Uses of the Litani and Awali Rivers (mcm per year)

		Observed Flow	Natural Flow
Litani River	Upper Valley Springs	325	325
	Upper Valley Runoff	210	210
	Irrigation Withdrawals	(120)	
	Return Flow	40	
Subtotal above Qirawn Reservoir		455	535
Qirawn Reservoir	Evaporation/Seepage	(26)	
	Mid-Valley inflows	121	121
	Markabah Diversions to Awali	(236)	
	Lower Valley inflows	264	264
	Qasmiyeh Irrigation (-RF)	(77)	
Total		501	920
Awali River	Local Flow	124	124
	Markabah Tunnel Inflow	30	(30)
	From the Litani	236	
Total		390	124 (154)

natural, annual flow of about 1.7 bcm (Kolars and Mitchell 1991, Chapter 9). Upstream from the Khabur, the Balikh contributes perhaps 300 mcm, and the Sajur (right bank) 138.6 mcm. Approximately 28 bcm enter Syria from Turkey in the main stream. It should be kept in mind, however, that all the Syrian tributaries mentioned above derive their flow from watersheds in Turkey. Thus, it is estimated that 98.6 percent of all the water entering the Euphrates River comes from Turkey (Kolars and Mitchell 1991, p. 191) (see also Table 2, this chapter).

The Jezirah and Mesopotamia of the twin rivers are bounded on the west by an extension of the Anti-Taurus range that parallels the Mediterranean southward. This extension consists of the Jebel Alawi in Syria, the mountains of Lebanon, including Mt. Lebanon and the Anti-Lebanon range east of the Bekaa, Mt. Hermon and the Golan Heights, and the hills of Palestine (the West Bank) and Israel. Precipitation upon the latter provides water for aquifers contested by Palestine and Israel (particularly the Yarqon-Taninim or Western Aquifer), while the watersheds of Lebanon and Syria produce the Litani River (Table 3) and the Orontes (Asi) River (Table 4), which flows north from Lebanon, through Syria, where it is joined by the Afrin, and then flows west to the sea through Hatay. The northern and eastern tributaries of the Jordan (the Hasbani and Yarmuk Rivers respectively) provide 60 percent of that stream's volume (derived from Lonergan and Brooks 1993, Fig. 5, p. 30; see also Table 5).

A narrow coastal plain with a matching Mediterranean climate provides some relief from the drier steppe climates on the lee side of the mountains, and the arid interior deserts of Jordan, southern Syria, and Iraq (Bakour

T A B L E   4

Sources and Uses of the Orontes (Asi) River (mcm per year)

	Observed Flow	Natural Flow
Headwaters (Lebanon)	420	420
Irrigation	(80)	
Return Flow	30	
Upstream Inflow (Syria)	370	370
Irrigation (Homs to Hama)	(200)	
Return Flow	70	
Subtotal	610	790
Irrigation (Ghab)	(580)	
Return Flow	230	
Inflows to Afrin River (est)	60	60
Inflows to Turkey (Afrin)	250	250
Turkish Withdrawals	??	
Total to Turkish Hatay	570	1,100

1991, pp. 6–10). The circle of the Mashrek is completed by the latter deserts, which extend eastward to the mouth of the Shatt al-Arab and the Zagros.

Precipitation diminishes from north to south across the entire Mashrek, both along the Mediterranean shore and in the interior. Southern Turkey enjoys an annual average of 686 mm of rain at Mardin and 550 mm at Gaziantep. Urfa, farther south, records 452 mm, but by the time Hasakah (300 mm) and Deir ez-Zor (200 mm) in Syria are reached the limits of rain-fed agriculture have been passed (Kolars and Mitchell 1991, Table 9.1).

T A B L E 5
Sources and Uses of the Jordan River* (mcm per year)

		Estimated Observed Flow	Estimated Natural Flow
North Jordan	Hasbani River (Lebanon)	125	125
System	Dan Spring (Israel)	250	250
	Banias River (Golan Heights)	125	125
	Local Runoff (Upper Valley)	140	140
	Irrigation–Return Flow		
	Huleh Valley	(100)	
Subtotal to Lake Tiberius		540	640
Lake Tiberius	Spring Flow (salty)	65	65
	Precipitation	65	65
	Local Runoff	70	70
	From Yarmuk	100	0
	Sub-total	840	840
	Evaporation	(270)	(270)
	To National Watercarrier	(500)	
Subtotal to Lower Stem of Jordan River (North Jordan)		70	570

		(Al Fatftah) (Salameh)	(Beaumont) (Gruen)	
The Yarmuk River	Flow from Syria	400**	495**	400
	Syrian Irrigation	(90)	(250)***	
	Syrian Return Flow	20	(est) 50	
	To East Ghor Canal	(158)**	(150)**	
	To Israel (via Tiberius)	(100)**	(80)**	
Subtotal to Lower Stem of the Jordan River (Yarmuk)		72	65	400

		Observed		Natural
Lower Stem of	Lower Jordan Spring Flow	185		185
Jordan River	Zarqa River and Wadis	322	539	322
	East Ghor Return Flow	32		
Overall Total		681****	674****	1,477

 * In an average year, climatic variations can change the values given by +/- 30%

 ** Conflicting sources of data account for these variations.

 *** Smaller values from the Johnson Plan; recent evidence indicates as many as 20 small diversionary dams have been built on the headwaters of the Yarmuk in Syria. Larger withdrawal values reflect such possible diversions.

 **** Once in the mainstream, this water is unusable due to high salt concentrations.

Baghdad receives 151 mm, Kuwait 111 mm, and Riyadh 82 mm. On the Mediterranean shore, Antalya, Turkey, averages 1,028 mm, Tripoli 745 mm, Beirut 517 mm, Jerusalem 492 mm, and Alexandria 169 mm. Eilat on the Gulf of Aqaba receives only 27 mm, Jiddah 25 mm, and Aden 39 mm. Sobering as such figures are, the previous discussion of variability should be kept in mind. The deserts are predictably dry; the northern areas predictably moist; but the average rainfall figures of the geographically transitional Mashrek region are made less meaningful by the area's great year-to-year unpredictability.

Groundwater, Surface Water, and National Needs

Groundwater, as mentioned previously, plays an important role in the provision of water for the countries of the Mashrek, but it is a very limited resource. The main feature of groundwater systems is that major aquifers in the Mashrek are carbonate aquifers, which are highly fissured and karstified. Their main advantage is that infiltration is high in recharge areas that receive high amounts of precipitation. However their drawbacks are numerous: high vulnerability to pollution, irregular flow, high losses to the sea, and an intricate surface-groundwater interrelationship. The latter feature favors a joint surface-groundwater development and management. With the exception of the Gaza where absolute supplies are at a dangerous low,[4] Jordan faces the most severe problem. Jordan has renewable groundwater resources of 275 mcm per year, while mining of nonrenewable aquifers amounted to an estimated 190 mcm in 1990 (Salameh and Garber 1990, pp. 10–14). Twelve basins provide the renewable flow while two are being mined. Of the latter, the Disi sandstone in southeast Jordan is being depleted at an alarming rate, in part the result of extensive pumping on the Saudi Arabian extension of the same formation where it is known as the Saq aquifer (Abu Rizaiza 1989, pp. 66–68). The Azraq Oasis east of Amman has suffered severe damage in recent years as the result of its being tapped for domestic use in the capital. This situation in combination with the limited supplies of surface water available from the Yarmuk and other streams (amounting to no more than 715.2 mcm per year) means that Jordan faces a critical water situation within this decade. Its present budget totals about 730 mcm annually, with an estimated need of 1.02 bcm by the year 2005 (Kolars 1992, p. 115).

Lebanon falls at the opposite extreme regarding water supply and demand. Groundwater consumption in 1975 (one of the few years for which data are available) amounted to 317 mcm (37 percent of all consumption), and surface water to 537 mcm (63 percent). Available surface water in major rivers amounts to 3.7 bcm per year (including 125 mcm in the Hasbani, which flows to Israel). The situation is complicated somewhat by the Orontes (Asi), which flows from Lebanon into Syria and thence into Turkey. However, the northern Bekaa that is the source of that river is of less agricultural significance than the valley south of the Beirut-Damascus road, which depends upon the Litani River. But the Syrian industrial utilization of the Orontes river waters is very important. The use of the Litani has been clouded by rumors of Israeli designs upon its waters, but no evidence has come to hand indicating that any significant amount has been removed to date.[5] Almost all aquifers within Lebanon are readily renewable. But capture of this flow is extremely difficult because it is a karstic system on the Mediterranean side.

Syria, with a 1990 population of 12.1 million people, growing at 3.8 percent per year, faces acute and immediate shortages of water for domestic use in its large cities. It also must provide water for village use and to expand its irrigated lands in order to meet the needs of its growing population. Thus, the problems of water in the Middle East come to a focus here (Tvedt 1992, pp. 14–19).

Syria retrieves approximately 3.5 bcm of groundwater from pumped wells and another 2.1 bcm from natural springs. At present, pumping of groundwater is from upper strata and the question of what deep usable reserves exist remains open.

Rivers are the main sources of surface water in Syria, and Bakour (1991) divides them into perennial and seasonal types. The perennials are further subdivided into local and international. The local rivers are the Khabur, Barada, and Sinn, which originate in Syria; they are generally small and of secondary importance, but they are utilized very efficiently before ending in the Euphrates, in internal sinks, or in the Mediterranean Sea. The international rivers are the Euphrates, Tigris, Sajour, Queiq, Asi (Orontes), Afrin and Jaghjagh, and Yarmuk.

The sharing of the Yarmuk's waters has a long history, including the agreement related to the yet to be realized construction of the dam called "Al Wahda." Jordan depends upon the river to supply its East Ghor Canal, vital to Jordanian agriculture. Israel diverts between 50 and 125 mcm annually into Lake Tiberius. Syria in recent years has built a series of retaining dams on the upstream tributaries of the Yarmuk and may be extracting more than 200 mcm annually for its uses. Certainly, the issue of this stream needs to be resolved, despite its small size.

Most significant for Syria are the Euphrates and the Tigris. Among the problems presented by these streams within Syria is their location far from centers of population. Furthermore, the Euphrates is incised in its valley, and until the building of the Tabqa (Ath-Thawra) Dam in 1974 its waters were available only along a narrow flood plain. The ambitious Syrian Euphrates Development Project hoped to irrigate as much as 620,000 hectares of land, although subsequent investigation of soil suitability indicates that the total eventually will be less. The Tigris River touches Syria for a scant 44 kilometers (39 shared with Turkey, 5 with Iraq) but its flow of 580 m³ per second (cumecs) (18.3 bcm/year) ranks second only to that of the Euphrates with 795 cumecs (25.2 bcm/year).[6]

Difficulty regarding these two rivers has developed with the creation of Turkey's Southeast Anatolia Development Project (Turkish acronym GAP), which may deplete the flow of the Euphrates at the point it leaves Turkey by half within the next 50 years. Recent negotiations between Syria and Turkey have ensured a cross-border Euphrates flow of 500 cumecs (15.8 bcm/year), which would suffice for Syria, but would leave far to small an amount to be passed on to Iraq, which also demands at least 500 cumecs in the Euphrates as it enters Iraq. But Syria and Iraq have so far agreed on continuing to share the currently available waters of the Euphrates river.

Iraq's population is projected to increase from 18,800,000 in 1990 to nearly 31,000,000 in 2010. Among the three riparians, it has the largest number of people actually living in the Tigris-Euphrates basin. Rapid population growth — a 300 percent increase since 1947 — has resulted in the spending of petroleum revenues to import food. Agriculture generates less than 10 percent of GNP, and as the second largest oil producer in the world, Iraq long depended almost completely on oil revenues for foreign exchange. The ravages of two wars have slowed Iraq's development and damaged numerous hydraulic projects. Iraq's hydraulic future depends heavily upon developments within Syria, Turkey, and Iran, since most of its waters, abundant as they are, are derived from international sources. This ominous situation has developed since 1960, a time when only Iraq used significant quantities of water from the Tigris and Euphrates (Tvedt 1992, pp. 14–19). There are 1,230,000 hectares of irrigated land between Hit and Hindiya at risk not only from salination but also from denial of necessary upstream waters. Turkey's promise to discharge 500 cumecs of water to Syria and Iraq, and the sharing agreement between Syria and Iraq were mentioned above.

Mention has scarcely been made of the crisis facing Palestine vis-à-vis Israel, and Israeli use of shared waters. All the nations of the Mashrek are involved in that confrontation, and will continue to be influenced by it. Numerous suggestions have been made to bring Turkish water to either

Jordan, or to the entire Mashrek and the Arabian Peninsula, or to Israel, in the latter case, by sea-borne Medusa bags. In the first two cases, the replacement of confiscated waters with resources from outside the immediate region implies acceptance of the status quo, and raises serious questions. In view of the recent breakthrough between Palestine and Israel, questions like these will undoubtedly be among those that will receive much attention as matters are worked out.

The Role of Water in Regional Politics

The politics of water

Politics and water are closely interconnected. Invariably the final decision to develop water resource schemes is very much part of a domestic political process. Planners in the field of water resources development decide the feasibility of projects, and politicians decide the implementation of the plans. Water affairs, especially investment decisions to develop water resources, can be very important economically and socially. However, such decisions are always out of the hands of those who have the technical know-how. Extreme financial losses may result for developing countries where political decisions are issued without technical support.

There are other political dimensions of water management. In federal countries where there are state or provincial governments, chances exist for conflict between the two levels of government. Conflicts between neighboring sub-units within a nation frequently occur over water resources and development projects. Although some Arab Mashrek countries are subdivided into provinces or regions, major issues are always in the hands of the central government's ministers. Hence, very few conflicts arise between political subdivisions, and if such conflicts occur, especially regarding natural resources like water, quick and convincing decisions can be made centrally.

The situation is much more difficult regarding international river basin disputes. Institutional arrangements and precedents to solve intra-country conflicts exist, but these are virtually lacking for international problems (Fox and Le Marquand 1978). Beyond the disapproval of the international community, there are few third party sanctions to apply against those countries that disregard internationally accepted rules of behavior and principles of sharing.

It is not surprising that water conflicts develop between countries, conflicts that have severely strained mutual relations. Many of these are yet to be resolved, like those between India and Bangladesh, Iraq and Syria,

Brazil and Argentina. Such international conflicts over shared water resources will continue to increase, especially since some writers have already suggested that water, rather than land, will be the major constraint or the critical factor for increasing world food production during the remaining years of this century (Brown 1976).

Studies are now available assessing the water situation on regional and global bases up to the year 2000 and to the year 2060. While the figures they present are general, their conclusions are remarkably similar (Table 6).

This table indicates that without extensive conservation and recycling processes, one-sixth of the water potentially available will be used by the year 2000. If current trends continue, water problems will be exacerbated throughout the world. With its basically dry and highly variable climate, there is no doubt that the Mashrek is on the forward edge of the problem, with far less unused water potential per person already than world averages, and with the problem continuing to intensify with population growth and economic development.

General international law defines as international 1) those rivers that separate two or more countries, i.e., rivers forming natural geographic boundaries, like the Rhine and the Senegal in Africa; 2) rivers that flow across two or more countries, like the Nile; or 3) rivers that have both the above two characteristics like the Rhine and the Tigris rivers. According to this definition the Euphrates River became an international river only after the fall of the Ottoman Empire.

In addition to the countries of the Mashrek proper, the non-Arab states of Israel, Turkey and Iran have riparian status on Mashrek watercourses. With the exception of Israel, they and the Mashrek countries share Islam as their predominant faith. Figure 3 shows surface and underground water sources in the Arab region. Table 7 shows traditional surface and underground water sources in the Arab Mashrek as well as the quantity of water consumed per person.

T A B L E   6
The Global Water Situation — 1973 and 2000

	1973	2000
1. Estimated population in millions	3,860	6,500
2. Potentially available water per person per year in m^3	10,400	6,200
3. Prospective demand per person per year in m^3	1,000	1,000
4. Ratio of 2 to 3	10	6

Source: de Mare 1976

The Euphrates is the longest river (2,700 km) in Southwest Asia. The Tigris (1,840 km), however, has greater volume (49.2 bcm/yr vs. 32.7 bcm/yr). Because of ambitious development plans being carried out by Turkey and Syria, the Euphrates has assumed greater importance. When the Turkish projects are completed there will be seven sub-projects on the Euphrates River including 15 dams, 14 hydroelectric power stations and 19 irrigation projects totalling 1,350,000 hectares (Kolars and Mitchell 1991). Syria, in turn, has completed the Tabqa (Ath-Thawra) Dam. Iraq usually irrigates about 1,550,000 hectares with Euphrates' waters.

With regard to the Tigris River, Turkey is now establishing three main dams that will secure large quantities of water for future agricultural expansion (Arab League of States 1993).

The waters of the Euphrates and Tigris are very important for both Iraq and Syria. Both countries in order to meet the needs of their increasing populations have almost fully utilized their available waters other then those from the basins of the two rivers. Both Syria and Iraq respect the rights of Turkey to withdraw water from the Tigris and Euphrates, and have had various discussions with Turkey on rights to river waters. Syria

FIGURE 3

Water Sources in Arab Region

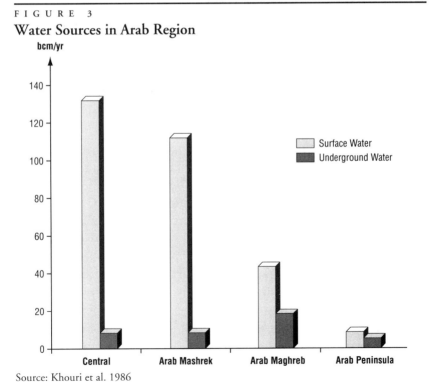

Source: Khouri et al. 1986

T A B L E   7

Traditional Water Sources in the Arab Mashrek (mcm)

Countries	1 Surface Water	2 Underground Water Recharged	 Stored	1 plus 2 Renewable Water	Pop. in millions	Water m³/cap/yr
Jordan	900	590	12,000	1490	2.645	563
Syria	22,100	2,935	—	25,035	10.600	2362
Iraq	80,000	1,000	—	81,000	15.601	5192
Palestine	4,000	950	—	4,950	4.360	1135
Lebanon	4,800	3,000	1361	7,800	3.435	2271
Total	111,800	8,475	13,361	120,275		

Source: Khouri et al. 1986

and Iraq recognize Turkey's point that storage of flood waters in Turkish reservoirs, with subsequent release during low water periods, could ensure predictable supplies downstream, which is a benefit to the lower riparians, especially for their hydropower operations.   However, Iraq and Syria believe that it is very important to have a signed tripartite agreement among themselves and Turkey that specifies each country's share. This has not yet come into existence.

The other main rivers in the region are the Jordan, the Litani, the Yarmuk, and the Orontes (Asi). Rational management of the Jordan River is now more critical than that of the Tigris and Euphrates, since Israel, Jordan and the inhabitants of the West Bank all have vested and immediately critical interests in the water of the Jordan and its tributaries. Syria and Lebanon are also actively concerned with the management of the Jordan.

The Yarmuk, a tributary of the Jordan, provides water to Syria, Jordan and Israel. An agreement on the joint development of the basin has been signed by Syria and Jordan, but no agreement covers actual distribution of Yarmuk water. The Litani River, unlike the others in the region, rises and flows entirely within the borders of one country, Lebanon. Despite the Litani's being a Lebanese river, interest on the part of Israel and Jordan in use of the Litani waters is increasing. The Orontes rises in the Bekaa Valley, north of the headwaters of the Litani, and flows north through Syria and Turkey before reaching the Mediterranean Sea. Syrian industry is particularly concentrated in the Orontes basin, and half of the country's industrial water comes from the Orontes. Any plan to use its water, however small the quantity, may well cause tension among the people sharing it.

Regionally, as water becomes more scarce, a new paradigm is developing. Food production and security based on ample supplies of water are beginning to be as important as petroleum revenues.

Only Turkey, just outside the region, has surplus water. Before the Gulf War, Turkey had reached the penultimate stage in the construction of the Ataturk dam, and was faced with the need to cut off the flow of the river to begin filling the reservoir. In January 1990, river flows were stopped at the dam for 27 days. Iraq and Syria confronted Turkey on this issue; Syria's hydropower generation was affected, and the Iraqis blamed Turkey for significant water shortages.

The ongoing development of Turkish and Syrian hydrological resources raises critical issues for all three Euphrates riparians, particularly Iraq. In separate bilateral talks between Turkey, Iraq and Syria, the Turkish authorities promised a flow of 500 cumecs across the Turkish-Syrian border. Iraq subsequently claimed a right to 570 cumecs. In separate talks between Syria and Iraq, it was agreed that Syria receive 42 percent and Iraq 58 percent of the water available in the Euphrates.

These problems extend back to 1920, when a treaty between the mandated territories of Syria and Iraq was signed, limiting the extent to which Syria might alter the flow of the river into Iraq. In 1923, the Treaty of Lausanne, which made no mention of Syria, noted that Turkey should confer with Iraq before beginning any activities that might alter the flow of the Euphrates. Iraq raised the question of "acquired rights" to a share of the river in 1962, when it agreed with Syria to form a joint technical commission to exchange data on river levels and discharges, but because no major hydraulic works were undertaken then, there was little effort to specify the nature of the shares. By 1967, Iraq claimed rights to 59 percent of the natural flow of the river at the Syrian-Iraqi border. In 1980, Iraq, Syria, and Turkey established a tripartite technical commission.

Water is the key to sustainable development in the Middle East and to the well-being of the region's population. Although proven oil reserves in the area will last 100 years, water supplies are already insufficient throughout the region. The situation in the Tigris-Euphrates basin is serious enough to warrant the immediate attention not only of the three riparians, but also of neighboring Middle Eastern states and the entire world community.

Problems of bilateral hydro-diplomacy

In the present situation, in which successful cooperation remains elusive, the outlook is that the Turks will continue to develop the GAP as long as their financial situation allows. The Ataturk reservoir will be filled, and its water will be used for hydroelectric generation and irrigation according to Turkey's needs. Turkey will continue to take the largest share of the Euphrates River. The demand for energy will continue to increase, and petroleum costs may have an impact on the country's economy.

Syria will attempt to expand its irrigated lands to at least 600,000 hectares. Turkish return flows down the Balikh and Khabur will increase the volume of those streams, resulting in environmental imbalances. Syria's population will continue to increase with increasingly insufficient water supplies, and the country will focus more attention on the Yarmuk and the snow waters of the Golan Heights, which are managed by the Israelis as part of their own water supply.

Iraq's reconstruction will be hampered by decreased and polluted downstream flows. Scarce water supplies will inhibit efforts toward the desalination of soils and may seriously impede the natural flushing of waste and salts from the mainstream of the river. Water from the Tigris river may be transferred via the Tharthar depression to help offset shortages on the Euphrates.

As needs change over time in this setting of separate bilateral diplomacy (Syria and Turkey, Iraq and Turkey, Syria and Iraq), there could be a continuous shifting of alliances, possibly providing short-term solutions to problems related to the sharing of the Euphrates and Tigris waters. Syria may again have periods of détente with Iraq to bargain with Turkey, but tensions may increase as water becomes more scarce. Waters of the Tigris may be partly transferred to the western part of Iraq, but may still prove insufficient for increasing agricultural needs.

Although piecemeal solutions may be achieved, they would be unstable, doing little for the long-term needs of each nation. The Middle East would continue to be subject to the manipulation of world powers. Turkey would experience fragmentary development and slow progress on the GAP and agricultural expansion. Only a fraction of its targeted area would be irrigated. Although this would delay downstream water shortages, Syria would feel an increasing scarcity of water and may spend more revenue to meet domestic needs through the reuse of water. Overall, ephemeral separate bilateral arrangements among combinations of Turkey, Syria and Iraq are likely to prevent continuous and sensible long-term planning.

The Development of Activities in the Field of Water Resources

Information-led alternatives to the status quo; moves toward more than bilateral international cooperation

The rainfall of the Mashrek region is meager. As was raised as a key point at the opening of this paper, it is also most unpredictable with respect to quantity, distribution, and timing. The perennial rivers provide a water resource base that is only somewhat more stable, and the rivers of significance are transboundary in character, which means that reliance on them is conditioned by the problems of ongoing negotiations. In principle, any country that withdraws water from a transboundary river or aquifer should take into account the shares and interests of other riparians and users. In practice, however, how much weight a water-withdrawing country will actually give to the interests of others is the outcome of a complex pattern of needs, power, and short- and long-term perspectives. As pressure on the water resource intensifies with the growth of population, the Mashrek countries have much work ahead of them.

Such situations necessitate more data and information about existing water resources. Bakour (1991) emphasizes the need for detailed studies of existing water sources. Productive negotiations among countries depend largely upon the availability of complete and accurate information regarding the resources at issue. Such studies should concentrate on deep aquifers, the quantities of water available, and suggestions for alternative uses of water. Agreements between neighboring countries, or among several countries, indicating the entitlement shares of each country, for both surface and underground waters, are badly needed, and are the ultimate goal of study efforts. These could be accomplished at least cost by utilizing existing information and data handling systems. The output of such systems could be gathered and incorporated, for instance, into a regional information clearinghouse (RICH), located either within the region or elsewhere on "neutral" territory. To identify specific problems in their early stages, a RICH would aggregate pertinent data from satellites and individual nations and would tap into worldwide information networks to generate information on a real-time basis. Nations would be encouraged to participate by the prospect of having access to a timely, accurate, and politically neutral information system. A RICH would compile factual data bases, and refrain from using them for political analysis, which would be the prerogative of nations sharing the data bases.

Moving toward river basin management, data predicting future water availability might be the basis for committing resources to the development of interlocking systems, such as the Turkish Peace Pipeline Project. Early in 1987, Turkish Prime Minister Ozal suggested the idea of a peace pipeline as an answer to the escalating water shortages of the region south of Turkey. The pipeline could carry water from southern and western Turkey, and from the Tigris River in eastern Turkey, to western Saudi Arabia and the United Arab Emirates. The initial estimate of water transfer was 1.28 bcm per year, flowing south through a double western pipeline, and 0.91 bcm per year flowing east. Costs were estimated at US$17–20 billion. Technologically feasible, these seemingly expensive lines could deliver water, it is said, at one-third the cost of a similar desalinated quantity. The peace pipeline has attracted considerable academic and public interest. However, as yet no official interest has been shown, nor commitment made, nor has a feasibility study been undertaken, and the project is currently presumed dead. A smaller version of the peace pipeline carrying much-needed water only as far as Amman, Jordan, would be financially more attainable and politically less complex.

The water issue among the Mashrek countries is a complex and critical one that needs special and wise handling by the riparian nations involved. Water seldom stands alone as an issue, but is rather always mixed with ethnic and religious rivalries. Increasing populations and increasing development in the agricultural and industrial sectors necessitate proper management in order to conserve the watersheds upon which the region depends. Bilateral, or even better, multilateral water agreements must be signed by the Mashrek countries. Such agreements must clearly spell out the rights and shares of each country. If nations would share both water technology and resources, available water could satisfy regional needs.

Within-nation conservation and efficiency measures needed

A prerequisite of effective international cooperation is efficient use of water within each country in the several basins of the Mashrek. Conservation can be achieved in all sectors: domestic, agricultural, and industrial. Water quality can vary depending on the purpose for which the water is intended — irrigation water need not meet the standards of that used for domestic or industrial purposes. Conservation means increasing crop water-use efficiency as well as improving domestic delivery systems. Water-use efficiency in agriculture can be achieved by breeding early-maturing and drought-tolerant varieties of crops; determining the critical water deficit of crops and giving preference to those that are less water-demanding; using mulch and indoor agriculture to decrease evapotranspiration; following proper crop management practices, e.g., choice of the right sowing date, application of the appropriate doses of fertilizer, insecticides, water, etc.; constructing small terraces to reduce runoff; establishing or improving of extension services to lead these agronomic changes; using new technologies in irrigation systems; harvesting rain water to be used during dry spells; adopting water recharge systems for aquifers; and operationalizing the use of recycled and desalinated waters.

The Role of National and Regional Organizations in Developing and Managing the Water Sector

Water — that sacred material — which is mentioned in all holy books, plays a paramount role in the daily life of all peoples. A recent definition of both human beings, and animals and plants is that they are moving or standing columns of water. Because water's importance is so pervasive, when we speak of the roles of national and/or regional organizations in the development of the water sector, reference is actually made to the role of every individual. Major changes in this sector must have at their base information and mobilization campaigns that reach to the level of individuals.

On the other hand, this must be done through leadership structures. Official government bodies, such as ministries, as well as volunteer organizations, must play vital roles. For example, they can introduce adaptable technology for increasing water-use efficiency in agriculture, industry, or domestic use. These official or non-official organizations through efforts such as extension services, the press, radio broadcasts, or TV, can raise an awareness of the need for efficient water use among people in all walks of life. National and regional organizations can also introduce the idea of the importance of water into the curricula of educational institutions, from primary schools to universities.

Toward the regional development and conservation of watersheds and groundwater basins

The water situation in the Arab Mashrek imposes a regional approach on the development of all water resources. The people of the Arab Mashrek may adopt several approaches to achieve this goal. Since the increase in population is an important factor in the utilization of regional waters, measures must be taken to adopt birth control policies. Water-use efficiency with respect to all sectors must be achieved. Since agriculture is the largest user of water, previously mentioned approaches to conservation should be adopted in order to maximize regional water-use efficiency and development. Measures should be implemented to reduce run-off. The harvesting of rain water should be encouraged. The recharge of groundwater basins in different countries should be increased, and there should be renewed attempts to discover new sources of water in the region.

Water resource development nowadays plays an important strategic role in the social and economic development, as well as the political stability, of the water-scarce Arab Mashrek. The water situation has been aggravated during the last decade by frequent droughts that have affected almost the entire region, and it is exacerbated by continuous population increase. Serious measures have to be taken so that water demand is well managed, and the meager existing water supply can be efficiently used. This is made doubly challenging for the nations of the Arab Mashrek by the region's great variance in precipitation. A component of this management is being able to forecast available supplies with the greatest possible reliability, and to this end agreements specifying each country's share of transboundary water must be signed between contending nations. By the same token, all other sources of water must be maximized and made secure in order to avoid internal unrest stemming from the unmet hydrologic needs of the people of the Mashrek.

Appendix

The water balance (Figure 1) represents a systematic, month by month matching of the water needed for vegetative growth with available water provided by a combination of natural precipitation and soil moisture (Thornthwaite and Mather 1955). The water needed for plant growth depends upon the amount of energy in the environment (ambient air temperature) and the amount of water necessary to satisfy both evaporation and transpiration vital to plant metabolism. This quantity is known as potential evapotranspiration (PE). During the seasonal march of vegetative growth and precipitation, rain may at first satisfy a plant's need for water. As precipitation diminishes, the plant extracts more and more moisture held in the soil. Available rain plus available soil moisture comprise actual evapotranspiration (AE), which may or may not equal PE. A point may be reached where AE fails to satisfy PE. When this occurs, the wilt point is reached and drought sets in. The amount of water lacking that would have to be added artificially (i.e., by irrigation) to ensure successful plant growth is the annual deficit (Delta). Thus, a given amount of rain in an area with a low PE may satisfy plant growth, whereas the same amount in an area with higher temperatures may not suffice (Thornthwaite et al. 1958, pp. 26, 27).

Notes

[1]   See Appendix for a technical explanation of the water balance.

[2]   BP: Before Present, present being accepted as 1950 AD. BCE: Before Current Era. CE: Current Era.

[3]   Salination, then and now, has been a major problem along the twin rivers. The Nile, before the building of the high dam, did not face this problem, for the annual flood effectively washed accumulated salts from the soil while depositing fresh alluvium.

4   The condition of Arabs in the West Bank of Palestine remains critical, but the absolute amount of water available on the West Bank would be sufficient for their needs if an equitable distribution system were present.

5   See Kolars 1992b for a more complete discussion of this issue.

6   These flow figures differ from those derived by Kolars and Mitchell, but fall comfortably within a range acceptable for discussion.

6

Water Resource Development in the Maghreb Countries

Mohammed Jellali and Ali Jebali

Introduction

As is the case everywhere around the Mediterranean Sea, the population of the western south shore has historically been aware of the strategic role that water plays for human survival in arid and semi-arid zones. The siting of cities near water sources, the development of water springs and wells, the construction of galleries for groundwater drainage, the exploitation of floods for agriculture and the building of water conveyance and distribution networks for domestic and agricultural uses all illustrate the traditional relationship that North African peoples have had with water. Thus, for many past centuries, the population has lived in harmony with the water environment, knowing how to take advantage of wet years and how to deal with dry ones.

However, the socioeconomic conditions of the region have changed noticeably since the beginning of the present century due to population growth, the improvement of living conditions, accelerated urbanization, the introduction of large scale irrigation and the development of industry. This hastened socioeconomic evolution has generated increasing pressure on the water resources of the region resulting in unprecedented water needs, emerging geographic disparities, and acute water pollution problems.

To face this rising water demand, which has accelerated further in the second half of the century, the countries of the region have implemented specific policies and strategies for water resource development that are coherently integrated within the framework of national socioeconomic development policies. Considerable progress in the management of water resources has been accomplished in the region, but colossal efforts remain, and they will have to take into account a particularly delicate hydrological setting:

- All countries of the region must be prepared to face strong pressure on limited water resources.

- Situations of limited water availability are already observed in certain areas, and it is expected that the total use of the entire water resource will be attained in a more or less brief period, depending on each country's specific situation. Total use of water resources in the whole region will probably occur around the year 2020.

- Although dryness is already the major characteristic of the Maghreb hydrologic setting, its impact will be more and more accentuated with the increase of the water demand and competition for water access.

- Coming decades will be marked by an increasing need for efficient water resources management in both the quality and quantity contexts.

Demographic Characteristics in the Maghreb

The Maghreb has a population estimated at somewhat over 70 million inhabitants living in an area of 6 million km². Morocco, Algeria and Tunisia represent half of the Maghreb total area and nearly 90 percent of the population. The population is extremely concentrated in a coastal band north of a line extending from Agadir to Bechar to Gabes. Thus, most of the large cities, with the exception of a few in Morocco, are on the Mediterranean or the Atlantic coastline.

Since 1950, the Maghreb region, in addition to an increase in the standard of living, has experienced large-scale population growth due to improvement in health and sanitary conditions. This trend is expected to continue as demographic estimates foresee the population doubling in 20 to 30 years (Table 1), with the regional total exceeding 120 million inhabitants by 2025. Despite the clear tendency towards accelerated urbanization (Table 2), the rate of the rural population's growth is expected to remain relatively high.

Climate and the Regional Context of Water Resources

The climate of the Maghreb is subject to Mediterranean influences in its northeast, to Atlantic influences in its northwest, and to continental and Saharan influences towards the south. It has hot and dry summers, and mild to cold winters. During the summer, precipitation is nearly absent and the evaporation rate is extremely high.

TABLE 1

Population Growth in the Maghreb Countries (10⁶ inhabitants)

	1950	1960	1990	2000	2025
Mauritania	0.83	1.00	2.02	2.75	5.12
Morocco	8.95	11.60	25.06	31.35	45.65
Algeria	8.75	10.80	24.96	32.90	51.95
Tunisia	3.53	4.20	8.18	9.93	13.63
Libya	1.03	1.30	4.55	6.50	12.84
Total	23.09	28.90	64.97	83.43	129.19

TABLE 2

Percentage of the Population that is Urban

	1960	1990	2000
Mauritania	6	47	59
Morocco	29	48	55
Algeria	32	52	60
Tunisia	36	54	59
Libya	23	70	76

In the extreme southwest of the region, the climate of southern Mauritania is subject to intertropical influences and is marked by two seasons: dry from November to May, and wet from June to October. In general, the Maghreb climate is quite variable, which results in a complex hydrological regime at the seasonal, annual and inter-annual levels.

Precipitation

In the northern part of the region, precipitation is concentrated within at the most one hundred rainy days per year. Locally, rainfall intensity may exceed 100 mm in less than one day, and a large part of a year's rain can be concentrated within a few days of the year.

Snowfall occurs in the higher Atlas Mountains of Morocco and Algeria. In the northern zones, the annual average precipitation may exceed 500 mm, and can reach 1500 mm or even 2000 mm in certain high mountain areas. Precipitation decreases progressively towards the south to less than 100 mm per year in the pre-Saharan zones, and it approaches zero in the Sahara desert.

Along the southern edge of the Sahara, Mauritania experiences a different precipitation regime. Almost a quarter of the country, which is subjected to a Saharan climate, receives less than 200 mm on average, marked by irregularities and important over-year gaps in rainfall. The southern part of the country, along the Senegal River, records an average precipitation of about 600 mm during the period between June and October.

The Water Resources

The water resources of the Maghreb are marked by great spatial disparity in distribution, both within countries and between countries.

The presence of the Atlas Mountains, which cross almost the entire region along an east-west line, combined with diversified geological and geomorphological characteristics, create unique surface and groundwater resource systems located within limited geographic areas. The Atlantic and Mediterranean coastal watersheds, which represent about 12 percent of the region's area, produce approximately 80 percent of the total surface flow. Conversely, the Saharan zones of Algeria, Tunisia and Libya include large sedimentary basins that contain very important nonrenewable ground-water resources. South of the Saharan zones, Mauritania's water supply depends on a single source, the Senegal River, which is shared with Guinea, Mali and Senegal.

Table 3 summarizes for each country of the region the present and future per capita water availability. It also shows that Mauritania possesses valuable surface water resources originating from its share of the Senegal River flow. A part of Tunisia's surface water (14 percent) originates in Algeria, and 2 percent of Algeria's surface water originates in Morocco. A comparison of the 1990 water availability per capita with the values for the year 2025 clearly reveals the impending chronic water shortages.

The Role of Water in Economic Development

The Maghreb's traditional expertise in water management includes ground water development by digging to great depths, the use of galleries to drain base flow (khettaras or foggaras in Morocco and Algeria), slope

---

T A B L E   3

Water Resources in the Maghreb Countries

	Population (10⁶ inhab)		Precipitation (bcm)	Flow (bcm)			Per Cap Avail (10³ m³/inhab)	
	1990	2025		Surface Water	Ground Water	Total	1990	2025
Mauritania	2.02	5.12	—	7.00	0.40	7.40	3.66	1.44
Morocco	25.06	45.65	150	22.50	7.50	30.00	1.19	0.65
Algeria	24.96	51.95	65	12.40	6.70	19.10	0.76	0.36
Tunisia	8.18	13.63	33	2.70	1.80	4.50	0.55	0.36
Libya	4.55	12.84	—	0.20	3.63	3.83	0.84	0.30

works to drain the run-off (jessours in the south of Tunisia) and flood harnessing for seasonal irrigation (in the south of Algeria and Morocco).

This hydraulic experience has recently been strengthened in order to deal with drought periods and to take advantage of flood water supply. The introduction of modern structures began in the region in the 1920s with the construction of the first dams in Algeria and Morocco, some of which are still in operation. These dams were intended to help meet municipal water supply needs, and respond to the demand for irrigation water and electrical energy production. Beginning in the 1960s the tendency has been to develop large hydraulic structures to develop surface water for irrigation in order to provide for the nutritional needs of an increasing population. Massive water development efforts are necessary for agriculture as a great part of the region receives less than 400 mm of rainfall per year, and irrigation is required during a long portion of the year.

Beyond developing maximum groundwater production near the big cities, the need to augment the supply of drinking water for the growing population has made it necessary to construct reservoirs and to move water, often further than 100 km. Consequently, as early as the beginning of the 1990s the countries of the Maghreb have achieved high levels of water resource development, near or exceeding 50 percent of the developable potential (see Table 4).

In the southwestern part of the region, Mauritania has appreciable surface water resources that are already developed through the construction of two reservoirs: Manantali (upstream) and Diama (downstream) on the Senegal River. On the other hand, Libya is facing a great shortage of renewable water resources and is undertaking the exploitation of its large nonrenewable groundwater resources.

The state of water use

A main characteristic of the Maghreb is the large requirement of water for irrigation, which represents 70 percent of the total water used in

TABLE 4

State of Water Use in the Maghreb as of the Early 1990s

	Total Flow (bcm/yr)	Developable Potential (bcm/yr)	Developed Potential (bcm/yr)			% Developed
			Surface Water	Ground Water	Total	
Morocco	30.0	21.0	7.5	3.5	11.0	52.4
Algeria	19.1	8.4	2.0	1.7	3.7	44.6
Tunisia	4.8	3.8	1.4	1.5	2.9	75.6

Algeria, 83 percent in Tunisia and exceeds 90 percent in Morocco and Libya. Accordingly, domestic water supply represents 6 percent of the total requirement in Morocco and approximately 25 percent in Algeria, while the requirement for industry is low in the whole region. Constructed hydroelectric power capacity for Morocco, Algeria and Tunisia remains less than 1,000 MW, of which three quarters is concentrated in Morocco where it is used mainly to meet energy demand during the winter peak hours.

In general, the first priority in water allocation is given to domestic water supply, and the second to irrigation, while the production of hydroelectricity is subordinated to those two priority uses. Table 5 below recapitulates the water demand structure in the four northern countries of the Maghreb for the year 1991. The figures show water demands for a normal year, but in practice the years can fluctuate significantly. In Morocco for example, the satisfaction of irrigation demand may vary between 40 percent and 100 percent, and the water available for hydroelectricity can be from 25 percent to 100 percent of what the generating capacity could absorb, depending on the annual rainfall. This underlines the importance of the drought phenomenon for both the development and the management of water resources in the Maghreb region.

Although in general, with the exception of Libya, the overall ratio of water resources to water demand is still favorable, throughout the Maghreb chronic water shortages occur in many subregions. In these subregions the over-exploitation of nonrenewable groundwater resources is already very much practiced.

Water, a driving force for development and regional organization

The future socioeconomic status of the Maghreb is closely tied to the development of water resources. However, the high variability of the climate imposes a strong irregularity on water availability in time, as well

T A B L E   5

Water demand in the Maghreb in 1991

	Total water demand	Municipal and industrial water (mcm)	Irrigation		Installed hydroelectric power (MW)
			Area (ha)	Water demand (mcm)	
Morocco	11,000	1,000	880,000	10,000	622
Algeria	2,800 to 2,900	1,000	318,500	1,800 to 1,900	285
Tunisia	1,660	276	300,000	1,384	64
Libya	4,683	482	364,000	4,275	—

as from one region to another. Large storage reservoirs have been constructed, as well as tunnels and canals up to 300 km long to transfer large quantities of water, meaning that hydraulic infrastructure has an important role in the regional organization of the Maghreb. Centers of economic development have been established around irrigated areas, to counterbalance the expansion of urban centers, and to contribute to a balanced development of urban and rural zones.

Along with the progressive development and use of groundwater resources, attention has turned to surface water to meet both the accelerating increase in the municipal water requirement, and to increase agricultural production to ensure food security for the whole region. Overcoming unfavorable climatic conditions, the position of the agricultural sector among the national economies has been strengthened with the development of irrigation. In 1990 agriculture accounted for 13 percent and 15.7 percent of the gross natural product respectively in Tunisia and Morocco. In general, the irrigated areas in the three northern countries of the Maghreb (i.e., Morocco, Algeria, Tunisia) account for 30 to 40 percent of the national agricultural production. This development of irrigation has tended to reinforce the food security of the region and alleviate dependence on food importation. As an example, Morocco imported less than a quarter of its agricultural foodstuffs in 1992 and 1993, and less than 7 percent of its non-food agricultural requirements. To avoid regional disparities of development in rural zones, particularly between those that receive large scale irrigation and those that do not, small projects are designed and constructed in the areas outside major irrigation projects to provide some water through groundwater pumping and small dam construction for local irrigation.

The population of many small urban centers has increased, generating competition between these centers, and the large coastal cities and regional major towns, but also establishing an urban network over the region. Currently, the Maghreb has 10 large cities of more than 500,000 inhabitants and almost 80 cities of 100,000 to 500,000 inhabitants, but the region has been able to develop and expand water distribution systems for all of the urban centers. However, easy access to safe potable water for the rural population remains a subject of concern for some countries, where the nature of obstacles is more institutional than technical or financial.

Another major requirement for water is hydroelectricity production. With an installed capacity of 687 MW, amounting to 30 percent of total power capacity, the hydroelectric capacity of Morocco allows an annual production of 500 GWH (1983–85) to 1,500 GWH (1980) representing

10 to 27 percent of the total production of electricity. The annual average oil fuel saved is about 660,000 tons. The main value of this hydroelectric production is its flexibility, which makes it usable to meet a peaking need of 500 MW on average.

The Major Issues of the Future

Water: a limiting parameter of development

While the Maghreb countries have accomplished substantial progress in their water resources development, additional efforts must be mounted to meet the following anticipated stresses:

■ The current use levels are close to if not greater than 50 percent of the developable water potentialities, and already exceed the renewal rate in Libya. Moreover, disparities within countries are emerging, so that there are localities where overexploitation is occurring or will shortly do so.

■ The large increase of water needs caused by the expected growth of the population in the coming 30 to 40 years will require a particularly high level of water development, simultaneously with rising development costs.

■ The increasing technological complexity of the hydraulic infrastructure, combined with the rising volume of water demand, and the multiplicity of users and operators make the water management process extremely complex and laborious.

■ The apparent rising frequency of drought episodes during the last two decades is of concern. (See section on drought management below.)

All of these considerations create a scenario of continual and increasing pressure on limited water resources, and more and more intensive competition for them. The competition for water has moved water planning from a sectoral and project basis toward integrated planning, which takes into account the entirety of the water supply and water needs, along with the interrelationship between water and socioeconomic development. The main objective of such planning is to make water, its usefulness having been maximized by an optimal program of investments in hydraulic structures, play a leading role in the economic development of the region, in spite of the limited water supply and the great expansion of demand for water.

Integrated water resources planning: a strategic need

Beginning in the 1970s, planning procedures have been implemented within the countries of the Maghreb to ensure proper integration of water policy with the strategic planning of each country and its development policies.

- In Tunisia, the first water plan established master plans for the north, center, and the south of the country. The master plan for the northern part is particularly important as it covers one-quarter of the area and 77 percent of the country's total surface water. Its objective is to respond to increasing needs for potable and industrial water, and to provide water for urban areas as well as irrigation water for agricultural regions in the northern portion of Tunisia. The plan defined the main infrastructures for surface water development, and the dam and tunnel construction necessary for water transfers, covering the region's water needs until the year 2000.

  A new planning step is currently being implemented that establishes strategies to develop, stock, and transfer water, in addition to a strategy for economizing water consumption, and one for additional non-conventional water resources. Beyond the establishment of "water needs/water resources" ratios and the hydraulic infrastructure programs that are required by the year 2010, this Tunisian program has the merit of incorporating demand management. Nonrenewable and non-conventional water resources are considered as absolutely necessary aspects of water resources development strategy in Tunisia (Salim and Kedri 1992).

- In Morocco, the planning process includes a range of procedures from the identification of water supply problems to the implementation of water management. This planning process is currently applied to many hydraulic basins and is projected over a period of 30 to 40 years.

  Moroccan water planning is generally considered as a tool for achieving the national development objectives. It is intended: a) to make potable water generally accessible and b) to provide water for the irrigation of one million hectares by the year 2000, up from the 880,000 hectares irrigated in 1991. At the sub-unit level within the country, the main goal is to identify water needs, conceive and design required infrastructures and recommend an optimal allocation of water between the different users in accordance with national and regional development options.

These area plans are the basic elements for establishment of a national plan of water resources development, which will be launched in the very near future. Its objective is to ensure coherence among hydraulic basins plans, to promote a balanced water access for all regions of the country and to define common measures of administrative, legislative and economic parameters that are necessary to carry out proper water planning.

■ In Algeria, water resources master plans have been established, or are being written, for the main hydraulic areas of the country. The results of these plans have already been incorporated into the national territorial organization plan. A national master water plan is also foreseen in the near future.

All these activities demonstrate the strategic nature of water resource planning to ensure rational water resources development and permanent allocation between the Maghreb's total water needs (20.243 bcm) and available resources (33.2 bcm). The main objective is the avoidance of water shortages that could become an obstacle to the socioeconomic development of the region.

Interim Development Programs

The demographic challenge facing the region determines the strategies to satisfy increasing water demand. The primary effort has to be focused on conjunctive groundwater and surface water development, and on improved management in order to minimize the use of water by different present users.

■ In Algeria, the effort to be realized by the year 2010 is to ensure potable water to a population of approximately 40 million inhabitants and to irrigate 500,000 hectares. For municipal and industrial needs, it is mandatory to develop nearly 2.5 bcm per year while the irrigation of 500,000 hectares will require approximately 3 bcm. To accomplish this it will be necessary to complement the present hydraulic infrastructure with fifty additional dams and ten diversion structures and to develop nearly 5 bcm of nonrenewable groundwater in the Saharan region (Synthèse des rapports nationaux, Alger, 20-30 Mai 1990).

■ In Tunisia, the objectives for the year 2000 are to cover 100 percent of urban and industrial needs, and 90 percent of potable water needs of the rural areas, and to extend the irrigated area to 400,000 hectares. Thus, by 2000, Tunisia will have developed 90 percent of its surface water and 100 percent of its groundwater. This will be achieved by creating:

◆ 21 large dams to develop 739 mcm per year

◆ 203 small dams to develop 110 mcm per year

◆ 610 drilled wells to develop 288 mcm of groundwater.

(Ministère de l'Agriculture, Tunisie 1990)

■ In Morocco, the effort is also enormous since the objectives are to provide access to potable water for all the urban population by the year 2000 and for all the rural population by the year 2010. Irrigated areas will be extended from 880,000 to 1,000,000 hectares by the year 2000, and to the maximum of 1,350,000 hectares by the year 2020.

It will also be necessary to increase the controlled surface water volumes from 7.5 to 13.5 or 14 bcm and extracted groundwater volume from 3.5 to 4.5 or 5 bcm. This effort requires building:

◆ 60 large dams

◆ An annual average of 100 km of boreholes to exploit groundwater and in particular deep groundwater

◆ Regional water transfer facilities for a distance of approximately 280 km

◆ Potable water supply infrastructures for an approximate discharge of 100 m^3/s.

■ Libya's water resources balance shows an approximate annual deficit of 1.68 bcm by the year 2000. This deficit will be reached even though renewable water resources use increases from 210 to 350 mcm through seawater desalination (130 mcm) and reuse of reclaimed water (220 mcm). The water deficit projected for the year 2000 will be covered by mining nonrenewable groundwater resources in the south of the country. The large transfer of groundwater from the southern regions will enable 1,642 mcm per year to be provided in the first stage and 2,226 mcm per year later (Salim and Kedri 1992).

■ In Mauritania, the Senegal River drains a catchment area of 300,000 km^2, and has a very irregular river regime, being practically dry for half of the year. Two main projects, the Diama and Manantali dams, were initiated by the Organization for Development of the Senegal Valley founded in 1972 by Mali, Mauritania and Senegal. The Diama project is expected to increase Mauritania's irrigated areas from the existing 24,000 hectares in 1990 to 141,000 hectares by the end of the project. This dam, which is situated on a bend in the Senegal River, 23 km upstream of St. Louis and on the border between Senegal and Mauritania, will serve the following purposes:

◆ It will form a barrier against the salt-water wedge, which otherwise extends as far as 250 km upstream during low-flow periods. In this way, protection will be provided for the existing and projected upstream irrigation and domestic water supply intakes of the dam.

◆ It will create a reservoir which, also benefitting from the inter-annual flow regulation provided by Manantali, will provide irrigation water for an area of around 120,000 hectares, while reducing the delivery head of the pumping stations.

◆ It will enable effective regulation of inflow to the Guiers and R'kiz lakes, as well as to other depressions situated along the Senegal valley.

Supply and Demand Management

Projections of "water needs/water resources" balances foresee that the total use of the supply potentialities will be reached in some subregions of the Maghreb by the year 2000. At the same time, the magnitude of the investments to be made in the coming 20 to 30 years reinforces the regional awareness of the necessity to give efficient management of water a high priority. The objective is to find means to act on water supply and demand in order to master its evolution.

Increasing Water Availability

First of all, it is necessary to utilize flood discharges through promoting their infiltration into aquifers by appropriate and simple surface structures. Such hydraulic structures presently exist near the center of Tunisia on the Zeroud and Merguellil rivers, and in Morocco on the Souss, Draa and Ziz rivers. They are intended to impede flood water discharge and to provide water to satisfy irrigation needs and to recharge groundwater aquifers downstream. In arid zones, flood water diversion structures dispersing water over large areas will also provide groundwater aquifer recharge as well as supply water for use in farming flood areas.

Sewage water re-use after purification is mainly implemented in Tunisia (81.5 mcm) and in Libya (110 mcm). In Algeria, in spite of an installed capacity for sewage water treatment that was approximately 140 mcm in 1990, the use of reclaimed water remains low due to equipment exploitation difficulties.

Libya uses desalination to produce 100 mcm of fresh water for industrial needs. In other countries of the region, seawater desalination or brackish water demineralization are not widely used at present, although the future plans are for these technologies to supply potable water for small urban centers or to satisfy industrial needs. The main handicap for expanding the use of desalinization techniques is its cost, which remains higher than the conventional long distance water transfer methods. Artificial rain technology should also be mentioned; although in an experimental stage, it has been used in Morocco for more than a decade.

Demand Management

The management of users' water demand requires daily mastery of the water supply situation. This implies a thorough knowledge of water resources through the establishment of an operational hydrologic network that provides, in real time, sufficient and reliable information on qualitative and quantitative values. It implies also making use of decision-making tools, in particular simulation models for surface water and for groundwater conjunctive management scenarios.

Over the past years, the countries of the region have created national hydrologic networks and established management models for the largest hydrologic basins and groundwater aquifers. The objective has been a monthly management plan predicting reservoir storage of surface water for each hydraulic basin to guide the management of available water supplies. This approach enables water delivery scheduling and permits reservoirs to keep low water flows up to certain minimum levels in order to avoid water pollution problems.

The groundwater monitoring network and models have been used to get a thorough knowledge of the aquifer and to establish optimal exploitation scenarios to avoid over-exploitation.

The management of water demand is a complex process as it brings together many operators from many entities and different sectors, often with conflicting interests. It is mandatory to create a climate of confidence and dialogue with the different operators and to obtain their support for the national and regional objectives of water management. It is important that water planning results be discussed and agreed upon by all users, and that a consensus on priorities be reached. In the Maghreb, as mentioned previously, priority is generally given first to potable water supply, then to irrigation, and finally to hydroelectric production.

Water use conflicts are aggravated by both pollution problems and water shortage during drought periods. To face these situations, the national water management authority is responsible for coordination between all the concerned parties. In Morocco, for example, the present focus is on strengthening coordination between the different technical authorities and sectors dealing with water in different parts of the country. This approach reinforces team spirit within each hydraulic area.

Water demand management is a concept new to the region. It was introduced during the drought that affected the Maghreb countries in the 1980s. Facing water shortage problems has led to the initiation of the approach, stressing:

■ The necessity for users to verify and justify their water needs;

■ The search for means of more efficient water use either by operating entities or at the end-users' level;

■ The necessity to share drought-caused water shortage effects between all sectors, including municipal water supply;

■ The urgent necessity to conduct water quality preservation and protection programs.

As result of the drought experience, rehabilitation programs have been implemented to modernize water distribution networks and increase their output. In Morocco, for example, the experience gained during the last few years can be summarized as follows (Boutayeb 1985):

■ For potable water supply, improvement programs since 1986 have been concentrated on the efficiency of distribution networks, which have reached 44 urban centers with a total linear distance of 930 km. The repairs have allowed a cumulative gain of almost 450 liters per second, which represents the water supply required by a city of 120,000 inhabitants.

■ For irrigation, among the various possible methods, there is a clear trend towards water-saving techniques. Thus, 120,000 hectares are presently equipped with sprinkling irrigation systems. This is a quarter of the total area of main irrigation.

■ The use of micro-irrigation techniques (sprinkling, pivot, and drip irrigation) has also accelerated; they now equip 15,000 hectares, for the most part in the Souss region. The financial incentives that were granted to private promoters to acquire the necessary equipment, combined with the private operators' initiative, explain the development of micro-irrigation. Similarly, rehabilitation efforts have been undertaken in the main hydraulic projects as well as in small and medium projects. These programs have allowed a 6 percent increase in efficiency, representing a water saving of 270 mcm per year. Adjustments in water rates for both potable and irrigation water have undoubtedly also contributed to decreases in overall water demand during the last ten years.

Similar efforts have also been undertaken in Tunisia (Khazen et Abib 1991), where the complete development of potential conventional water will be achieved by the year 2010. Consequently, it is imperative to promote water use efficiency. It is with this aim that the study "Economy 2000" is currently being undertaken to define all available actions likely to contribute to a new strategy for water use economy. The general studies undertaken show that the improvement of water use efficiency will postpone recourse to desalination by almost ten years (Tunisia, Ministry of Agriculture 1994).

Water demand management strategies are still recent in the area and their implementation requires reinforcement. The human factor is a key issue, so explanation and public awareness efforts are required in order to convince the different users of the increasing water scarcity, and of the necessity to adopt water-saving techniques in order to ensure sustainable development. Therefore, the measures taken to organize farmers into users' associations are of great importance since they put the responsibility for irrigation network management on the beneficiaries. They also constitute a useful tool for the promotion of water-saving techniques.

Another dimension of this problem is related to the inadequacy of equipment, particularly in the field of drinking water supply. Little effort has been made to give the public, for example, adequate water-saving plumbing.

Drought management

Drought phenomena and their genesis, as well as the mastering of their impact, are the major current concerns in the Maghreb. With limited water resources and the early prospect that they will be entirely in use, the region's fragile hydrological situation is becoming more and more exposed to the risk of failure during drought periods. Furthermore, there are potentials for climate change that may have adverse effects, mainly in arid and semi-arid regions.

During the last 20 years, the Maghreb has known exceptionally long severe drought periods: in Tunisia from 1987 to 1989, and in Morocco from 1979 to 1984 and 1991 to 1993. Morocco has not experienced such intense periods of drought since those of 1943–45, 1937, and 1930.

In Morocco for example, the available rainfall observations indicate a significant decrease, around 40 mm per year, at some gauge stations. In fact, Tangier and Ifrane gauge stations have shown rainfall decreases respectively of 100 mm in 40 years and 400 mm in 30 years. Moroccan dendrochronological data, covering 1000 years, augment the modern hydrological measurement series, which generally does not exceed 40 years. The tree-ring analysis indicates that:

■ A drought occurs every 8 years on the average.

■ On the average, during the period 1000–1984, 116 dry episodes were recorded; duration was less than 2 years in 84 percent of the cases.

■ During the same period, the recorded dry episodes are:

  ◆ Seven with a duration of 3 years

  ◆ Three with a duration of 4 years

  ◆ Four with a duration of 5 years

  ◆ Three with a duration of 6 years

  ◆ One with a duration of 7 years.

■ Drought periods of 6 to 7 years, like the drought of 1979 to 1984, have a return period of 455 years.

During drought periods today, reservoir analyses, which are based on the reservoir water levels at the beginning of the hydrological year, always provide for the needs of a dry period. Management scenarios are established taking into account both the expressed water demands and various restrictive hypotheses for water supply. In general, the analysis is conducted for a period of more than one year in order to ensure the inter-year transition. This analysis allows the identification of the geographic areas and the user sectors that could be influenced by the supply restrictions, and it is the basis for a dialogue with users, aimed at obtaining their adherence to the proposed drought management plan. In case of conflicts among uses, governmental authorities are requested to arbitrate.

The current drought that has affected Morocco since 1991 has brought a surface water supply deficit of 75 percent on the average. A drought management plan has been established, with the following points:

■ No restriction on domestic water supply, with the exception of Tangier where the restriction rates are 30 and 50 percent respectively during winter and summer seasons;

■ Irrigation water supply is subjected to restrictions ranging from 30 to 55 percent of the scheduled deliveries at the beginning of the agricultural season;

■ Hydroelectric energy production is limited to water releases available after priority needs have been met;

■ For domestic/potable water, committees are established at local levels to guard against all forms of water waste at both individual and collective levels, and programs are intensified against leakage in distribution networks;

■ For irrigation water, farmers are organized through both professional and water users' associations in order to improve the efficiency of irrigation networks. Irrigation is scheduled according to the following hierarchical priorities:

◆ Priority to arboriculture and semi-perennial farming

◆ Priority to industrial farming

◆ Reduction in irrigation application rates

◆ Reduction or suppression of new farming programs

◆ Banning of heavy water-consuming crops.

Water savings are also realized through the construction of small dams and temporary dikes to facilitate flow management on streams. Finally, recourse to groundwater, when it is feasible, is also highly recommended. Drought impacts on the rural population are reduced in large part by supplies from groundwater. In fact, drilling wells and boreholes are among the main strategic components of drought management plans.

Environmental impacts of economic development on water

The Maghreb region faces two major water resource development problems: erosion and soil loss leading to reservoir silting, and water pollution. Both contribute to an increasing fragility of the national hydrologic picture.

Soil conservation

The northern fringe of the Maghreb region has a strong problem of soil degradation, which often exceeds 2000 tons per km² per year. This loss is linked to human pressure exerted on mountain soil through deforestation and poor farming methods, and linked to unfavorable natural factors such as vulnerable geology and high intensity rainfalls. The resulting soil erosion affects the hydrological regime and imperils surface water distribution networks.

■ In Morocco, with a total storage capacity of almost 10 bcm from nearly 80 dams, an average yearly water volume of 50 mcm of storage space is lost to siltation. Reservoir capacity already lost to silting is on the order of 800 mcm at the end of 1990. With new dam construction, annual lost volume is expected to increase to 100 mcm by the year 2000. In general, for Moroccan reservoirs, we see that:

◆ Dams that were constructed 20 years ago have lost more than 10 percent of their initial storage capacity through silting; three reservoirs have lost 50 percent of their initial capacity through silting, and seven reservoirs have lost their dead storage.

◆ It has been necessary to raise four dams to increase storage.

■ In Algeria, for the 37 dams in operation in 1990, with an initial total capacity of 3.9 bcm, the accumulated siltation loss is estimated to be 430 mcm (11 percent). This lost volume is expected to reach 930 mcm (24 percent) by the year 2010.

Apart from the agricultural consequences of soil loss, the alarming problems that erosion brings for water management can be summarized as follows:

■ The reduction of reservoir capacity due to the silting means that there is less water under control.

■ Providing for siltation complicates structural design and increases the costs of dams.

■ The decrease in structural performance is translated into economic losses at the users' level. It forces a search for new reservoirs on more and more difficult sites at higher costs.

■ The sediment burden also complicates the operation of downstream hydraulic structures (pumping stations, canals, etc.), increasing the costs of their operation and maintenance.

For all these considerations, soil conservation is becoming an important line of strategy in the Maghreb. The objective is to promote a strong solidarity between the upstream and downstream portions of watersheds in order to ensure water resources development on a sustainable basis.

Water pollution problems

Problems created by water pollution are recent and have accompanied rapid urban, industrial, and agricultural development. Water quality conservation programs have not kept up with the increasing pace of water resources development in the region. The seriousness of water pollution problems has become manifest during recent droughts, when drops in surface and base flows aggravated water quality deterioration, and highlighted the fragility of the hydrologic systems. The principal observed pollution is that of large cities, as well as industrial discharges into the hydrological network, and recently, forms of pollution generated from agricultural activities, mainly by the leaching of fertilizers and plant control products.

Tunisia has realized the most significant progress in pollution abatement through the Office National de l'Assainissement (ONAS), created in 1974. With 24 operational sewage water treatment plants in 1990, ONAS is treating the pollution generated by three quarters of Tunisia's urban population. These stations allow the yearly collection and treatment of 176 mcm of used water, reflecting the connection of 70 percent of the urban population to the sewage network. The ongoing sewage treatment program, which includes the building of 20 new treatment plants, bringing treated water production to 200 mcm, will facilitate the use of treated water for irrigation.

In Algeria, 77 water sewage treatment plants have a capacity of 140 mcm/year but management difficulties are handicapping water quality preservation. Libya treated an estimated 100 mcm of sewage water in 1990.

In general, there is a very great need for a strategy of quality management in the Maghreb countries in order to avoid the degradation of the developed water resources, and to incorporate water reuse into "water needs/water resources" balances for the future decades.

The institutional framework

Legislative and regulation aspects

The objectives of the legislation in effect are to allow rational usage of water and to ensure its protection against degradation. Most legislation provides for preservation of public rights with regard to water, and for water quality conservation. In a planning context, however, specific legislation measures have not been promulgated in Morocco, Algeria nor Tunisia. However, the draft Moroccan water law that is now under study dedicates a full chapter to water resources master plans.

In spite of numerous legal measures defining conditions of withdrawal, usage, and discharge, implementation remains limited for many reasons:

■ Water control is ineffective in practice, and in certain cases almost non-existent. This situation contributes to inadequate water management.

■ In certain cases, some regulations are overly severe and consequently unenforceable. This has an adverse effect on the management of water.

In addition, associated measures to make these rules effective, such as financial incentives for water conservation and protection, remain insufficient.

Economic and financial tools

The imposition of charges on drinking water supplies in the region is intended to fulfill the following objectives:

■ To allow domestic water access for the entire population, with the intention of facilitating water access for the low-income population

■ To save water, through increasing the charges for high consumption

■ To promote the drinking water sector by allowing operators to develop financial resources for new investments.

To resolve these problems, the adopted tariff system, in Morocco, Algeria, and Tunisia, is based on a progressive selling price for drinking water. Thus, in Morocco:

■ The first portion of consumption is billed at a rate intended to cover basic, irreducible needs.

T A B L E   6
Average Price for Domestic Water in Some Countries of the Maghreb in 1992 ($/m³)

	Selling Price($/m³)
Mauritania	1.35
Tunisia	0.56
Morocco	0.36
Algeria	0.14

Source: National Office of Potable Water, Morocco

■ The second portion is generally billed at a tariff close to the cost price.

■ The third level is billed at a rate that compensates for the deficit generated by the first portion.

A list of tariffs charged by each country of the Maghreb is shown in Table 6. Note that disparities exist in the applied tariffs in the region.

Table 7 shows the charges for domestic water, by quarterly consumption bands, for urban centers in Morocco.

Considering the high cost of hydro-agricultural investments and in order to lighten the farmers' charges, irrigation tariffs are designed to charge agricultural users for only a part of the government public expenses:

■ In Morocco, farmer payments are fixed at a level of 40 percent of the equipment costs after deduction of the energy production costs; the first part of this participation is fixed at 30 percent as a direct contribution to the development of the irrigated lands; the second part is an annual tax for water usage fixed at 10 percent. Table 8 shows tariffs in some irrigated areas of Morocco in 1992.

T A B L E   7
Prices for Urban Water by Consumption Levels in Rabat and Casablanca (Morocco) in 1992 ($/m³)

	Quarterly Tariff for Domestic Use ($/m³)			
	First Level 0 – 24 m³	Second Level 25 – 60 m³	Third Level > 60 m³	Industrial Tariff
Rabat	0.174	0.490	0.676	0.375
Casablanca	0.147	0.405	0.588	0.361

Source: National Office of Potable Water, Morocco

TABLE   8

Irrigation Water Charges in Some Areas of Morocco in 1992 (US $/m³)

Area	Selling price ($/m³)
Gharb	0.020
Haouz	0.018
Tadla	0.018
Doukkala	0.019
Low Moulouya	0.020

Source: Directorate of Planning and Economic Studies, Ministry of Agriculture, Morocco

■ In Algeria, the adopted fee schedule for water irrigation aims to cover full operation and maintenance expenses. Thus, there is a single tariff applied in Algeria. In addition a tax is applied on a maximal subscribed discharge, which varies according to the specifics of each irrigated tract.

Generally, water charge policies do not pass along to the farmers or consumers all water development costs. They are worked out from financial considerations that are specific to the operators who are in charge of water services. The applied rates reflect neither the total water development costs nor, in some cases, the full expenses of the infrastructure's operation and maintenance. Water cost recovery also remains weak in relation to the scale of future water development that must be carried out in the mid and long terms, and also relative to the funds needed for water conservation and protection of water quality.

Water administration

In all countries of the region, the responsibility for conducting national water policies is in the hands of a central governmental department, which is in charge of water assessment, and planning and management of both quantity and quality aspects. These departments are, in Mauritania, the Ministry of Hydraulics and Energy; in Morocco, the Ministry of Public Works, Vocational Training, Managerial and Staff Training; in Algeria, the Ministry of Equipment; in Tunisia, the Ministry of Agriculture; and in Libya, the Secretariat of Dams and Water Resources.

Other ministries as well inevitably take part in water resources management, such as the ministries of Agriculture, Industry, Energy, and Interior. They implement the policies of their respective sectors and ensure the supervision of water users in those sectors. These departments intervene in water resources management through public administration channels at the national or regional level or through organizations that act as water users' representatives. Other departments also intervene at a horizontal level such as the departments of Health, Environment, Economy and Finance.

Certain water activities are attributed to special public departments, such as the National Office of Potable Water and the National Office of Electricity in Morocco, and in Tunisia, ONAS and SONEDE, which is a company for water distribution. In Algeria, network management is ensured by 9 regional firms and 26 state-owned firms for drinking water, 4 offices covering broad areas and 8 county offices for irrigation. All these public departments are endowed with autonomy in order to act flexibly. To ensure a rational management of water, and in view of the numerous participants, coordination of different actions is indispensable at national and regional levels.

In Morocco, the National Water and Climate Council was created to set the general orientation of the country's water development. Although recent, the experience this body has gained in coordination is promising since it gathers together administrative representatives, water users, local councilors and local community representatives as well as professional organizations and experts.

Education and public awareness

In the education and public awareness fields, programs are intended to develop a general awareness of the importance of water and the danger of its scarcity. However, these programs, which concern only drinking water supply, have been only occasionally implemented, during either a water shortage crisis arising for technical reasons or a drought period. Public awareness activities suffer from a lack of effectiveness and do not fulfill the desired objectives. To have positive impacts on water management, they must be planned as part of the whole water resources management strategy, and should take fully into account the human factor. Creating public awareness of the need for efficient water use has to be consciously targeted, and then implemented. Appropriate awareness campaigns have to be conducted in order to inform users and to enable them to contribute to participatory water resources management.

Conclusion

Despite the unfavorable natural context, the Maghreb region has appreciable water resources that are indispensable for its economic and social development. The expected strong population growth in the region requires a high level of water resources development, which is, however, confronted by various physical, technical and economical constraints.

The strong hydrologic variability requires large capacities of surface water storage in order to regularize the large supplies of wet years and to conduct multi-annual management. The high cost of system investments mandates an active water conservation and protection policy, for both quantity and quality aspects. The following points are essential:

■ A better integration of water resources development and watershed anti-erosion management is required to ensure the sustainability of infrastructures in the face of siltation.

■ The conservation of water quality is essential from two points of view: the regeneration of water quality and water conservation create, in effect, additional invaluable water resources on one hand, and on the other hand, they contribute to the improvement of the water resources environment, mainly in the rural area.

■ The implementation of rational and economic water management is also an imperative strategy in a fragile hydrologic system marked by increasing scarcity of water. In the future, attention must be focused on both the management and the development of water resources.

■ Water management has to be decentralized at a local watershed level in order to develop more participation, coordination and solidarity, and to allow a better involvement of users in the decision-making process.

The current experience of the Maghreb region demonstrates similarities among its countries with regard both to shared problems, and to similar approaches seeking adequate solutions. We see a broad scope of cooperation, which allows the countries to unify their efforts in order to deal with the water scarcity management challenge of the year 2000.

7

Water in the Arabian Peninsula: Problems and Perspectives

Jamil Al Alawi and Mohammed Abdulrazzak

Overview and Introduction

Economic development and increasing population worldwide are placing great demands on existing water resources and are reducing water quality. Water supply problems are particularly pronounced in arid regions, notably in the countries of the Arabian Peninsula (shown in Figure 1): Saudi Arabia, Kuwait, Bahrain, Qatar, the United Arab Emirates, Oman, and Yemen — a region that covers 3.11 million square kilometers and has a total population of 31 million people (Al-Gazi 1990).

These nations are in general devoid of reliable surface water supplies and depend entirely on groundwater and desalination to meet their water requirements. Rapid population growth (as shown in Table 1), along with the expansion of irrigated farming, urbanization, increased economic activities, and improvement in the standard of living have brought about substantial water demand increases. In particular, ambitious irrigated agriculture programs and extensive urban expansion, coupled with the absence of conservation measures, are causing significant overdraft of water resources beyond their natural renewal capacity. Total annual demand has increased from 6 to 22.5 billion cubic meters (bcm) in the ten year period 1980–1990, and the agricultural sector is now taking substantial volumes from fossil groundwater reserve. If current water use practices continue, annual water demand may reach 35 bcm by the year 2010. This would require construction of additional desalination plants and even more intense mining of the non-renewable groundwater.

Set against the background of basic dryness, the region's water resource problems include the difficulties of capturing very intermittent surface runoff, over-consumption, depletion of fossil groundwater, saltwater

FIGURE   1

Geographical Location of Countries of the Arabian Peninsula

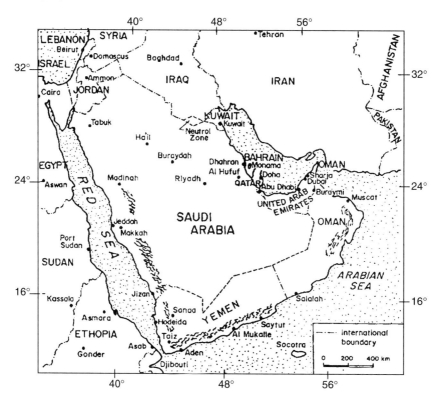

TABLE   1

Population Distribution of Countries of Arabian Peninsula

		Population in Millions					
		1970			1990		
	Area (km²)	Urban	Rural	Total	Urban	Rural	Total
S. Arabia	2,149,690	3.767	3.973	7.74	9.188	2.592	11.78
Kuwait	17,818	0.581	0.179	0.76	1.942	0.169	2.11
Bahrain	652	0.213	0.069	0.282	0.408	0.102	0.51
Qatar	11,610	0.137	0.034	0.171	0.378	0.042	0.42
UAE	77,700	0.29	0.215	0.505	1.409	0.331	1.74
Oman	300,000	0.112	0.638	0.75	0.153	1.237	1.39
Yemen	550,000	0.98	6.5	7.48	2.592	10.367	12.96
Total	3,107,470	6.08	11.608	17.688	16.07	14.84	30.91

intrusion, pollution of shallow aquifers, deficient institutional arrangements, lack of trained personnel, and poor management practices. With most renewable water resources fully developed, both shallow and deep fossil groundwater aquifers are being depleted at an accelerating rate to meet the rapidly rising water demand of irrigated agriculture.

Water is provided free of charge for agricultural development. Domestic water supply, especially for major urban centers that depend primarily on seawater desalination, is heavily subsidized; the consumer pays only about one-tenth of the actual production costs. These non-existent and low water charges have encouraged over-consumption, not only leading to extensive mining of the groundwater reserve but also stimulating production of the excessive wastewater that is contaminating some groundwater sources.

Despite the natural supply limitations and over-consumption, comprehensive water policies, and plans to assess, develop, and manage water resources on national or regional levels are lacking. In the past, each country's fragmented policies stressed development of one economic sector without regard to effects on other sectors. Agriculture was greatly expanded to establish food self-sufficiency; cereal production has now reached export level. Water supplies were developed without concurrent conservation programs. Ambitious wastewater re-use programs are being developed without provision for the enforcement of necessary regulations. Water resource programs usually focus on specific areas such as water supply, flood control, and wastewater re-use, without comprehensive or integrated approaches. Most of the peninsular countries do not have water plans in place, and the plans that do exist need to be reevaluated and updated. The weakness of existing institutional arrangements is a major constraint. Water agencies have been multiplied, resulting in an overlap of responsibilities and duplication of efforts. Lack of cooperation and coordination among agencies on local and regional levels, as well as the absence of clear legislated guidance for their administrative work gives rise to conflicts and to gaps in activities.

The solution for such impending water crises lies in the formulation and implementation of a long-term management plan that emphasizes conservation, recycling, pollution control, appropriate institutional arrangements, and capacity building. Regional cooperation and coordination between the countries of the peninsula is also needed, particularly for the management of the shared groundwater resources.

In this paper the water resource problems, as well as water requirements, for all countries of the Arabian Peninsula are addressed, and management alternatives to cope with the imbalance between water supplies and demand are suggested.

Water Resources

Climate

The countries of the Arabian Peninsula have similar physiographic, social, and economic characteristics including extremely arid climates, sparse natural vegetation, and fragile soil conditions. The peninsula is largely desert with the exception of the coastal strips and mountain ranges. The climate is characterized by long, hot, dry summers and short, cool winters for the interior regions and hot, somewhat more humid summers, and mild winters for coastal regions. Hydro-meteorological parameters exhibit great variation: seasonal temperatures may range from -5° to 46°C in the north, central, and eastern parts of the peninsula. The coastal areas and mountainous highlands have lower and less extreme temperatures, ranging from 5° to 35°C. Humidity is generally low in the interior, ranging from 10 to 30 percent, while in the coastal areas it may range between 60 and 95 percent. The low percentage of cloudy days and high solar radiation over the region result in high evaporation rates. The total annual potential evaporation ranges from 2500 mm in the coastal areas to more than 4500 mm inland.

The Arabian Peninsula generally has scanty and irregular rainfall and is devoid of rivers or lakes. The average annual rainfall ranges from 70 to 130 mm, except in the mountain ranges of southwestern Saudi Arabia, Yemen, and southern Oman, where rainfall may reach more than 500 mm. Average rainfall has little meaning, however, since many desert areas receive no rainfall for months or years due to extremely random storm patterns. Rainfall throughout the area is generally governed by regional Mediterranean and Indian subcontinental air circulation patterns. Cyclonic lifting associated with the eastward passage of depressions from the Mediterranean region causes winter rainfall in the months of November and April over the northern and eastern parts of the peninsula. Spring rainfall sometimes occurs over the central and southwestern parts of Saudi Arabia, Yemen, and Oman. The Indian subcontinent monsoons influence the weather over the southern region of the peninsula, resulting in summer rainfall that occurs mostly over the southern parts of Saudi Arabia, Oman, the United Arab Emirates, and most of Yemen. All of these circulation patterns are modified by local topographic relief and distance from the sea. In general, rainfall amount decreases sharply with distance from the sea and in a northern direction. The steep relief of the mountain ranges of the Asir highlands in southwestern Saudi Arabia, Sarat mountains in western Yemen, and the Hajar and Dhofar mountains in the southern regions of Oman that run parallel to the Red Sea and the Gulf usually causes orographic rainfall that frequently causes flash flooding, particularly in the summer months. Most rainfalls in these areas are of high intensity and short duration, producing a large volume of surface runoff that gathers in wadis, stream beds that are normally dry. From there, such sharp rainfalls are utilized in flood irrigation, and impounded behind dams, and they also usually recharge the alluvial aquifer beneath a wadi.

Physiographic and geological features

Water availability in the region is governed by rainfall distribution, which is related in turn to topographic and geological features. The main topographic features are the western, southwestern, and southeastern mountain ridges, as well as the central plateau of the peninsula, which bring orographic rain as discussed above. Surface runoff is more abundant in these regions than in the rest of the peninsula. The mountain ridges create and divide the large number of small drainage basins that empty toward the Red Sea, Arabian Sea, and the Gulf of Oman, as well as the larger dryer basins that drain toward the central plateau, some of which continue eastward toward the Gulf. Generally, the coastal drainage basins have steep relief and narrow coastal plains, compared to the mild slope and large size catchment of the central region. The remainder of the peninsula is characterized by low relief and poorly defined drainage patterns.

The other major features that influence the availability of groundwater resources are the peninsula's igneous and metamorphic basement rock known as the "Arabian Shield," and the sequences of sedimentary layers known as the "Arabian Shelf," shown in Figures 2 and 3. The shield, which covers one-third of the peninsula, consists of an outcrop of hard rock that begins in the western part of Saudi Arabia and extends from the Gulf of Aqaba in the north to the Gulf of Aden in the south. The shield has limited groundwater stores in the alluvial deposits of wadi channels, and geological joints and fracture zones.

The dependable groundwater reserves are those stored in the thick extensive sequences of sedimentary formations of the Arabian Shelf, underlying two-thirds of the peninsula (Figure 2). The outcrops of these formations where recharge may take place are located in the western part of the peninsula and the formations slope gently with increasing thickness as they extend eastward under the Gulf, to the northeast into Jordan and

F I G U R E   2

General Geological Map and Aquifers of the Arabian Peninsula

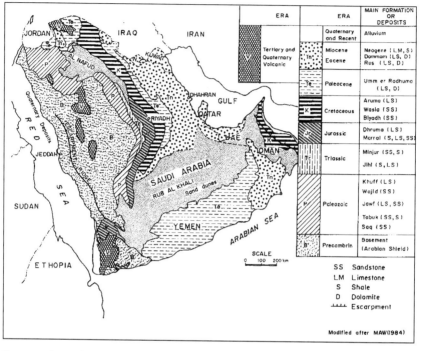

Iraq, and southwest to Yemen. In combination with rainfall distribution, these topographic and geological features control surface and groundwater availability and use in different parts of the peninsula. The water resources of each country of the peninsula are shown in Table 2.

Surface Water

Due to infrequent and low amounts of rainfall, surface runoff is small over most of the Arabian Peninsula. In the northern and central parts of Saudi Arabia, most of Kuwait, Bahrain, Qatar, and the northern part of the United Arab Emirates, runoff is generally unavailable. Ephemeral flow occurs only from localized storms, but still constitutes the main source of groundwater recharge to the shallow alluvial aquifers. The rainfall runoff coefficient is very small, ranging from 2 to 15 percent for most parts of the peninsula due to the nature of the soil cover and drainage characteristics. The only areas receiving enough rainfall to generate reasonable amounts of runoff are the southwestern parts of Saudi Arabia, most of Yemen, the southern part of the United Arab Emirates, and southern parts of Oman.

T A B L E 2

Water resources of Arabian Peninsula Countries (1992)

	Annual Rainfall (mm)	Annual Evaporation (mm)	Runoff (mcm)	Shallow Ground-water Reserve (mcm)	Ground-water Recharge (mcm)	Per Capita Consump-tion (L/D)	Ground-water Use (mcm)	Desalination (mcm)	Waste-water Re-use (mcm)
S. Arabia	75	3500–4500	2230	84,000	3850	252	14430	795	217
Kuwait	70	1900–3500	0.10	182	—	400	80	240	83
Bahrain	70	1650–2050	0.20	90	100	455	160	56	32
Qatar	67	2000–2700	1.35	2,500	45	407	144	83	23
UAE	89	3900–4050	150	20,000	125	81	900	342	62
Oman	71	1900–3000	918	10,500	550	146	645	32	10.5
Yemen	122	1900–3500	3500	13,500	1550	27	1200	9	6
Total	—	—	6800	130,772	6220	206	17559	1557	433

Source: BAAC 1980, De Jong 1989, Shahin 1989, Khoury et al. 1986, Danish 1990 and Mohamed 1986.

Annual surface runoff volume in the peninsula is estimated to average 6.8 bcm (Abdulrazzak 1992); its inter-year variation, however, is very pronounced. Almost half the annual runoff is generated in southwestern Saudi Arabia and Yemen, estimated at 1.45 bcm and 2.1 bcm, respectively. The national totals for Saudi Arabia (Authman 1983; BAAC 1980) and Yemen (Mohamed 1986; Al-Fusail et al. 1991) are estimated at 2.23 bcm and 3.5 bcm, respectively. Amounts of surface water available in Oman (El-Zawahry 1992) and the United Arab Emirates (Al-Asam 1992; Uqba 1992) were estimated at 918 million cubic meters (mcm) and 150 mcm, respectively. The remaining countries have only negligible amounts of surface runoff.

In general, surface runoff is used for flood irrigation and the recharge of alluvial aquifers. Approximately 185 dams of various sizes have been constructed in Saudi Arabia for flood protection and groundwater recharge; they have a combined storage capacity of 475 mcm. Thirty dams have been or are being constructed in Yemen, the United Arab Emirates, and Oman.

Shallow alluvial aquifers

Alluvial deposits along the main wadi channels and the flood plains of drainage basins make up the shallow groundwater system in the peninsula. Groundwater in the shallow aquifer is the only renewable water source for these countries. The shallow aquifers in the eastern part of the peninsula, particularly in the United Arab Emirates and Oman, are generally thicker and wider than in the west, and alluvial thickness in the inland basins is greater than those of the coastal basins. Alluvial aquifer thicknesses generally range from 20 to 200 meters, with the exception of the coastal areas of Oman where thicknesses may reach 400 meters. The width of these alluvial aquifers may range from a few hundred meters to several kilometers. The widths of the aquifers decrease in a southerly direction for basins on both the western and eastern coasts. The coastal alluvial aquifers are subject to saltwater intrusion, especially on the Gulf, due to extensive groundwater withdrawals. Shallow aquifer water quality is generally good with total dissolved solids ranging from 300 ppm to 3000 ppm. Combined reserves of the alluvial aquifers shown in Table 2 are estimated at 131 bcm (Abdulrazzak 1992; Shahin 1989; Khoury et al. 1986), with the largest single reserve, that of Saudi Arabia, estimated at 84 bcm (BAAC 1980; MAW 1984; Ukayli 1988). Groundwater from the shallow alluvials is used mainly for domestic and irrigation purposes.

Fossil groundwater aquifers

The other main source of water for the countries of the Arabian Peninsula is the non-renewable fossil groundwater stored in the sedimentary deep aquifers. The Arabian shelf formations of sandstone and limestone, shown in Figures 2 and 3, store significant amounts of groundwater that is thousands of years old (Burdon 1973, Edgell 1987).

The sedimentary aquifers in the Arabian Peninsula have been classified as either primary or secondary based on areal extent, groundwater volume and quality, and development potential (MAW 1984). The primary aquifers are the Saq, Tabuk, Wajid, Minjur-Druma, Wasia-Biyadh, Dammam, Um er-Radhuma, and Neogene. The latter two are carbonate aquifers while the remainder are sandstone. Secondary aquifers are the Aruma, Jauf, Khuff, Jilh, Sakaka, the upper Jurassic, the lower Cretaceous, and Buwaib. These aquifers cover two-thirds of Saudi Arabia and some of them extend into Kuwait, Bahrain, Qatar, the United Arab Emirates, Oman, and Yemen as well as into Jordan, Syria, and Iraq. A typical cross-section of aquifer sequences is shown in Figure 3.

FIGURE   3

Schematic Geological Section of Deep Aquifers

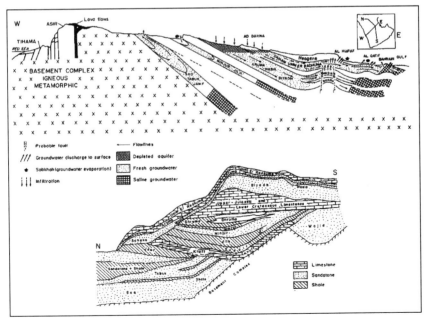

Vast amounts of groundwater stored in the primary deep aquifers serve as the only dependable source of water for Saudi Arabia and, to a lesser extent, the other countries of the peninsula. Deep groundwater reserves for the aquifers in the peninsula, shown in Table 3, are estimated at 2175 bcm with the major portion (1919 bcm) located in Saudi Arabia. Recharge for all the deep aquifers is estimated at a very limited 2.7 bcm per year. This reserve represents groundwater exploitable by lowering the water level to 300 meters below the ground surface, the maximum depth currently possible with modern pumping technology.

Although water in the deep aquifers is large in quantity, the quality varies greatly, being suitable for domestic consumption in only a few areas. Total dissolved solids range from 200 to 20,000 ppm. Good quality water is stored in the Saq, Tabuk, and Wajid aquifers in Saudi Arabia, and the Neogene in Bahrain and Kuwait. Brackish water from the Minjur, Wasia-Biyadh, Dammam, and Um er-Radhuma aquifers usually requires treatment in most of the countries for hardness and high temperature. Water temperatures vary between 40° and 65°C depending on the depth of extraction. Water from the deep aquifers tends to be saturated with calcium and magnesium salts, and to have high concentrations of sulfate and chloride ions, and there are relatively large quantities of hydrogen sulfide and carbon dioxide gases. The brackish water from some of these deep aquifers is used without treatment for

T A B L E   3

Deep Aquifer Groundwater Reserves to a Depth of 300m below Surface

Aquifer name	Reserve (bcm)	Recharge (mcm)	Water quality (ppm)
Saq	280	310	300–1,500
Tabuk	205	455	200–3,500
Wajid	225	104	500–1,200
Minjur-Dhruma	180	80	1,100–20,000
Wasia-Biyadh	590	480	900–10,000
Um er-Radhuma	190	406	2,500–15,000
Dammam	45	200	2,600–6,000
Khuff & Tuwail	30	132	3,800–6,000
Aruma	85	80	1,600–2,000
Jauf & Sakaka	100	95	400–5,000
Jilh	115	60	3,800–5,000
Neogene	130	290	3,700–4,000
Total	2,175	2,692	

Source: MAW 1984; Khoury et al. 1986; Edgell 1987; BAAC 1980; Lloyd 1990; and Abdulrazzak 1992.

agricultural purposes, and for domestic purposes in some locations in Saudi Arabia, Bahrain, Qatar, and the United Arab Emirates. However, before use in the remainder of the peninsula, brackish water requires treatment such as cooling, aeration to remove hydrogen sulfide and carbon dioxide gases, and lime soda processing.

Shared Water Resources

Regional water resources management requires careful consideration because sharing water among countries of the peninsula is a delicate matter. As a result of the infrequent rainfall and intermittent nature of runoff over the shared drainage basins, the resource is basically scarce and unpredictable. Most of the shared drainage basins are between Saudi Arabia and its neighbors. The shared catchments include Hamadah and Sirhan among Saudi Arabia, Jordan and Iraq; Al-Batin among Saudi Arabia, Kuwait and Iraq; Najran, Liyah and Khulab between Saudi Arabia and Yemen; and Burami, Al-Ayn among Saudi Arabia, Oman, and the United Arab Emirates. Surface water availability in these shared basins is negligible; what is important is that each represents a shared alluvial (shallow) aquifer. Developing each requires an agreed plan for pumping in different locations and understandings on groundwater recharge. Achieving both shared use and optimum use of trans-border shallow groundwater resources presents a major challenge to decision-makers in the region.

Some of the deep fossil aquifers, which are the most dependable sources of water for urban expansion and extensive agricultural activities, also extend under national borders. Therefore they, too, raise the issue of sharing and cooperative management. Shared deep aquifers cover most of Saudi Arabia and extend from Saudi Arabia into Kuwait (Dammam, Aruma, and Neogene), into Iraq, Bahrain, Qatar, United Arab Emirates, and Oman (Dammam, Aruma, Um er-Radhuma, and Wasia-Biyadh), into Yemen (Wajid, Um er-Radhuma, and Wasia-Biyadh) as well as the neighboring countries of Jordan and Syria (Saq and Tabuk) (FAO 1979). The approximate areal extent of these aquifers is shown in Figure 4, while their vertical relationship is indicated by Figure 3.

The Saq and Tabuk deep aquifers, which underlie the central and northern part of Saudi Arabia and southern Jordan and Syria (MAW 1984, Edgell 1987), support extensive agricultural development in these regions and provide water for many towns and villages. The Wajid lies below the southern part of Saudi Arabia and the northern part of Yemen. Its water is used mainly for agriculture in southern Saudi Arabia, but is the main water supply for the Yemen's urban region of Sanaa. The Wasia-Biyadh

FIGURE   4

Shared Groundwater Aquifers

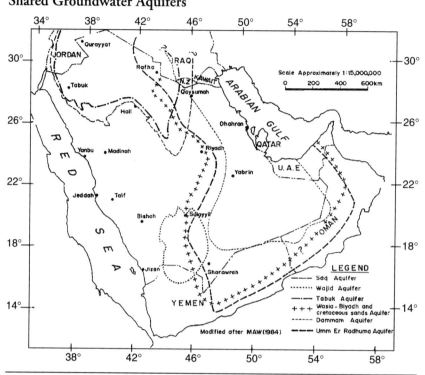

Modified after MAW (1984)

LEGEND
— - — Saq Aquifer
· · · · · · · Wajid Aquifer
——— Tabuk Aquifer
+ + + Wasia - Biyadh and cretaceous sands Aquifer
- - - - - Dammam Aquifer
━ ━ ━ Umm Er Radhuma Aquifer

aquifer is one of the main sources for Saudi Arabia's capital, Riyadh, and a large number of towns and villages, and agriculture activities in the central regions of Saudi Arabia. The Wasia aquifer also supplies water to Sanaa in central Yemen and a number of villages in Oman. The Um er-Radhuma is tapped to meet a portion of irrigation requirements in eastern Saudi Arabia, Bahrain, Qatar (Al-Mahmoud 1992), and the United Arab Emirates. The Dammam aquifer covers most of the eastern part of Saudi Arabia and territories of Bahrain (Al-Mansour 1986), Kuwait, Qatar, the United Arab Emirates, and western Oman and its water serves both domestic and agricultural purposes. The Neogene aquifer satisfies domestic and agricultural requirements in eastern Saudi Arabia, Kuwait, and Bahrain.

Desalination

During the last twenty years, the countries of the Arabian Peninsula have become increasingly dependent on desalination to meet their water supply requirements. They have become, by necessity, world leaders in desalinating sea water or brackish groundwater for domestic consumption. The present annual designed desalination capacity of the seven countries has reached 2.02 bcm, compared with a worldwide capacity of 5.68 bcm (Bushnak 1990, Wagnick 1992). These capacities cover all desalination plants and include numerous units in private sector ownership for industrial or other purposes. Particularly Saudi Arabia, Kuwait, and the United Arab Emirates rely on large-scale plants capable of producing up to 500 mcm per year. In 1992 about 80 percent of the desalination capacity in the region used the multi-stage flash (MSF) process and most of the remaining large-scale plants are reverse osmosis (RO).

Desalination production ranges between 70 percent and 85 percent of designed plant capacity. Total regional volume of actually produced desalinized water in 1992 was estimated at 1.56 bcm (Al-Mugran 1992, Al-Sufy 1992), as shown in Table 2. Desalinated water met over 51 percent of urban and industrial water demand in 1992. The major producers of desalinated water are Saudi Arabia (51 percent), United Arab Emirates (22 percent), Kuwait (15 percent), Qatar (5 percent), Bahrain (4 percent), Oman (2 percent), and Yemen (1 percent).

The total number of desalination plants in operation as of 1992 was 45: 23 in Saudi Arabia, 8 in the United Arab Emirates, 6 in Kuwait, 3 in Bahrain, and 2 each in Oman and Qatar, and 1 in Yemen (Al-Sufy 1992). In Saudi Arabia 17 plants are located on the Red Sea coast and 6 on the Gulf. Three large-scale MSF plants are located at Al-Jubail, Jeddah, and Al-Khobar with annual production capacity of 394 mcm, 217 mcm, and 83 mcm, respectively. The locations of desalination plants in the peninsula are shown in Figure 5.

Renovated Wastewater

Existing wastewater treatment facilities in the Arabian Peninsula face difficulties in handling the ever increasing volumes of wastewater generated by increased water consumption and urbanization. Wastewater discharge from major urban centers is polluting shallow alluvial aquifers and the coastline, and has caused urban water tables to rise. The main emphasis to date in these countries has been on simple disposal of wastewater, rather than on treating and re-using effluent. Planning for the full utilization of treated effluent remains in the early stages, and the regional treatment capacity is sufficient to handle only 40 percent of the domestic wastewater generated. The total volume of renovated wastewater used in the Arabian Peninsula is estimated to be about 433 mcm. The re-

FIGURE 5

Geographical Location of Desalination Plants in the Arabian Peninsula

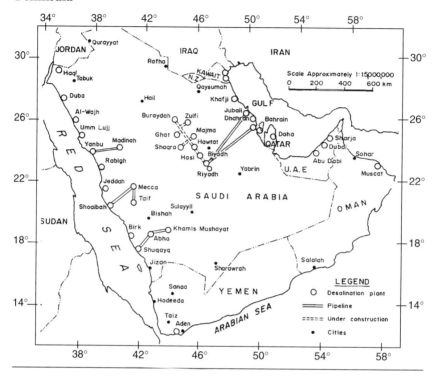

use volumes, which represent approximately 25 percent of the available treated wastewater, are shown in Table 2.

Wastewater re-use ranges between 217 mcm in Saudi Arabia and 6 mcm in Yemen. In the region as a whole, renovated wastewater meets about 2 percent of total water demand or 14 percent of domestic and industrial demand. In Saudi Arabia, reclaimed wastewater is used for irrigation of non-edible crops, landscape irrigation, and for industrial cooling. In Kuwait, Bahrain, the United Arab Emirates, and Oman it is used for municipal irrigation of landscaped areas, while in Qatar it is used to irrigate animal food crops.

Water Demand

Current and future regional requirements

During the last decade, extensive development, rapid population growth, and substantial improvement in the standard of living in the

T A B L E   4

Water Demand of Countries of the Arabian Peninsula, 1980-1990 (in mcm)

	1980		1990			Total Demand 1980	Total Demand 1990
	Domestic & Industrial	Agri-culture	Domestic	Industrial	Agri-culture		
S. Arabia	502	1,860	1,508	192	14,600	2,362	16,300
Kuwait	146	40	295	8	80	186	383
Bahrain	46	92	86	17	113	138	216
Qatar	50	60	76	9	109	110	194
UAE	229	560	513	27	950	789	1,490
Oman	15	650	81	5	1,150	665	1,236
Yemen	98	1,600	144	72	2,500	1,698	2,716
Total	1,086	4,862	2,703	330	19,502	5,948	22,535

Source: MOP 1985, Al-Fusail et al. 1991, Al Mahmoud 1992, Bushnak 1990, Uqba 1992, El-Zawahry 1992.

countries of the peninsula have all intensified an imbalance between rising water demand and very limited existing water resources. Most of the countries have experienced a 20 to 30 percent annual increase in domestic and industrial water demand over the last 10 years. Substantial increases in agricultural water use, as shown in Table 4, were also experienced, particularly in Saudi Arabia (Abu Rizaiza 1989, Al-Ibrahim 1990).

Domestic and industrial water requirements are satisfied through desalination and by a limited portion of the groundwater pumping. Agricultural requirements are met through abstraction of water from shallow alluvial aquifers located in the coastal strips and inland basins and, more recently, are satisfied mainly from the deep aquifers covering most of the Arabian Peninsula. In Saudi Arabia, the rapid expansion of agricultural activities encouraged by the government during the last 15 years has caused substantial increases in water demand, leading to extensive mining of the deep aquifers. Likewise, agricultural water demand has sharply increased in Bahrain, Oman, and the United Arab Emirates, similarly as a direct result of government policies encouraging self-sufficiency in food production. Government incentives and subsidies have made it possible for large areas to be cultivated, placing great strain on the existing groundwater resources. High population growth, further improvement in the standard of living and continuation of current agricultural policy will lead to increases in water requirements. Projected demand for the periods 2000–2010 is shown in Table 5, assuming an agricultural water consumption growth of 1 percent per year.

TABLE   5

Projected Water Demand for the Years 2000 and 2010 (in mcm)

	Population (in millions)		Domestic and Industrial		Agriculture		Total Demand	
	Year 2000	Year 2010	Year 2000	Year 2010	Year 2000	Year 2010	Year 2000	Year 2010
S. Arabia	15.553	19.315	2,900	3,600	20,211	21,700	23,111	25,300
Kuwait	1.511	1.710	530	650	110	121	640	771
Bahrain	0.654	0.981	155	180	130	135	285	315
Qatar	0.425	0.525	140	184	185	204	334	388
UAE	1.922	2.104	832	911	1,400	1,545	2,232	2,456
Oman	1.826	2.262	147	270	1,270	1,403	1,417	1,585
Yemen	17.75	23.45	360	640	3,250	4,000	3,610	4,572
Total	39.641	50.347	5,064	6,435	26,556	29,108	31,629	35,395

Country demand situations: Saudi Arabia

In 1992, water demands for different sectors in Saudi Arabia were satisfied through desalination (795 mcm), surface water (900 mcm), shallow alluvial aquifers (950 mcm), reclaimed wastewater (217 mcm), and non-renewable deep aquifers (13.48 bcm). As of 1992, approximately 47 percent of the domestic demand was satisfied through desalination from 23 plants located at the coastal regions. The major cities that rely mainly on desalination are Jeddah, Yanbu, Dammam, Jubail, and Al-Khobar. A large number of smaller coastal townships also rely on desalination for their water supply. Ambitious pipeline schemes were implemented to deliver desalinated seawater from the coast to the large inland cities of Riyadh, Mecca, Medina, Taif, and Abha, as well as a number of towns and villages, to meet growing demands. Desalination projects are being implemented to increase the annual supply from this source by 339 mcm.

Domestic demand experienced a three-fold increase from 500 mcm in 1980 to 1.7 bcm in 1990, due to urban expansion, improvement of the standard of living, lack of conservation measures, and over-consumption. Future demand is expected to reach 2.9 bcm and 3.6 bcm in the years 2000 and 2010, respectively. Since the greater part of domestic water demand has traditionally been satisfied through the use of desalinated seawater, the current production of 795 mcm will have to be increased three-fold in order to meet the projected demand for the year 2010.

The major water consumer in Saudi Arabia during the last decade has been the agricultural sector (MOP 1990, Al-Tokhais 1992). Water consumption in this sector increased from 1.86 bcm to 14.6 bcm during the period 1980–1990, an eight-fold increase made possible mainly through mining of the deep aquifers of Saq, Tabuk, Minjur-Dhruma, Wasia-Biyadh, Um er-Radhuma, Sakaka, Neogene, Dammam, Jauf, and Aruma. Agricultural demand is forecast to reach 21.7 bcm in the year 2010. Groundwater abstraction for agricultural purposes is expected to reach 20.31 bcm and 22.2 bcm in the years 2000 and 2010, respectively.

Kuwait

Water demand is satisfied mainly through desalination, supplemented by a limited amount of groundwater and treated wastewater in amounts of 240 mcm, 80 mcm, and 83 mcm, respectively (Country Report 1986, Bushnak 1990). Desalinated seawater is blended with small amounts of groundwater from the Neogene and Dammam aquifers to meet domestic and industrial requirements. Total water demand, including irrigation, during the last 10 years has increased from 186 mcm to 383 mcm. Demand is expected to reach 640 mcm and 771 mcm in the years 2000 and 2010, respectively. The major water consumer is the domestic sector, which relies mainly on desalinated seawater. Water requirements for the relatively modest amount of irrigated farming and urban landscaping are met mainly through the use of groundwater and treated effluent.

Bahrain

Water demand for the islands of Bahrain is satisfied mainly through deep fossil groundwater sources supplemented with desalinated water and limited use of treated effluent. Water from Neogene, Dammam, Rus, and Um er-Radhuma aquifers is used to satisfy most of the domestic, industrial, and agricultural requirements (Al-Mansour 1986, Danish et al. 1992). The total annual water requirement increased from 138 mcm to 216 mcm from 1980 to 1990. It is expected to reach 250 mcm and 315 mcm in the years 2000 and 2010, respectively. Desalination now contributes approximately 56 mcm while groundwater abstraction of 53 mcm meets domestic water requirements. The installed desalination capacity nowadays is inadequate to cope with the domestic demand. To bridge the gap, groundwater is continuing to be abstracted at an alarming level.

Qatar

Water requirements are satisfied through groundwater from the shallow alluvial, and the Dammam and Um er-Radhuma deep aquifers and from desalinated seawater. During the period 1980–1990 annual water demand increased from 128 mcm to 194 mcm (Al-Kenson 1986, Bushnak

1990). Domestic and industrial water requirements during the period 1980–1990 were satisfied mainly from a blending of 83 mcm of desalinated water with pumpage from the shallow aquifer. Agricultural water demand in 1990 was estimated at 109 mcm and will reach 185 mcm in 2000. Total demand is expected to reach 388 mcm in 2010.

United Arab Emirates

Total demand is satisfied through a combination of runoff diversion (75 mcm), groundwater exploitation mainly from shallow aquifers (900 mcm), desalination (342 mcm), and reclaimed wastewater (62 mcm) (Country Report 1986, Uqba 1992). Domestic water demand is being met through seawater desalination and pumpage from shallow aquifers. Demand for water has increased substantially during the last ten years, from 789 mcm to 1.49 bcm. This rise can be attributed to increased agricultural activities encouraged through government-sponsored financial incentives and to urban landscaping. Agricultural demands are met largely through groundwater abstracts from alluvial aquifers supplemented by pumping from the Dammam and Um er-Radhuma aquifers. Agricultural demand was estimated at 950 mcm in 1990 and is expected to reach 1.4 bcm and 1.55 bcm, respectively, in the years 2000 and 2010. Total demand is expected to reach 2.23 bcm and 2.45 bcm, respectively, in the years 2000 and 2010.

Oman

Oman's water requirement during the last ten years has been satisfied through flood water diversion (275 mcm), spring discharge (375 mcm), shallow alluvial (645 mcm), desalination (32 mcm), and reclaimed wastewater (10.5 mcm). Domestic water requirements are met mainly from shallow aquifers, supplemented by desalination. Agricultural water demand is satisfied mainly through exploitation of groundwater from shallow aquifers supplemented from the Neogene, Taqa, Aruma, Wasia, and Saiq deep aquifers. Traditional infiltration gallery systems known as qanats (aflaj), numbering more than 6000, provide 50 percent and 70 percent of irrigation and water supply requirements, respectively (Abdelrahman 1992). Water demand for various purposes in 1980 in Oman was estimated at 665 mcm and increased two-fold to 1.24 bcm in 1990. Agricultural water demand was estimated at 650 mcm in 1980 and 1.15 bcm in 1990. Demand for agricultural purposes in the years 2000 and 2010 is expected to increase to 1.27 bcm and 1.40 bcm, respectively. Total demand is expected to reach 1.42 bcm and 1.59 bcm in the years 2000 and 2010, respectively.

Yemen

Water demand for the now unified Republic of Yemen is met through the use of surface runoff (1.45 bcm), groundwater (1.7 bcm), and a limited amount of desalinated sea water (9 mcm) (Khoury et al. 1986, Mohamed et al. 1986). Domestic and industrial requirements are met largely through the use of groundwater from shallow alluvial and deep aquifers of Wajid, Wasia, Um er-Radhuma, Amran, and Kohlan. Domestic and industrial demand, while relatively small in comparison to the rest of the countries in the peninsula, has increased from 98 mcm to 216 mcm during the period between 1980 and 1990 (Al-Fusail et al. 1991). It is expected that such demand will reach 360 mcm in the year 2000 and 640 mcm in 2010. Agricultural water demand is met largely through diversion of flood water and groundwater exploitation from shallow aquifers. Frequent rainfall produces a relatively dependable source of surface water. However, increased use of modern pumping technology has resulted in uncontrolled drilling, causing water consumption for agricultural purposes to rise from 1.6 bcm to 2.5 bcm in a ten-year period (1980–1990). At the current rate, agricultural requirements are expected to reach 3.2 bcm by 2000 and 4.0 bcm by 2010.

Projected water demands for domestic, industrial, and agricultural sectors in each country of the peninsula, for the period 2000–2010, are

T A B L E   6

Projected Water Availability for the Years 2000 and 2010

| | Water Sources (mcm) | | | | | | | |
| | Year 2000 | | | | Year 2010 | | | |
	Surface*	Re-claimed	Desali-nation	Ground-water**	Surface*	Re-claimed	Desali-nation	Ground-water**
S. Arabia	900	710	1,289	20,212	900	1,000	1,300	22,100
Kuwait	—	80	428	132	—	106	428	237
Bahrain	—	42	115	93	—	53	141	121
Qatar	0.4	43	216	75	0.4	43	216	129
UAE	75	200	772	1,185	75	250	772	1,359
Oman	227	50	68	1,072	227	61	68	1,229
Yemen	1,450	36	10	2,105	1,450	57	10	3,055
Total	2,652.4	1,161	2,898	24,909	2,652.4	1,570	2,935	28,230

*   Diversion of surface runoff
**   Mainly deep aquifers

shown in Table 5, while Table 6 indicates estimated availabilities by water sources for the same period.

Imbalance between demand and supply

Combined water demand in 1990 for all the countries of the Arabian Peninsula was estimated at 22.5 bcm. Consumption, by sector, was broken down as follows: agriculture (86 percent), municipal (12 percent), and industrial (2 percent). In 1990 the agricultural sectors consumed an estimated 19.5 bcm with Saudi Arabia using the largest share, mainly from mining of deep aquifers. The estimated agricultural water demand is expected to reach 26.57 bcm in the year 2000 and 29.11 bcm by 2010, assuming an annual growth rate of only 1 percent. This can be set against a projected total water requirement for the peninsula for the years 2000 and 2010 of 31.6 bcm and 35.4 bcm, respectively, as shown in Table 5. The major consumer, it is clear, will continue to be the agricultural sector. Based on current trends, future programs, and projections for individual countries, renewable water resources such as surface runoff, desalination, rechargeable alluvial aquifer and reclaimed wastewater are already greatly insufficient to meet the expected demand. Supplies available from renewable sources for the years 2000 and 2010 are estimated at 6.71 bcm and 7.16 bcm, respectively, as shown in Table 6, which is far less than the projected water demands of 31.63 bcm and 35.4 bcm for the same years. It is expected that the deficit of 24.91 bcm for the year 2000 and 28.23 bcm for the year 2010 will be offset through the use of groundwater reserves, especially from the deep aquifers, as shown in Table 6. Thus, by the year 2010, a total of 550 bcm will have been withdrawn, mainly for irrigation, reducing the total estimated groundwater reserves of 2175 bcm by approximately 25 percent.

Expected domestic and industrial demand increases in the next 20 years will also necessitate the large scale construction of additional desalination plants unless strict water conservation measures are implemented and good quality groundwater is used solely for domestic use. The countries on the eastern part of the peninsula, namely Kuwait, Bahrain, Qatar, the United Arab Emirates, and some parts of Saudi Arabia, are expected to face water shortages for domestic supply by the year 2000, when existing desalination plants will reach the end of their operational lives. The groundwater in the deep aquifers under this area is usually highly saline and unfit for domestic consumption. If present domestic consumption patterns continue unaltered, countries in this region will be required to invest at least $30 billion in the construction of new desalination plants and support facilities capable of handling future demands. A large number of waste treatment plants will also be required to handle the resulting wastes. This huge investment may result in tremendous economic strain, especially in those countries with limited oil income.

Future desalination

The present installed capacity of government-owned desalination plants in the Arabian Peninsula is 1.9 bcm. Dependency on desalinated water has substantially increased during the last 15 years in most countries. Kuwait and Qatar seem to have sufficient desalination capacity to meet their entire current domestic and industrial water demand, while in Saudi Arabia, the United Arab Emirates, Bahrain, and Oman, desalination accounts for between 47 percent and 75 percent of domestic and industrial public supply. Only Yemen, with more rainfall, places little reliance (5 percent) on desalination at the present time.

Additional desalination tends to be regarded as the easiest means of meeting the ever-increasing demand for water in the region. Most of the countries are constructing and planning significant expansion of their desalination capacity to meet future water requirements, as shown in Table 7. Saudi Arabia, the largest producer of desalinated water in the region, is constructing additional plants with a capacity of 126 mcm for the major cities of Medina, Yanbu, Jeddah, and Jubail. Future plans call for additional capacity of 213 mcm for the cities of Jeddah and Al-Khobar. The United Arab Emirates is also expected to increase its desalination capacity by 270 mcm in the near future for its urban centers at Abu Dhabi and Dubai. Kuwait is planning to increase its desalination capacity by 110 mcm by 1997. In Qatar, the present desalination capacity of 112 mcm at Ras Aba Fontas will be raised by 16 mcm in the near future, and an additional capacity of 88 mcm will be added to the system by the end of this century. In Bahrain, desalination capacity reached 75 mcm in 1993 and 25 mcm of further capacity is proposed for Maharraq city, while a further 40 mcm of capacity is being studied as part of a privatization scheme for power and water. In Oman, the present desalination capacity is 55 mcm and will be increased by 13 mcm with the construction of several smaller units by 1995. Present and future capacities for countries of the peninsula are shown in Table 7. They will reach 2.92 bcm by the year 2000.

Desalination now contributes 51 percent of the region's domestic and industrial water, and is expected to contribute 58 percent by 2000. However, many existing desalination plants are heading toward the end of their economic life. In addition to the costs of new capacity, estimated at

T A B L E   7

Desalination Projects in Arabian Peninsula Countries

	1990				2000			
	Installed Desal'n Capacity (mcm)	Desal'n Production (mcm)	Domestic/ Industrial Demand (mcm)	Desal'n to Demand Ratio (%)	Planned Desal'n Cap. (new) (mcm)	Total Desal'n Capacity (mcm)	Domestic/ Industrial Demand (mcm)	Desal'n Ratio (%)
S. Arabia	950	795	1,700	47	339	1,289	2,900	44
Kuwait	318	240	303	80	110	428	530	81
Bahrain	75	56	103	54	65	140	155	90
Qatar	112	83	85	98	104	216	140	>100
UAE	502	342	540	63	270	772	832	93
Oman	55	32	86	37	13	68	147	46
Yemen	10	9	216	4	0	10	360	3
Total	2,027	1,557	3,033	—	901	2,923	5,029	—

at least $30 billion, a huge investment, estimated at $20 billion, will be needed after the year 2000 to replace the large number of aging plants constructed during the period 1975–1985. The capital cost of new units is increasing, and operation and maintenance costs for desalination plants are very high as well. The average cost of desalinated water reaching the end user in the region ranges from $0.5–$3.5 per cubic meter (Bushnak 1990). Higher costs are expected for locations dependent upon small-scale desalination plants. Budget allocations for desalination have been adversely affected by falling oil revenues and by increased expenditure on regional defense. Many proposed new desalination projects are being delayed by budget restrictions, and new approaches, such as privatization plans, are being given serious consideration by several of the Arabian Peninsula countries.

Environmental Impact of Water Development

Water quality

Deterioration of groundwater quality is being experienced in many regions of the peninsula, especially in coastal zones. Pumpage from groundwater aquifers in excess of natural recharge, mainly along the eastern perimeter of the peninsula, is causing an advance of salt intrusion wedges, a very serious and damaging effect. Many water wells in Bahrain, Qatar, the United Arab Emirates, and Oman have been abandoned as a result of seawater intrusion. Higher pumpage in the eastern parts of Saudi Arabia and Bahrain has also disturbed the dynamic equilibrium between aquifers, leading to leakage of poor quality water from one aquifer into others. Over-irrigation, especially with water of poor quality, and uncontrolled use of fertilizers and pesticides, has generated large volumes of saline drainage water that increases the pollution levels of shallow aquifers. Municipal wastewater from urban centers is discharged either into the sea for coastal cities or into the alluvial channels for inland cities and towns. This contaminates groundwater resources as a result of the low wastewater treatment level. High water consumption in unsewered areas, and sewer network leakage in urban areas of Riyadh, Jeddah, Mecca, Qatf, Kuwait, Doha, and Dubai, have resulted in groundwater table rises that have caused damage to the foundations of buildings and created human health hazards. In rural areas, improperly constructed and maintained septic tanks and cesspools have led to percolation of effluent to shallow aquifers and to contamination of water supply. Little is known about groundwater contamination in most countries of the peninsula due to lack of monitoring procedures and pollution prevention measures.

The major causes of water quality deterioration in the region are: the over-consumption of water, especially in the agricultural sector, increases in the volume of saline drainage, the lack of wastewater treatment capacity reaching at least to the secondary treatment level, inadequate sewerage coverage, and an absence of monitoring and enforcement legislation.

Agriculture

The needs felt for food security or self-sufficiency in essential items such as wheat, meat, and poultry have prompted decision-makers in the region, especially in Saudi Arabia, to encourage agriculture. Successful subsidy and incentive programs have resulted in a large scale expansion of farming activities with substantial water requirements. These are satisfied mainly by mining deep aquifers, as shown in Table 4. Deep aquifer water levels are declining, with a consequent increase in pumping cost, saltwater intrusion, and disturbance of the dynamic equilibrium between aquifers. These have led in turn to the abandonment of farm land, a decline in the agricultural labor force and urban migration.

In some parts of the peninsula, pumping programs from shallow and deep aquifers in excess of natural recharge are being practiced widely. Over-pumping for irrigation means that wells must be further deepened, especially in the central and northern regions, but to some degree throughout the peninsula. Present groundwater law regulates drilling, but once a well has been drilled, gives a landowner the right to unlimited extraction, without being liable for damage to the aquifer, to his neighbors, or to the environment. This unregulated pumpage, along with lack of enforcement of rules against unlawful drilling, and poor irrigation practices, has resulted in substantially excessive agricultural water consumption. The free availability of groundwater has encouraged its depletion, and even brought about localized water logging with decreases in soil fertility, especially in the eastern part of the peninsula. Since food is now being exported, there is a need to evaluate agriculture policy to assure that food production is limited to that necessary for self-sufficiency, and to reduce agricultural water consumption through conservation measures.

Industrial

Industrial and commercial purposes in the Arabian Peninsula represent 2 percent of total water demand, or 11 percent of domestic and industrial demand combined. Water for industrial and commercial use is thus a relatively insignificant proportion of total water demand, although in Kuwait, Bahrain, and Qatar it is slightly above the regional norm. Industrial development in the region has had a lower than expected impact on general water availability, largely because the industrial sector is generally responsible for meeting its own water supply requirements by desalination or other means, and has not been permitted unrestricted access to groundwater. This has provided a degree of self-regulation that has avoided the excesses and wastage of water that exist in other sectors.

Desalination

The building of a large number of desalination plants may have caused negative effects on the surrounding ecosystem. Early desalination plant construction emphasized water and power production objectives with little emphasis on environmental impact assessment. Seashore or inland desalination plants, especially large ones, can cause air and water pollution problems as a result of the combustion they entail, and because the separation of desalinated water from intake sea or brackish water leaves a brine with concentrated salts that must be disposed of.

Emission into the air of oxides from a desalination plant's stacks affects urban air quality. Water pollution may result when rejected brine is discharged back into a water body that was the source of feedstock. Marine life may be affected when raw water needed for desalination purposes is screened to remove aquatic organisms, which are damaged in the process, or when the filtered small organisms are exposed to high desalination temperatures and pressure conditions. Both of these effects may increase the mortality rates for plants of all types as well as fish.

Rejected brine causes chemical, physical and biological changes in the surrounding environment into which it is discharged, due both to its increased salt concentration and to residual treatment chemicals and trace elements it may have picked up within the desalinating plant. The degree of brine salinity depends on the original salinity of the feed water and the amount of desalinated water that has been extracted. Recovery from seawater produces intensifications ranging from 20 to 68 percent depending on the type of desalination process used; this means increases in the brine's concentration ratio of 1.25 to 3. Reverse osmosis and electrodialysis recovery range from 50 to 90 percent with corresponding salinity concentrations of 2–10 times as much as the original feed water. Treatment chemicals and internal plant corrosion cause release into the brine of heavy metals and chemicals such as copper, iron, nickel, molybdenum and others that may pose environmental problems. The amount of these ions is directly related to the water's pH and the construction material of the desalination plant, and these contaminants can be reduced by various pre-treatment steps.

Desalination brine is discharged at an elevated temperature, which can be environmentally significant, as can biological effects related to changes in biological oxygen demand that may be caused by the prevailing condition of the desalination plant.

All these factors tend to interact to form environmental problems at a seashore or inland brine discharge site unless consideration is given to minimizing their effects in the plant's design and operation. Recently, desalination decision-makers have shown keen interest in addressing environmental concern through limited environmental impact studies. Further plant construction should require prior environmental impact study.

Management Options

Development of additional sources

All countries in the Arabian Peninsula are actively trying to develop additional sources of water to meet their spiraling demands. Options under consideration include desalination of seawater and brackish groundwater, wastewater re-use, groundwater recharge, water importation, and weather modification.

Desalination of sea and brackish water has provided additional supplies for most countries of the peninsula. Desalination schemes have cost governments in the region huge sums as a result of the low water prices charged to consumers. Desalination of brackish groundwater (3000–10,000 ppm) may provide a viable alternative to seawater desalination for many inland sites due to low desalting and transportation costs, and reduced vulnerability to pollution and sabotage. However, the cost of brine disposal to avoid pollution may be prohibitive. A regional research center on desalination is needed to evaluate alternative means of selecting the best technologies and reducing production and maintenance costs.

Increasing the volume of groundwater recharge from surface runoff, especially for Saudi Arabia, Yemen, and Oman, can provide additional water for times of need, since much surface runoff is now being lost to the sea from coastal drainage basins. Evaporation from inland drainage basins can also be reduced, yielding water for recharge. There is need to increase the efficiency of recharging dams that have been build in different regions of the peninsula through better dam operation and the removal of silt and clay deposits. Increases of aquifer recharge using surface runoff, surplus desalinated water, and imported water from neighboring countries can provide a strategic reserve of potable groundwater, held as a standby for an emergency. Such schemes in the United States and elsewhere indicate that, with appropriate design and operation, a volume of water equal to the volume of water stored can usually be recovered when needed. A series of recharge dams or injection well fields can utilize recaptured excess runoff, especially in the southwestern and southern parts of the peninsula, while imported water can be used to build a series of water reserves elsewhere through recharge dams, spreading basins, or injection wells.

There are a number of proposals for augmenting the water supply of the Arabian Peninsula by the importation of fresh water from outside the region. The best known is the "Turkish Peace Pipeline" scheme. Through the construction of two pipelines, it would transfer water to the Arabian Gulf states from rivers of Turkey that flow toward the Mediterranean. The proposed projects will move 2.2 bcm per year with approximately half of the volume for Syria and Jordan and half to the Arabian Peninsula. The cost of the project is estimated at 20 billion dollars, and its construction time at 8 to 10 years. The large western pipeline would pass through Syria and Jordan and terminate at Mecca in western Saudi Arabia. The smaller eastern pipeline would cross Syria and Iraq and then pass down the west side of the Gulf, supplying water to Kuwait, the Eastern province of Saudi Arabia, Bahrain, Qatar, United Arab Emirates, and Oman. Saudi Arabia and Kuwait would be supplied 840 mcm and 220 mcm, respectively. Although this importation of water would ease the shortfall situation in the region, the Arabian Peninsula countries are concerned about the political implications of becoming dependent on upstream states for the security of their water supply as well as the potential vulnerability of the pipelines to sabotage or attack, and therefore this importation proposal is not now an active one.

Two other importation proposals involve pipelines under the Arabian Gulf: one from the Garon river of Iran to Qatar and the other from Pakistan to the United Arab Emirates. The Iran-Qatar scheme involves a 1.5-meter-diameter gravity flow pipeline extending 770 kilometers, of which 560 kilometers would be within Iranian territory. It would provide Qatar with an estimated annual volume of 135 mcm. The cost is estimated at $1.5 billion, and completion time at three years. Other proposals for importing water to the region include towing icebergs from the Arctic and utilizing the empty holds of incoming petroleum tankers. Feasibility studies were undertaken but such plans are not currently under consideration. Weather modification is said to hold the potential of increasing the availability of surface water that could be used for flood irrigation or for increasing the magnitude and frequency of groundwater recharge. This alternative would be particularly applicable in the southern region of Saudi Arabia, Yemen, and the Dhofer mountains in Oman. Preliminary investigations in the southwestern region of Saudi Arabia show potential, and a detailed study is planned.

Demand management

A long-term solution to the water deficit problem requires the conservation of water resources. It is essential that each country develop a water management plan that includes programs to reduce water waste as one of its major components.

The major conservation effect should be concentrated on the agricultural sector where the current consumption is six times the water used for domestic and industrial purposes. Government incentives and subsidies in agriculture can be used as leverage to implement conservation measures such as improved irrigation efficiency through sprinkler and drip systems, laser leveling, canal lining, and farmer education. Irrigation efficiency can be increased from its current low level of 30–45 percent to 60–70 percent. Since exporting surplus food production is tantamount to exporting water, a critical balance between agricultural production and water consumption should be examined carefully and governmental policies adjusted accordingly. A significant reduction in overall water consumption in the region could be achieved by a reappraisal of agricultural policies, reflected in comprehensive legislation on groundwater use. New laws would provide for monitoring and for the improvement of extension services.

Desalinated water is used almost exclusively for domestic purposes and reduced consumption in this sector would have a corresponding impact on the need for future desalination and wastewater schemes. The high per capita water consumption at present (see Table 2) shows the need for urban water conservation. Reduction in domestic consumption can sometimes be achieved through increasing public awareness of the value of water as a scarce resource, but reliance on voluntary compliance is rarely effective.

Water metering and charging for water has been widely adopted throughout the peninsula region, but has not lowered water consumption because water tariffs are so low that costs to the consumer do not effectively discourage excessive water use or waste. Tariffs per cubic meter of potable

T A B L E   8
Water Tariffs

	Monthly Tariff *	Type/Use	$/m³
Saudi Arabia	0.15 SR/m³ 1-100 m³	potable	0.04
	1.00 SR/m³ 101-200 m³	potable	0.27
	2.00 SR/m³ 201-300 m³	potable	0.53
	4.00 SR/m³ >300 m³	potable	1.07
Kuwait	0.800 KD/1000 gal	potable domestic	0.58
	0.250 KD/1000 gal	potable industrial	0.18
	0.100 KD/1000 gal	brackish domestic	0.07
	0.100 KD/1000 gal	brackish industrial	0.07
	0.200 KD/1000 gal	brackish agricult	0.15
Bahrain	0.025 BD/m³ <60 m³	potable domestic	0.07
	0.080 BD/m³ 61-100 m³	potable domestic	0.21
	0.200 BD/m³ >100 m³	potable domestic	0.53
	0.300 BD/m³ <450 m³	potable industrial	0.80
	0.400 BD/m³ >450 m³	potable industrial	1.06
	0.002 BD/m³ <60 m³	brackish	0.01
	0.025 BD/m³ 61-100 m³	brackish	0.07
	0.085 BD/m³ >100 m³	brackish	0.23
Qatar	4.40 QR/m³, free for citizens	potable	1.21
UAE	15.00 DH/1000 gal	potable	0.90
Oman	2.000 OR/1000 gal	potable domestic	1.14
	3.000 OR/1000 gal	potable industrial	1.71
Yemen	5.40 YR/m³ <10 m³	potable	0.50
	6.90 YR/m³ 11-20 m³	potable	0.64
	9.50 YR/m³ 21-30 m³	potable	0.88
	12.60 YR/m³ 31-40 m³	potable	1.16
	15.50 YR/m³ >40 m³	potable	1.43

* SR, KD, BD, QR, DH, OR and YR are local currencies.

water in the lowest charge band range from $0.04 in Saudi Arabia to $1.21 in Qatar (although water is free for Qatari citizens) as shown in Table 8. The highest charge rate per cubic meter in any of the countries is $1.71 for industrial usage in Oman. Some countries have a separate charging structure for brackish water; the monthly cost per cubic meter is in the range of $0.01 to $0.23. Since desalinated water production costs are currently on the order of $0.50 to $3.5 per cubic meter, it can be seen that very few consumers in the region are paying a commercially viable rate, and that subsidies of over 90 percent are common. Conservation needs to be encouraged through price incentives mandated through regulations and enforced. Charging unsubsidized rates for monthly water consumption in excess of 50 cubic meters per household should lead to reductions in water demand. Billing for consumed water should be done monthly, and collection of charges needs to be enforced.

There is also potential to cut water demand through delivery network analysis in conjunction with leak detection and pipeline rehabilitation. In Bahrain, leakage levels of 24 percent were assessed in 1985, and in areas where repairs were carried out water consumption was reduced by 19 percent. In Qatar, leakage from the system is currently assessed at 15 percent, which is equivalent to losing 13 mcm of desalinated water. The governments of the Arabian Peninsula countries are not allocating sufficient budget funds for comprehensive leak detection programs. Maintenance programs should be a part of every long-term planning effort.

Any reductions in demand achieved to date are small in comparison with the potential overall water savings that could be realized in the region by full implementation of a range of conservation measures (Akkad 1990). A few conservation measures in the domestic sector in the eastern region of Saudi Arabia (ARMOS communities) and in Bahrain have demonstrated their effectiveness in reducing consumption in the range of 15–20 percent. Voluntary or compulsory water conservation should be incorporated as a basic component in all water agency management work. Components should include public education, the use of water-saving technologies for distribution systems and households, rebates for retrofitting, modification of existing building codes to promote use efficiency, gray water use for landscaping, water-conserving plant species, and import restrictions on non-conserving devices. Existing housing subsidies and loan programs that have been practiced in most countries of the region can be used as incentives, and a system of restrictions and/or penalties to enforce compliance with regulations is also required. A high priority should be given to implementation of a conservation program on a continuous basis, with regular evaluation of its effectiveness and frequent updating.

Planning

The major and first step for the peninsula countries in responding to imminent water shortages is the formulation and implementation of comprehensive national water management plans within each country, and on a regional basis.

Most countries of the peninsula have not yet worked out, much less implemented, water plans. Even for those countries that have a plan, effectiveness is limited due to lack of comprehensive coverage. In addition to being comprehensive, national or regional plans should be flexible in order to accommodate social and economic changes. Sustainability of the existing water resource should be a main goal. Plan design should emphasize the long-term involvement of local professionals and researchers to ensure continuous updating and improvement. The collection and archiving of reliable data and the dissemination of information must be improved in most countries. Formulation of an integrated and flexible water management plan begins with a reliable quantitative and qualitative estimation of water resources within each country as well as of the sources shared by a number of countries, particularly the deep aquifer reserves. Data banks and data quality control on consumption in all sectors are essential for planning.

Comprehensive plans within each country should offer viable alternatives for the reduction of imbalances between supply and demand, and should include both short- and long-term policies and goals. The development programs for water resources and for agriculture currently being promoted in most of the peninsular countries lack clear directives and goals. Water resources are developed haphazardly with no long-range planning, inter-sectoral coordination, or management guidelines. The five-year planning cycle is far too short to be adequate, since experience has shown that fragmented five-year development plans in some countries abruptly shift from, for example, an emphasis on groundwater development for agricultural expansion to a stress on increased desalination. Such an erratic path fosters widespread misuse and abuse of water in all sectors.

Policy

Since all water projects in the region are supported through governments, either in the form of financial or administrative support or monetary incentives, directive policies and actions should come from the governments. These policies should include laws and procedures for reducing water consumption, such as water pricing mechanisms, modification of building codes, establishing and respecting safe yields of groundwater, and limiting agricultural acreage and government subsidies to growers. Domestic and industrial water pricing policies should reflect at least the cost of operation and maintenance of the desalination plants. Short- and long-term policies on groundwater development must emphasize sustainability of resources. Policies should be formulated to manage groundwater reserves, wastewater renovation, and seawater desalination in order to meet future urban, industrial, and agricultural requirements. Agricultural development plans should delineate limits on agricultural food production, outline market pricing mechanisms, and require irrigation efficiency. Such policies will contribute to better management of water resources and help curb rising demand.

Weaknesses in institutional arrangements are one of the major constraints on the management of water resources in each country of the peninsula. Comprehensive legislation should define the responsibilities of each organization, and require exchange and dissemination of information among them on a local, provincial, and national level, and among countries as well as within them. Each country should set up a single coordinating body such as a national water resource committee or council. This organization, with powers delegated to it by the government, would ensure timely coordination of activities within a country, and between it and its neighbors. It would be responsible for determining a national position on optimal water resource development and management. Such a position is especially needed for water resources shared between countries, a subject on which each country's council would have exchanges with its adjacent or regional opposite numbers.

Summary and Conclusion

The countries of the Arabian Peninsula are located within an extremely arid region. They have limited renewable fresh water supplies, which have been nearly fully developed, and their only dependable source is fossil groundwater reserves. In some regions, depletion of these nonrenewable groundwater resources is taking place at an alarming rate due to over-pumping in order to meet agricultural requirements.

Nonetheless, agricultural water demands have increased almost eight-fold during the period 1980–1990. Interest-free loans and the desire to become self-sufficient in certain foodstuffs, as well as other incentives for agricultural development, have resulted in substantial increases in irrigated acreage, which has caused, in turn, extensive mining of groundwater reserves. If agricultural activities are allowed to continue in this manner, 25 percent of the existing deep fossil groundwater reserves will be exhausted by the year 2010.

Improvement in the standard of living and urban migration, coupled with the absence of conservation programs, have brought about high domestic water consumption, which itself tripled from 1980 to 1990. Programs currently in force in many of the countries stress only the increase of urban water supplies through building desalination plants.

Water shortages are already a reality because of limited supply of desalinated water, and wasteful and inefficient water consumption patterns. Conservative forecasts of demand during the period 1990–2010 for the domestic and industrial sector expect a two-fold increase, while water demand for agriculture will increase by half, assuming a low annual growth of 1 percent.

To cope with future water demands, emphasis must be placed on efficient management of water resources in the region. It is essential that each country of the region establish an up-to-date water plan that emphasizes not only the sustainability of water resources, but promotes a number of important policies, such as optimum allocation of water in accordance with market values, conservation, pollution control, and improvement of the coordination of efforts between water institutions. Technical research and development, as well as manpower development and training, are also essential aspects of any water program. Key policies should address short- and long-term programs for agricultural development, capacity building, review of water-pricing subsidies, development and application of appropriate technology, institutional arrangements, and water importation.

An institutional infrastructure capable of coordinating and managing complex water policies within each country, and between countries, must be recognized as a fundamental element in resource management. Finally, utilization of water resources, especially groundwater, which are shared between most of the countries of the peninsula requires comprehensive regional coordination and cooperation in order to assure optimum development and management and to avoid future conflict. Development of a regional water plan would lead to the efficient management practices that are needed to reduce the imbalance between demand and supply.

8

Desalination, an Emergent Option[*]

Taysir Dabbagh, Peter Sadler, Abdulaziz Al-Saqabi,
Mohamed Sadeqi

Introduction

Desalination is usually neglected as an option for the provision of
potable water owing to its high cost. While cost cannot be overlooked,
there is compelling evidence that desalination will have to be adopted as the
only acceptable way to provide potable water in some developing countries, particularly in the Arab world (Dabbagh et al. 1989). Reducing the
cost of desalinated water, therefore, is becoming a pressing issue that has
to be explored. It depends not only on improving desalination technology,
but also on improving the overall management of the use and conservation
of desalinated water.

The technology of desalination can no longer be regarded as new since
it has been used on a commercial basis for nearly 40 years. In the 1950s
Kuwait embarked on desalination as the most practical solution to its water
supply problems, and has used it successfully ever since. In fact, Kuwait,
small country that it is, financed research to improve and advance the
budding technology when it was seriously disputed as a viable main source
of potable water. The original work by Professor R.S. Silver at Glasgow
University, which spanned many years, provided a foundation for the
development of desalination in Kuwait and, indeed, for much of the
progress that has since been made in desalination as a major source of
potable water.

Today, however, the momentum Kuwait gave to commercial desalination has somehow been lost while other technologies born during the same
era or later have flourished and become widespread: computers, electronics

* The views expressed in this paper are not necessarily those of the Kuwait Fund
for Arab Economic Development.

and undersea oil exploration, not to mention power generation and space technology, have leaped forward while desalination has been neglected in favor of other sources of water.

Possible Reasons for the Neglect of Desalination

Shortages of water are becoming a vital issue, not only since water is an essential commodity for human survival and a major factor in economic development, but also because it is a possible cause for conflict between neighboring countries. It is therefore fitting to consider the reasons underlying the relative lack of major progress in desalination technology, since it has the potential to help solve the water supply problems of future generations while at the same time safeguarding the environment.

A number of possible factors may have resulted in the neglect of the development of desalination:

- *The location of need:* The countries in need of desalination have been mostly developing countries, while the industrial countries — the seat of technology — have not until recently suffered from water shortages.

- *The market:* Competition between the major manufacturers has been limited and has had a negligible effect on market prices, especially since by far the largest group of customers for desalination capacity over the last two or three decades has been the Arab oil-producing states. Since their demand has been backed by plentiful funds, there has been little incentive to minimize the cost of the process.

- *The research centers:* The public sector has carried out little research (U.S. Office of Technology Assessment 1988). Even the collation of international data has been carried out on the initiative of private sector researchers. The development of desalination techniques has been largely left to the main manufacturers, who have a vested interest in existing technology and usually undertake research "in-house" and keep the results secret (Dracup and Glater 1991).

- *The universities:* Opportunities for post-graduate studies or courses at universities are extremely limited as there is a lack of suitable staff available to supervise research in desalination. At the undergraduate level desalination is given little attention, but if the fundamental importance of water is considered it would surely merit a place alongside such technologies as computing, electronics or petroleum engineering.

■ *Training:* The training required for operating and managing a desalination plant is substantially different from that required for other water treatment plants. The adoption of desalination by well-established water authorities has been resisted as it would necessitate the introduction of new training procedures and expertise.

Options for Alleviating the Present Water Shortage

Current shortages of water, coupled with water's unequal distribution, have led to the need to choose between two major solutions to the problems of localized supply shortage: long distance conveyance, and desalination and/or recycling (Khouri, N. 1992). Storage is also a "major solution" in this supply-side framework; but it is largely exhausted as a measure in the Arab world. Water can be conveyed by canal, pipeline, tunnel or tanker, either between different regions in the same country or across country borders. Examples of major national water transport projects and proposals are Libya's artificial river (1900 km), Tunisia's Majradah canal (about 160 km), the study for Senegal's Cayor canal (230 km), Turkey's Melen project for Istanbul (180 km), and Egypt's interest in the conveyance of water from the Nile to the Sinai and the Red Sea. International projects include the proposed "Peace Pipeline" from Turkey to the Levant and the Gulf (over 3000 km, $20 billion in cost; 8–10 years construction) (Starr 1992), and the existing pipelines from Malaysia to Singapore (30 km), and from China to Hong Kong (60 km).

The advantage of transporting water is that the technologies of canal building and pipeline construction are well-established and can utilize expertise already developed by other industries such as oil production and transport. As for tanker use, the technology required for transporting water is far less complex than that required for oil. However, it is not always the case that water is cheaper to transport than oil. For instance the price the United States would pay to ship water to its troops in Saudi Arabia is at least 10 times the price of transporting oil (Starr 1992). In fact, before the introduction of desalination, water was conveyed to Kuwait from the Shatt-al-Arab in specially constructed dhows.

The disadvantages of transporting water, however, include the following:

■ A very high level of initial investment is required.

■ Long lead times are involved in planning.

■ Once schemes are implemented countries are committed to dependence on a particular water source for many years.

■ Water coming from a different country is strategically vulnerable.

■ A degree of inflexibility is involved in that the maximum supply is determined not only by the source, but also by the capital available initially for the construction of pipelines, storage capacity, etc.

Only the richest countries can make provision for their requirements many years in the future.

Desalination presents a quite different picture from the conveyance of water. Its advantages can be summarized as follows:

■ Capital expenditure on desalination can be carried out gradually in stages.

■ Plant output is more flexible within capacity limits.

■ The sources of input are less strategically vulnerable as they are normally sited within national borders.

■ Cost reductions in the future seem likely as the increasing use of the techniques means they will almost certainly be improved upon.

The relative disadvantages of desalination mainly relate to operating costs and can be summarized as follows:

■ The personnel required for operation and maintenance can be costly as the technology is not widespread and expertise is scarce. In the Arab setting, it may be necessary to employ expatriates.

■ Spare parts and materials can be expensive, especially as they usually need to be imported.

■ Energy requirements are high.

Principal Desalination Processes

Two main types of desalination processes are now available commercially: distillation processes and membrane processes. The most important distillation methods are multi-stage flash distillation (MSF) and multi-effect distillation (MED). Both involve the evaporation of saline feed water and its condensation back into fresh water, leaving dissolved substances in the waste brine. Both methods are improvements on the original sub-merged tube process, developed commercially in the 1950s, which in-volved heating stagnant brine.

In multi-stage flash (MSF), a stream of brine flows through the bottom of up to 25 stages or chambers (Figure 1). The pressure in each chamber

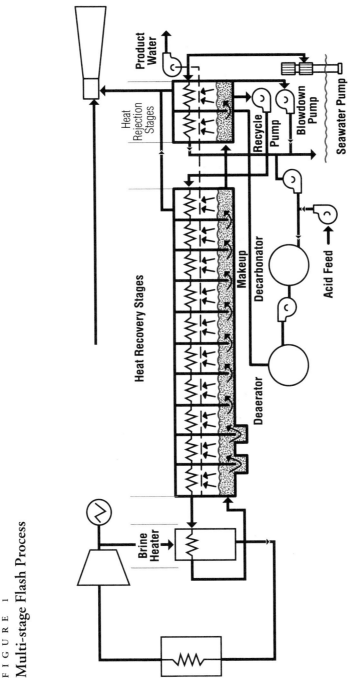

FIGURE 1
Multi-stage Flash Process

is maintained at a lower level than the saturation vapor pressure of the water and a proportion of it "flashes" into steam and is then condensed.

In multi-effect distillation (MED), evaporation takes place as a thin film of feed water passes over a heat transfer surface. This is usually the outside of horizontal tubes (MED/HTE) but in some small-scale operations is the inside of vertical tubes (MED/VTE) (Figures 2, 2A). The vapor formed in each effect condenses in the next, providing a heat source for further evaporation. Energy savings are made if the vapor from the last effect is re-compressed thermally or mechanically (MED/TVC or MED/MVC). Fewer stages are involved than in MSF.

The two membrane processes are reverse osmosis (RO) (Figure 3) and electrodialysis (ED). In RO, a pressure greater than the osmotic pressure of the feed water is applied to the feed-water side of a semi-permeable membrane producing a flow of fresh water through the membrane. In ED feed water flows through a stack of membranes to which an electric voltage is applied. Ions migrate to the charged electrodes formed by the membranes so that an ion-depleted product and concentrated reject brine stream are formed in alternate spaces between the membranes. While the feedstock generally assumed is either seawater or saline groundwater, the use of sewage effluent as a feedstock, suitably treated to remove harmful bacteria, is becoming widely acceptable.

Comparison of Desalination Processes

Table 1 shows the state of the art of desalination, but the main features of each process can be pointed out as follows:

- *Multi-stage Flash* (MSF) is the method most widely used on a large scale and there is still room for its improvement by better utilization of computer modeling and using high grade alloys to extend the plant's life span. It can make use of low grade heat produced during electricity generation — which would be otherwise wasted — thus considerably reducing the energy demands of the process. For this reason it is usually installed as part of a dual-purpose plant, along with the power generation function. MSF has the advantage over RO that it requires less specialized technical expertise and is much more robust. It is particularly competitive for desalinating sea water. The product is usually very pure with less than 30 ppm total dissolved solids (TDS), and may need post-treatment to provide drinking water and to protect the distribution network.

F I G U R E   2

Conceptual Diagram of Horizontal-Tube Multi-Effect Distillation Process

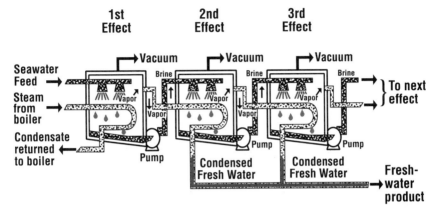

F I G U R E   2 A

Conceptual Diagram of Horizontal Multi-Effect with Vertical Stacked Effects

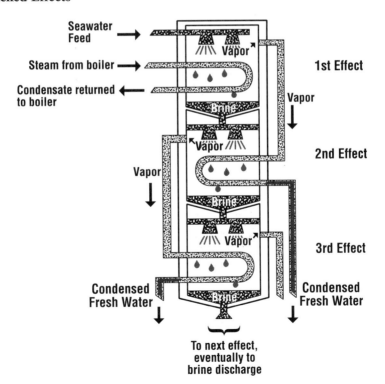

FIGURE 3
Flow Diagram of Reverse Osmosis Process

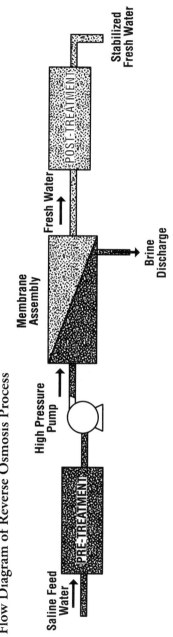

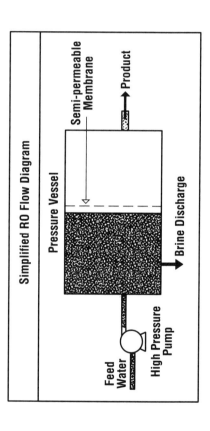

■ *Multi-effect Distillation* (MED), which produces a similar product to that of MSF plants has not been used on such a large scale as MSF. However, it promises certain advantages. The process takes place at a lower temperature than MSF thus reducing corrosion and scaling and thereby also reducing the demand for additives, maintenance and expensive corrosion-resistant materials. Moreover, fewer stages are necessary so less plant is required. The capacities of MED plants continue to increase and MED is on the point of competing with MSF for all but the largest installations. Vapor compression can reduce energy requirements, independent of electricity generation, to levels approaching those of RO, but MED/VC has not yet been used in large-scale plants.

■ *Reverse Osmosis* (RO) was introduced in the 1970s and is the most widely used process after MSF. It is particularly well suited to the desalination of brackish water — including waste waters — but in the 1980s was also developed for seawater desalination. The latter requires much more energy, since pressures of 65–70 bar are needed compared with 25–30 bar for brackish water. RO can be carried out independently of electricity generation — though it does of course use electricity — since it cannot use waste heat. The amount of desalination plant required is very much smaller than for MSF. The main disadvantage of RO is the sensitivity of the expensive membranes to fouling and the need for very careful operation. It is likely, however, that improvements can be made in process efficiency as well as the sensitivity and cost of membranes. The purity of the product is only fair. It can be 500 ppm TDS from the one-stage treatment of sea water, and may not need post-treatment. To obtain products with the desired qualities for a public water supply, various hybrid plants have been proposed, such as RO with MED/VC.

■ *Electrodialysis* (ED) is only used for relatively small units such as those required by hotels. It is suitable only for brackish waters of low to medium salinity, which must be given careful pre-treatment. Non-ionic substances in the water pass through unchanged, and ED is uneconomic for the production of water with less than 250 ppm TDS. Product quality is much more affected by variations in feed quality than in the case of RO.

T A B L E   1

Main Characteristics of In-use Desalination Processes

	Membrane Processes			Multi-Stage Flash Distillation (MSF)	Distillation Processes — Multiple Effect Distillation (MED)			
		Reverse Osmosis (RO)			Horizontal Tube Evaporation (HTE)			Vertical Tube Evaporation (VTE)
					With Vapor Compression		Without Vapor Compression	
	Electrodialysis (ED)	Brackish Water (RO)	Seawater (SWRO)		Thermal Vapor Compression (TVC)	Mechanical Vapor Compression (MVC)		
Principle	Feed water flows through a stack of alternate cation and anion exchange membranes to which an electric voltage is applied. Ions migrate to the electrodes and an ion-depleted product and concentrated reject brine stream are formed in alternate spaces.	A pressure greater than osmotic pressure is applied to the feed side of a semi-permeable membrane producing a flow of fresh water through the membrane.		A stream of heated brine flows through the bottom of up to 25 stages at successively reducing pressures. Vapor "flashes" from the brine and condenses or heat exchanger tubes through which recycled brine flows.	Evaporation takes place as a thin film of brine passes over a heat transfer surface. The vapor formed in each effect condenses in the next, providing a heat source for evaporation. Part of vapor formed in last effect is recompressed and returned to first effect to save energy.		Seawater is evaporated outside horizontal tubes	Seawater is evaporated inside vertical tubes
Required Pressure		Usually 25–30 bar	Usually 65–70 bar		Thermally	Mechanically		
Feed Water	Low to medium brackish up to 3,000 ppm TDS	Brackish water 5,000–6000 ppm TDS	Seawater					
			Usually 35,000 ppm TDS ; 45,000 ppm TDS in Arabian Gulf; 100,000 ppm TDS possible					
Pre-Treatment	To avoid particulates, and counter scaling, organics, iron, and metallic cations.	Careful pre-treatment required to avoid membrane fouling. Must be adjusted to water source. To avoid particulates, and counter scaling and biological fouling. Chlorine damages some membranes		Screening and straining to remove particulates. Anti-scalant according to operating temperature, anti-foaming agents, de-aeration and chlorination against biological fouling.				

TABLE 1

Main Characteristics of In-use Desalination Processes (continued)

	Membrane Processes			Distillation Processes				
		Reverse Osmosis (RO)				Multiple Effect Distillation (MED)		
						Horizontal Tube Evaporation (HTE)		Vertical Tube Evaporation (VTE)
						With Vapor Compression	Without Vapor Compression	
	Electrodialysis (ED)	Brackish Water (RO)	Seawater (SWRO)	Multi-Stage Flash Distillation (MSF)		Thermal Vapor Compression (TVC) / Mechanical Vapor Compression (MVC)		
Product	Less than 250 ppm TDS. Non-ionic species uncharged. Feed variations affect quality more than with RO	Purity fair. Can produce less than 500 ppm in one stage. May not need post-treatment or can be combined with MED/VC plant product.	Less than 500 ppm TDS in one stage. May not need post-treatment or can be combined with MEDC plant product	Post-treatment required to protect distribution network and for drinking water. Brackish water may be added or bicarbonate to increase calcium.		High Purity Less than 30 ppm		
Conversion	80% of feed	30–40% of feed		30–40%				
Energy Source	Usually electrical power. Others possible, e.g., steam or diesel engine.			Mainly low grade waste heat from electricity generation used in dual purpose installation. Some electric power for pumps.				
Factors Affecting Efficiency	Efficiency maximized by larger membrane area, current low enough to avoid polarization and scale, and good distribution of feed water.	At low salinities, the operating pressures are lower and the conversion rates higher, so energy recovery not worthwhile compared with SWRO.	At higher salinities operating pressures are higher and the conversion rates lower, so energy recovery turbines worthwhile.	Efficiency affected by scale control and heat transfer techniques, the use of corrosion resistant materials, and the mechanical design of larger structures and pumps. Higher temperatures increase efficiency but also increase scaling. Energy input largely independent of dissolved solids level.		Vapor compression reduces energy consumption for a given number of effects		

TABLE 1

Main Characteristics of In-use Desalination Processes
(continued)

	Membrane Processes			Distillation Processes				
		Reverse Osmosis (RO)			Multiple Effect Distillation (MED)			
					Horizontal Tube Evaporation (HTE)			Vertical Tube Evaporation (VTE)
					With Vapor Compression		Without Vapor Compression	
	Electrodialysis (ED)	Brackish Water (RO)	Seawater (SWRO)	Multi-Stage Flash Distillation (MSF)	Thermal Vapor Compression (TVC)	Mechanical Vapor Compression (MVC)		
Capacity of Existing Commercial Plants	Relatively small. Used for small commercial undertakings, e.g., hotels. Large outputs require a modular approach.	45,000 m³/day — Using modules 56,800 m³	56,800 m³/day	34,000 m³/day, but 45,000 m³/day planned	10,000 m³/day	1,500 m³/day		Small. (A large plant is under consideration for California).
First Commercial Use	Mid-1960s	1970s	1980	1960	late 1970s — These processes are derived from the submerged tube multiple effect distillation used in 1950s in Kuwait.			
Operating Temperatures	Ambient temperatures (under 400°C).			Maximum in first stage 110°C or 90°C according to anti-scalant used	Around 65°C in first effect			High temperature
Capital Cost	Less plant, materials, civil work and land required than for distillation. Depends on quality of product required.			Much larger, heavier plant required than for RO with extensive materials, civil work and more land. High temperature necessitates expensive corrosion-resistant alloys, e.g., titanium.	Fewer stages so less plant required than for MSF. Lower temperatures, therefore less corrosion.			
Operation and Maintenance	Reasonably simple. Can be highly automated. Main cost is membrane replacement. Normally extensive pre-treatment essential for successful operation and long life of membranes.			Minimal skill required by operator. High grade materials reduce maintenance time but are expensive. Labor intensive: all water boxes have to be opened for cleaning. Consumes chemicals (to use these costs much less than the energy).				

T A B L E   1

Main Characteristics of In-use Desalination Processes
(continued)

	Membrane Processes			Distillation Processes				
		Reverse Osmosis (RO)			Multiple Effect Distillation (MED)			
					Horizontal Tube Evaporation (HTE)			Vertical Tube Evaporation (VTE)
					With Vapor Compression		Without Vapor Compression	
	Electrodialysis (ED)	Brackish Water (RO)	Seawater (SWRO)	Multi-Stage Flash Distillation (MSF)	Thermal Vapor Compression (TVC)	Mechanical Vapor Compression (MVC)		
Choice	Operating costs for ED and RO lower than for MSF if salinity lower. Favored in USA and Europe where energy costs are high and expertise easily available.		Along with MED/MVC best for small-scale seawater desalination in remote areas.	Competitive in Middle East as it is linked to electricity production and cheap labor.	Major contender for all but largest installations.	Along with RO best for small-scale seawater desalination in remote areas.		Mixed results so far.
Prospects		Vast expansion last decade. Prospect of decreasing cost by improving process efficiency and membrane performance.		Limited potential for improvement except for increasing life-span substantially by using high quality materials and alloys.	Scope for development in unit size and reducing energy consumption. On point of competing with MSF as it is less complex requiring reasonable expertise.			

World Volumes of Desalinated Water

The growth in the world quantity of desalinated water can be seen from Figure 4, which also shows changes in the proportion of the totals contributed by MSF, RO and "other" methods. The proportion of these totals contributed by the Arab world is appreciable, and of this the majority is to be found in the Gulf. Of the 15,582,000 m³/day capacity world-wide at the end of 1991, 7,744,552 m³/day (or 50 percent) was in the six Gulf Cooperation Council (GCC) countries.

Graphs in Figures 5 and 6, showing the annual capacity built in the Arab world and the GCC countries for the period 1965 – 1991, illustrate the predominance of MSF in the area, where it accounts for 80.25 percent of the capacity. Of this, all but an insignificant proportion is provided by dual purpose, (i.e., electricity and water) plants. RO accounts for 16.25 percent only, and all other methods account for the remaining 2.5 percent. However, these proportions differ significantly between the six states. The total capacity in Saudi Arabia at the end of 1991 was 3,811,952 m³/day, and of this MSF represented 2,777,896 m³/day or 70.25 percent, RO 963,424 m³/day (25.25 percent) and others the remaining 4.5 percent. However, in Kuwait, of a total capacity of 1,393,190 m³/day, MSF represented 1,350,546 m³/day (97.1 percent), RO 29,098 m³/day (2.1 percent) and other methods the remaining 0.8 percent.

These figures also show significant differences when compared with figures for the world as a whole. Of the world-wide total capacity of 15,582,000 m³/day, some 7,994,500 m³/day or 51.3 percent is provided by MSF; 5,080,200 m³/day or 32.6 percent by RO; and the remaining 16.1 percent by other methods.

Impact of Local Conditions on Process Choice

The impact of local conditions on the choice of process can be illustrated by considering the factors which affect the GCC countries.

■   *The nature of the feed water to be treated:* Higher salt concentrations are best handled by distillation, while osmotic processes have traditionally found their advantages when used for water with lower concentrations of salt. Sea water, which has 35,000 ppm of salt or more, is always of higher salinity than the waters referred to as brackish, which typically run between 5,000 and 10,000. In GCC countries, often the only water available in abundance is sea water, so the only solution is to desalt it and even to pipe part of the desalinated water over long distances in order to serve many inland areas. Although improvements

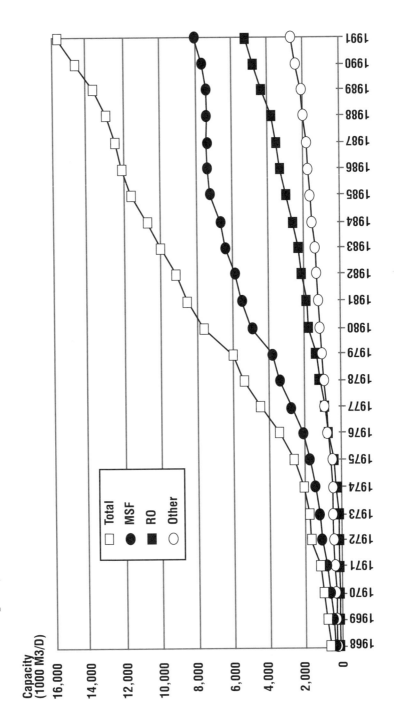

F I G U R E 4
Cumulative Capacity of All Land-Based Desalination Plants of Capacity 100m³/d or More (By Contract Year)

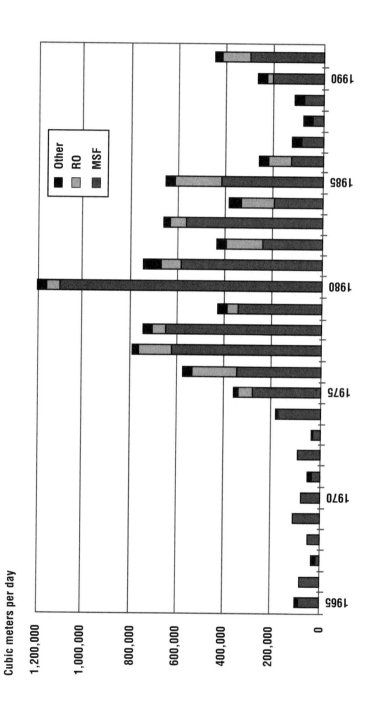

FIGURE 5
Annual New Capacity Contracted — Arab World

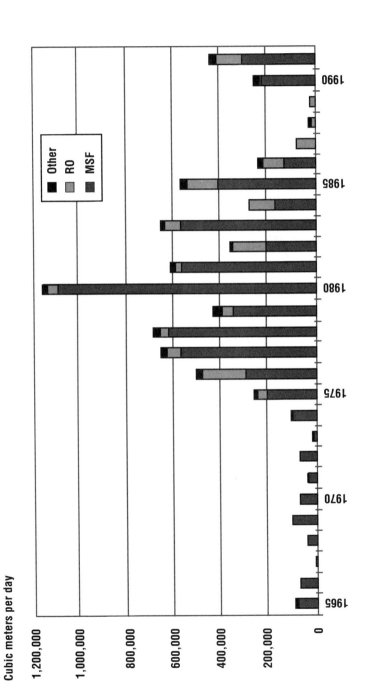

FIGURE 6
Annual New Capacity Contracted — Gulf

in RO membranes have narrowed the gap between MSF and RO for seawater desalination, MSF has maintained an advantage over RO, especially where the ocean's salt content is very high, as it is in the Gulf.

World-wide, however, RO has been gaining ground on MSF, with the demand increasing in inland areas where fresh-water supplies from natural sources have been diminishing. This is especially so in the United States where attention has turned to the purification of brackish groundwater or municipal wastes, for which RO is very suitable.

■ *Scale economies:* The demand for water in the GCC countries rose rapidly during the "oil boom" years of the 1970s due to a sudden regional increase in wealth and in population through the immigration of expatriate workers, often accompanied by their families. This situation favored the construction of large-scale plant at a time when MSF was the only method available offering large economies of scale. Much MSF plant in the Gulf has been in existence since the seventies, having been established when RO was still in the early stages of development.

■ *Co-production of electricity and fresh water:* The rapid increase in population and wealth had other ramifications, notably in the growth in demand for power for industrial, domestic and municipal use. MSF fitted in admirably as part of a dual-supply system, where waste heat from electrical generation is used for distillation, with cost advantages accruing in the production of both the water and the electricity.

■ *Availability of technical expertise:* The demand for technically qualified labor in the Gulf area increased far beyond the supply available in the booming seventies. Consequently, there was a great reliance on expatriate labor for most industrial activity. The electricity and water industries were no exception, but whereas there was a large pool of expertise concerning electricity supply available in the industrial world that could easily be adapted to Gulf conditions, RO was a new technique for which the expertise had yet to be developed. This gave MSF a further advantage during the early years.

■ *Simplicity of operation and availability of spare parts:* MSF is easier to operate, and pre-treatment is less sensitive than it is for RO. Whereas MSF spare parts are widely available and some can be produced locally, RO membranes are expensive, easily fouled, and must be imported from a limited number of manufacturers.

■ *The proximity of the sea:* The actual preferences reflect the proximity of the sea to population centers, which in turn determines the feed water to be used. Both Libya and Iraq are significant oil producers. The former has the capacity to desalinate 633,517 m³/day and the latter 333,540 m³/day, but whereas Iraq with a limited coastline has shown a preference for RO, Libya, with its major population centers near an extensive coastline, has favored MSF.

Turning to the Arab world in general, it can be seen that advances in the adoption of desalination have been very sporadic, mainly for reasons which, as suggested above, make the GCC countries a special case (Figure 5). Only Algeria of non-major oil producers has a desalination capacity in excess of 142,216 m³/day, but a number of other Arab countries have significant capacity.

It is obvious that one must guard against extrapolating only from the desalination experience of the GCC countries, yet their results and the adaptations of desalination processes that they have made provide a pool of knowledge of great potential value to the rest of the world, especially states with similar climatic conditions.

True Cost of Water Provision

Almost all modern developments in water provision involve an increase in the input of resources, and recently such increases have generally been of a capital nature. This is a way of saying that water is increasingly becoming a produced commodity requiring the input of scarce resources that have alternative uses. The price for the use of these resources is their "opportunity cost," i.e., the value of the opportunities lost by diverting the resources from use elsewhere. Consequently, the problem of costing of water is becoming a complex economic problem as well as an accounting one. Unfortunately, even the accounting framework has been sadly neglected in many parts of the world, so that when the costs of desalination are compared with the costs of water production by more traditional methods the traditional costs, in most cases, are grossly underestimated. Experience at the Kuwait Fund has shown that a wide variety of practices are used in costing existing supplies. Often, for example, no account is taken of depreciation. Replacement costs or expatriate fees are sometimes neglected as these are normally covered by technical assistance or grants. Where existing plant is very old, and no sinking fund has been created, the latter omission is crucial. Such poor accounting methods need improvement before valid cost comparisons of water supply by alternative methods can be made.

When economic costs are included the problem becomes even more complicated. As well as the financial costs of extraction and distribution other less tangible costs need to be considered. There are often costs, and sometimes benefits, that accrue to third parties and that do not appear in the accountant's profit and loss sheet, but that nevertheless must be reckoned, whether in money, time, or convenience.

Water has a number of uses, and when scarce it has to be allocated among them. If left to market forces, the price system would achieve this, but according to economic criteria only. When the socio-political system is involved, as it is when non-economic goals are sought, society has to make such decisions consciously. For instance providing water for agriculture may mean a choice between not providing water for industry or increasing total supply of water at a further cost.

Modern methods of cost-benefit analysis attempt to encompass these problems, but one of the great difficulties is the need to reduce all costs and benefits to a single index, and then to compare competing projects with each other in order to maximize the net benefit according to the index. The index to be maximized is ultimately, although not always explicitly, the social benefit which it is presumed arises from the project. This may be easily calculated for an export-producing project whose output can be measured in foreign exchange earnings, but for water with its many different uses, its health implications and its industrial and agricultural applications, the use of customary techniques for the economic analysis of its cost must be undertaken with great caution.

Incorrect pricing leads to the misuse and misallocation of scarce resources. For example, when water is supplied at far below cost, a false conception of true costs results. A rigorous method of costing water for developing countries should be established in order to support a sound water policy and also to allow valid comparisons to be made between the costs of different methods of supplying water. Although it may not be desirable to charge full costs under all circumstances, when a water policy is being formulated the costs of allocating water according to non-economic criteria should always be made clear.

Comparison of Costs of Water Supply Methods

Making valid comparisons between the cost of desalination and other forms of water supply is rendered difficult by the frequent lack of rigor in the costing of conventional methods. In 1987 the World Health Organization published the results of a world-wide survey on the costs of water (World Water/WHO 1987). Table 2, based upon that survey, shows the

T A B L E   2

Unit Cost of Water Production

Country	Cost $/m^3	GNP per Capita $	$Cost per Capita %
Cape Verde	4.65	317	1.5
Cayman Islands	2.75	13,000	0.02
Cameroon	2.0	800	0.25
Mexico	1.5	2,080	0.07
Argentina	1.5	1,929	0.8
Netherlands	1.25	9,290	0.01
Zambia	1.05	390	0.3
Saudi Arabia	1.0	8,850	0.01
Sierra Leone	0.9	200	0.45
Tonga	0.8	354	0.23
Botswana	0.75	840	0.09
Togo	0.66	300	0.22
Surinam	0.6	3,030	0.02
Seychelles	0.6	2,250	0.03
Malawi	0.6	170	0.35
Pap New Guinea	0.55	649	0.09
Tunisia	0.5	1,277	0.04
Cook Islands	0.4	7,170	0.006
Cyprus	0.4	3,572	0.01
Djibouti	0.4	480	0.08
Rwanda	0.4	280	0.14
Bahamas	0.37	7,556	0.01
Laos	0.35	100	0.35
Barbados	0.34	4,889	0.01
Ghana	0.35	420	0.08
Burundi	0.35	230	0.15
Mali	0.33	142	0.23
Switzerland	0.33	14,764	0.002
Afghanistan	0.3	163	0.18
Mauritius	0.29	1,020	0.03
Burma	0.25	188	0.13
Singapore	0.24	7,420	0.003
Spain	0.22	4,256	0.005
Vanuatu	0.22	529	0.04
Zaire	0.22	271	0.08
Hungary	0.23	1,909	0.01
Finland	0.21	10,531	0.002
Thailand	0.21	729	0.03
Honduras	0.2	720	0.03
Rep of Korea	0.19	2,032	0.009
Haiti	0.18	320	0.06
Malaysia	0.18	2,033	0.009
Costa Rica	0.17	1,300	0.013
Madagascar	0.17	240	0.07
Angola	0.15	560	0.03
Nicaragua	0.14	770	0.02
Morocco	0.14	512	0.03
Chile	0.12	1,430	0.008
Peru	0.12	1,010	0.012
Iraq	0.118	2,964	0.004
Bangladesh	0.09	136	0.07
Ecuador	0.09	1,160	0.008
Western Samoa	0.09	660	0.013
Panama	0.07	2,100	0.003
Philippines	0.05	585	0.008

disparity of the values quoted by those countries that responded (Dabbagh et al. 1989).

There seems to be no correlation between the costs quoted and any single factor. Singapore, with one of the lowest costs, $0.24/m³ relies heavily on water imported by pipelines from Malaysia, while Cameroon, with abundant surface water and rainfall, has one of the highest costs, $2.0/m³. It is interesting to note that an efficient water management, such as Singapore's, which keeps accurate records and includes all the basic costing components in calculations, can keep the cost of water reasonably low in spite of an adverse water resources situation, whereas some developing countries quote a high cost for water production and distribution based on calculations that exclude basic costing components. This indicates that efficient management can not only reduce the cost of water but also provide financial rewards that can be attractive to the private sector.

The Netherlands quotes a figure of $1.25/m³, which is almost four times that of Switzerland, at $0.33/m³, yet both are countries with high national incomes per capita. The difference may be due to the quality of their respective water sources, so that Netherlands with its low-lying land, potential seawater intrusion, and very high population density has to bear a water cost almost as high as that for desalinated water in spite of its recognized efficient management. Unfortunately, as can be seen from the table, many countries are facing more difficult situations than the Netherlands with the added disadvantages that they are countries with low national incomes per capita and often inadequate water management. The cost of some conventionally treated national water supplies is indeed approaching that of desalinated water.

Factors other than management efficiency and local conditions contribute to the disparity between the figures cited for water costs in different parts of the world, and the following are among those that have been noted from independent experience:

■ Existing infrastructure costs may have been ignored, or when capital has been completely written down, no allowance has been made for capital replacement. (It is worth noting that the cost of construction work for a water supply scheme in Africa can be three to five times the cost of a similar scheme in Europe.) (Dabbagh et al. 1991)

■ The price given may be that charged to the consumer and may bear no relationship to costs.

■ Estimates may have been made "off the cuff" because of a lack of data, or may have been based on poor or inefficiently gathered information.

■ Inputs — especially of labor — may have been calculated in local currency, which may not reflect costs accurately owing to artificial rates of exchange.

■ National estimates are extremely difficult to make and can be almost meaningless where a large number of independent authorities are responsible for supply and GNP per capita varies greatly between regions.

Allowance has to be made for these possibilities when costing conventional water supply schemes for comparison with other methods. It is also imperative to determine the cost of increasing capacity by further use of the existing method of supply, as this is often very expensive and may even be impracticable. Thus, desalination has not only to be compared with conventional methods of supply, but also with the actual alternative methods available in a particular concrete situation, which may include long-distance pipelines and tanker transport if conventional local water has been already been fully exploited.

Whether pipeline or tanker transport is the more economical again depends on many factors such as the terrain over which the pipelines need to be laid and the nature of the sea-leg of the journey. In addition to ships, tanker transport also requires much pipeline and other infrastructure to convey water to the loading ports and from the port of discharge to its destination for use in the receiving country or region.

When pipeline and tanker transport were compared for one particular developing country in Southeast Asia (unpublished reference) tanker transport only became the cheaper at distances greater than 200 to 300 miles, when the price was around $2.0/m^3$ at present prices, and then only when most of the distance was over the sea and little land-side infrastructure was required. Beyond that point, however, tanker transport became much more economical and the journey could be doubled without an appreciable increase in costs.

Desalination was not considered as an option in the above study, but current costings show that as the distances to water grow longer, desalination can become cheaper than pipeline transport well before tanker transport does. Indeed, it is fast becoming a serious contender world-wide as a solution to water shortages especially where it is produced as part of a dual process with the production of electricity.

A case in point is the proposed "Peace Pipeline," which is intended to supply 15 million people with approximately 350 liters per person per day. At a cost of $20 billion, a construction period of 10 years, and a suggested pipeline life of 50 years, the initial capital costs alone would represent between $0.735 per m^3 and $1.758 per m^3 depending on what interest rate/discount factor between 5 percent and 10 percent is chosen (See also Section 1 of Annex). Other costs must be added to these initial capital costs. These are the initial charges for water provided by the supplying country; way leave or toll charges for countries through which the pipeline passes; the costs of operation and maintenance (including pumping and other energy-linked costs); the replacement of pumping and other equipment at intervals; and the cost of treatment before the water enters the final distribution systems. Using equivalent discount rates, it is suggested that desalination would be at least as economic as the "Peace Pipeline."

The True Cost of Desalination

Comparisons between the costs of established and innovative methods of water production can be influenced by inefficient costing, and the validity of the costing methods themselves may be disputed.

An examination of a series of costings of desalination processes (Wade 1991; Sadukham et al. 1991; Leitner 1991) has been made. All exhibit the major elements of normal accounting procedures, in that costs are detailed for the initial inputs such as site preparation, building works, machinery and start-up period expenditure, together with labor and capital carrying charges. These are shown generally as capital costs. Detailed running or operational costs are shown on an annual basis. Capital costs are then transformed into annual charges. Briefly, an amount is calculated that, if invested annually, would reach, by the end of the life of the installation, a sum that would have been reached had the initial capital costs been invested in the same manner. When this sum is added to the annual operating costs, the whole is treated as total annual costs. (Where water is produced jointly with electricity, a "shadow" amount, equal to the cost of producing electricity independently by the most efficient means available, is first deducted to obtain the cost of producing water.) The result divided by the intended annual output, gives the unit cost of the production of water.

In all cases where such comparisons of costs are made, a number of difficulties arise, especially where the time configurations of the costs being compared are noticeably different. For example if the length of life of capital in a process were different from that assumed, then the allocation of capital costs to annual charges would be wrongly stated. Or, if process A had high initial capital costs and comparatively low operating costs, and process B the opposite, then the difference between the two costings will be affected by the choice of interest rate used to calculate the return on alternative investments.

Often a 20-year lifespan is assumed for capital in all processes being considered, but some MSF plants have been operating satisfactorily in Kuwait for as long as 26 years. Shuaibah South MSF desalination plant, which has served satisfactorily for the last 21 years, is expected to continue to do so for at least another 10 years under routine maintenance procedures. Lifespan is materially affected by standards of maintenance, the type of materials and alloys used in plant construction and the complexity of the moving parts. Discussions with specialized consultants in desalination indicate that a 20 percent to 30 percent increase in initial investment to provide better materials for the construction of some MSF plants could result in doubling their lifespan. Indeed, MSF could become the modern equivalent of the slow sand filter that lasted for decades, since the bulk of the capital cost of the plant is in the non-moving parts — the shell and pipework — which can last for a very long time indeed if current improvements in corrosion protection are used.

The interest rate (or discount rate) used needs careful selection. An oil-rich country might legitimately use the rate available on the world currency market, while a developing country, with a grave shortage of investment capital, might use the rate of return expected on an investment in foreign exchange-earning industry. Differing conditions demand different rates. It is also not possible to compare one process with another on an abstract basis to provide universally applicable answers, since these do not exist. Comparisons must be made using criteria applicable in each case where desalination is proposed as a solution to the shortage of water.

The following Tables 3, 4, and 5, based on figures provided in Wade (1991), illustrate the effect of changes, all within reasonable ranges, in three of the parameters (i.e., discount rate, oil prices and lifespan of capital) used in calculating the costs of water.

In Tables 3, 4 and 5, each variant has a 27,240 m^3 daily capacity, at 85 percent efficiency.

■ Project 1 uses MSF with a back pressure steam turbine

■ Project 2 uses MSF with a gas turbine and waste heat boiler

■ Project 3 uses MSF and is part of a plant co-generating electricity

TABLE   3

Variation in Unit Water Cost with Discount Rate ($ per m³)

Discount Rate	Project Category				
	1	2	3	4	5
4%	1.4	1.2	1.0	1.2	1.4
6%	1.5	1.3	1.0	1.2	1.5
8%	1.6	1.4	1.1	1.3	1.6
10%	1.7	1.4	1.2	1.4	1.7
12%	1.8	1.5	1.2	1.5	1.8
14%	1.9	1.6	1.3	1.6	1.9

TABLE   4

Variation in Unit Water Cost with Oil Prices ($ per m³)

Price in US$ per Barrel	Project Category				
	1	2	3	4	5
0.0	1.1	0.9	1	1.3	1.5
8.0	1.4	1.2	1	1.4	1.6
18.0	1.7	1.4	1.2	1.4	1.7
29.5	2	1.8	1.3	1.5	1.8
35.0	2.2	1.9	1.3	1.6	1.9

■ Project 4 uses RO with a single-stage system

■ Project 5 uses RO with a double-stage system

In Table 3, variations in the discount rate over the range shown result in changes of 36 percent, 28 percent and 34 percent in the unit price of water produced by the MSF processes, and 43 percent and 41 percent by those using RO.

In Table 4, the price of oil, presumed to be the source energy used in each case, has been varied over reasonable ranges, depending on the availability to the user. This results in increases of 100 percent, 109 percent and 40 percent over the three MSF processes, but only 24 percent and 23 percent over the two using RO. In some countries in the Middle East, associated gas is available as a co-product of petroleum, and its price may, in many cases, make its use a cheaper alternative to oil.

The effect on the unit cost of water of extending the lifespan of capital — without incurring extra costs — from the base case of 20 years to 25 and 30 years has been calculated in Table 5 for each of the five production projects illustrated in Tables 3 and 4. These calculations have been made for specimen discount rates of 10 percent (base case), 6 percent and 4 percent. The table also shows the savings in total annual costs that would

T A B L E   5

Variation in Costs per m³ with Life of Capital for Various Discount Rates

Capital Life (Yrs)	Project Category				
	1	2	3	4	5
Reduction in Total Annual Cost in $					
Case I : Discount Rate 10%					
20 Yrs	1.662	1.438	1.147	1.427	1.688
25 Yrs	1.589	1.349	1.024	1.387	1.6
30 Yrs	1.548	1.299	.956	1.365	1.618
20–25 Yrs	(618,102)	(753,577)	(1,041,461)	(338,686)	(381,022)
20–30 Yrs	(965,256)	(1,176,935)	(1,617,227)	(524,964)	(592,701)
Case II: Discount Rate 6%					
20 Yrs	1.5	1.3	1.0	1.2	1.5
25 Yrs	1.4	1.2	0.9	1.2	1.4
30 Yrs	1.3	1.1	0.8	1.2	1.4
20–25 Yrs	(753,577)	(922,920)	(1,270,074)	(414,891)	(474,161)
20–30 Yrs	(1,227,738)	(1,490,220)	(2,065,987)	(668,906)	(762,044)
Case III: Discount Rate 4%					
20 Yrs	1.375	1.243	0.959	1.148	1.369
25 Yrs	1.280	1.127	0.798	1.096	1.312
30 Yrs	1.217	1.052	0.695	1.063	1.272
20–25 Yrs	(804,380)	(982,190)	(1,363,212)	(440,292)	499,562)
20–30 Yrs	(1,337,811)	(1,617,227)	(2,235,330)	(719,709)	(821,314)

be occasioned by each extension from the 20 year lifespan base case. Thus reduction in production costs could be made if an extension in the life of capital to the levels shown could be brought about by increased expenditure, either on an annual basis as part of operating costs, or as an increase in initial capital amortised according to the appropriate discount rate, or by using a combination of both.

It is evident from the table that, due to the high capital investment and lower running costs of MSF compared with RO, reductions in unit water costs from extending the life of capital in an MSF plant are greater than the corresponding reductions for an RO plant, and that these differences increase as discount rates fall.

For the above illustration, however, variations have been made in only three of the parameters used, and then only on the basis of "other things being held equal." Comparison of five methods was made for plants of equal capacity, although it is realized that economies of scale, especially in those using the MSF method, would have enhanced the results if optimum sizes of plant had been chosen in each case. Also, the changes in the parameters themselves would encourage shifts in the proportions of the factors utilized. A lowering of interest rates would not only reduce the annual cost of capital, but it would encourage the substitution of capital for labor, thus enhancing the extent of the savings illustrated. Similar shifts would result from other changes of the same nature. It is obvious that severe difficulties arise when using currently accepted procedures to decide between methods of desalination when all possible variations are introduced.

Costing the Provision of Desalinated Water

Desalinated water has generally been costed "from plant intake to output," that is, according to the costs of the process used to transform saline water into water of a potable quality. However, this is not entirely valid when results are used for comparison with other methods. Seawater is taken in raw for desalination, while, for example, water drawn from a river for conventional treatment has often been used several times upstream and treated before being returned to its source and would thus require less subsequent treatment at a desalination plant.

Also, when costings are compared, it is generally assumed that all modes of treatment require a similar distribution system, and therefore distribution does not affect subsequent costs. This is not necessarily so, and changes in the means of distribution may be required to accommodate the water produced in each case. Differently produced waters have different properties, which have different effects on various parts of the distribution system. Desalinated water, for example, is very aggressive when passed through asbestos cement pipes, and leads to their rapid deterioration.

It is necessary, therefore, to examine the full cycle of use of desalinated water from the intake of seawater to after-use discharge at the sewage plant, and to examine the main components of the cycle to see whether savings can be made, or whether extra expenditure is needed to reduce costs elsewhere in the system. Components worthy of examination are:

- ■ *The distribution network:* Considerable savings are often possible by reducing unaccounted-for water, i.e., the difference between water produced and water sold. In virtually all the Arab countries, data on this problem are insufficient, but it is generally thought that the percentage of unaccounted-for water varies between 30 percent and 50 percent. Some is used in fountains and public places without being paid for, but much is lost through leakage or the overflow of reservoirs. Advances in leakage detection and control can reduce unaccounted-for or "missing" water to about 5 percent or 10 percent (Reed 1980).

■ *Storage:* The considerable variation in demand between summer and winter restricts the efficient use of the installed capacity of the desalination plant. This is basically due to the insufficient capacity for water storage. For this reason the desalination plant should be operated at its most efficient capacity and when excess water is produced it should be diverted to suitable aquifers and retrieved later (Pyne 1988). This should result in appreciable savings in plant capacity.

■ *Sewage effluent:* The collection and treatment of sewage effluent is of particular importance when desalination is being considered. Irrigation has been the most common use made of treated sewage (Khouri 1992), but the economic benefits of such use should be carefully examined. It may prove to be more beneficial to import agricultural products and to use the treated effluent for blending with and diluting salty water at the point of intake for desalination, thus reducing desalination costs. Alternatively, under careful supervision, the effluent could be used for recharging aquifers, in which case its quality should be improved by infiltration over time.

These are but examples of the many problems faced in the type of analysis proposed. In practice, each case and each location must be treated separately, and water provision, use, and subsequent disposal or re-use must be seen as part of a circular process. The costing of water needs to be carried out in such a way that comparisons can be made that permit the selection of the system that results in the cheapest cost of water on delivery to the point of use.

The Modeling Approach

To calculate the lowest cost of delivered water — that is, the optimum price obtainable — all the contributory costs must be determined simultaneously and then fed into a computer model of the relationship between the costs. To this end, a model has been constructed that, although simple, is capable of testing some of the hypotheses in this paper as well as being expanded later for application in more specific situations.

The main components of the model are illustrated in Figure 7. There are four sub-areas that require individual cost inputs: production, distribution, sewage collection and treatment, and disposal. These are assessed as follows:

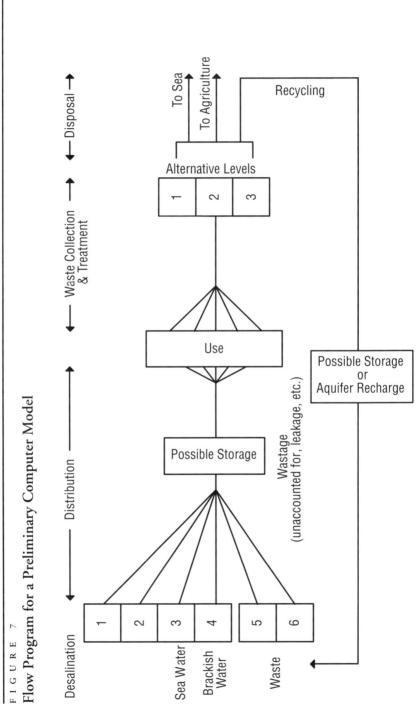

F I G U R E 7
Flow Program for a Preliminary Computer Model

■ Production costs depend on those for the individual desalination systems available. These are inserted into the model and the costs are calculated per m³ of output.

■ The costs of water at the point of use are determined by adding the costs of the distribution system to the costs of production, allowing for a percentage of unaccounted-for water, especially leakage, and then calculating the initial cost per m³ of delivered water.

■ Sewage treatment costs depend on the percentage of the delivered water that is collected as wastewater. The level of treatment required varies according to the subsequent disposition planned for the effluent.

■ Disposal costs or potential gains depend on whether sewage effluent is dumped into the sea, used for agriculture, or recycled through the desalination system.

A situation is hypothesized in which the delivered quantity of water required is 350,000m³ per day, (or 128 mcm per annum), using seawater as feedstock. This is nearly the volume of water used by the city of Tunis. In the base case there is an initial leakage factor in the delivery system of 30 percent and no recycling. Even though the initial cost of produced water is $1.66/m³, leakage and other costs raise the delivered cost of water to $3.39/m³. Table 6 illustrates: (1) how the cost per m³ falls as the quantity of sewage recycled is increased, thereby allowing for a reduction in the cost of sewage treatment. The increased quantity of sewage recycled into the system does not require the same expensive treatment as the saltwater intake it would replace and permits a lower intake of sea water for desalination; and (2) how a reduction in leakage lowers delivered costs as a result of lower production levels. These two improvements in combina-

TABLE 6

Cost Reduction by Leakage Control and Waste Recycling ($ Cost per m³)

Unaccounted-For Water	Volume Recycled		
	0.0	50%	100%
30%	3.39	3.26	3.13
25%	2.96	2.83	2.71
20%	2.60	2.48	2.37
15%	2.30	2.19	2.09
10%	2.05	1.95	1.85
5%	1.84	1.75	1.65

tion seem able to reduce the cost of delivered water by over 50 percent under the conditions postulated.

However, such results cannot be obtained at nil cost, and savings must always be balanced against the costs of achieving them. In the case illustrated, if maximum leakage reduction and full recycling of waste (taken as 50 percent of delivered water) were achieved, the saving of $1.74 per m³ delivered would represent a gain of over $222 million a year. A sum of this order would, if amortized, support an appreciable quantity of capital expenditure to be devoted to the system's improvement.

Although the model used is essentially very simple, it is being developed to test alternative approaches to the problems of water shortages in individual locations. By including costs of each alternative available (e.g., leakage reduction, extension of supply by MSF or RO, disposal or recycling of water) and treating these according to the parameters applicable to that location (e.g., feedstock locally available, fuel availability and co-generation possibilities, discount rates, size of market, anticipated life of capital) then the optimum points of investment in each location's supply system can be identified. Given the rising cost of water from both desalination and the more traditional sources, it is no longer valid to assume that impending shortages in any situation should automatically be met by additional capacity in the existing sources of supply.

Impact on the Environment

The use of desalination has a positive impact on the environment. It helps to conserve water resources by enabling water that is otherwise useless for human consumption to be purified and also by providing a means for recycling wastewater.

On the negative side, desalination processes produce reject brine with a very high concentration of salts. In the case of seawater production facilities, this brine can usually be returned to and dissolved in the sea without much adverse effect, although a temperature differential between intake water and reject brine may affect marine organisms. To protect the environment against harmful effects, various measures can be adopted, such as careful placement of intake and discharge outlets with regard to the local marine environment as well as to the turbulence, temperature, and major constituents of the reject brine, which vary according to the process used.

If reject brine from an inland desalination plant is deposited haphazardly it can have considerable adverse effects and pollute groundwater irreversibly. To avoid this the brine can be deposited in lined ponds, spread out over the desert, or injected into the ground at great depths where it is not expected to harm the environment. Disposing of reject brine safely can cost between 5 percent and 30 percent of the total cost of plant installation, depending on its type and location (Mandeel 1992).

Environmental protection measures necessary when using desalination are attainable, although there is still room for improvement in this field. There are even prospects for utilizing beneficially the reject brine that is the main pollutant of desalination.

All in all, the adverse impact on the environment when desalination is adopted as a major source of potable water is minimal in comparison with the harm that has already been done to conventional sources of water through overdraft of aquifers and other forms of mismanagement. Under these "conventional" abuses, some aquifers have been deteriorating so much in quantity and quality that they have often reached a state where they cannot be replenished, let alone protected from further deterioration. The misuse of natural water resources can lead to permanent damage. In contrast, desalination avoids such misuse and may indeed rectify it in certain circumstances, such as making usable the water from a brackish aquifer that would otherwise have to be abandoned due to deterioration of water quality as a result of over-pumping.

Future Growth of the Desalination Industry

It is expected that the desalination industry will continue to grow. The International Atomic Energy Agency in a comprehensive report (IAEA 1992) anticipates that the demand for desalinated water will double in each decade over the next 20 to 25 years, implying an annual growth rate of 7.18 percent. This figure is used as a guide for what follows.

Various cost estimates for the installation of desalination equipment (see, inter alia, Wade 1991; Sadukham, et al. 1991; Leitner 1991) have been analyzed. Allowing for the current proportion of new plant using the two main desalination methods, the cost of new plant at 1989–1991 prices is estimated to be around $1687 per m^3/day capacity (see Annex, Section 2). Using the same sources, it can be estimated that the 1989–1991 annual operating costs attributable to the equipment industry's manufacturing output, such as spares, membranes and, say, 10 percent of general repairs and maintenance, would be about $105 per m^3/day.

Based on the world capacity in 1991 of 15.6 mcm/day (Wagnick 1992) and taking into consideration the above figures, Table 7 has be drawn up to compare the capacity and costs for manufacturing plant and machinery for desalination for the years 1991 and 2016 using 1989–91 prices.

T A B L E   7

Comparison of Desalinating Capacity and Costs of Manufacture of the Desalination Equipment Industry's Output for the Years 1991 and 2016

		Capacity (mcm/day)		Cost (billion US$/year)	
	Year:	1991	2016	1991	2016
World capacity		15.6	88.3	—	—
Total O&M cost at $105 m³/day capacity		—	—	1.64	9.27
Additional annual capacity		0.25	6.34	—	—
Cost of additional investment at $1687 m³/day capacity		—	—	0.42	10.69
Replacing capital (20 year life assumed)		1.2	22.1	—	—
Cost of replacing capital at $1687 m³/day capacity		—	—	2.03	37.22
Total				4.09	57.18

While a complete analysis would require more data and resources than are available, the above analyses show that the total call on the desalination equipment industry for specialized manufacturing output destines it to become one of the world's major industries, growing from about $4 billion turnover in 1991 to nearly $57 billion in 2016. This is an annual rate of increase of 12.2 percent to cope with an annual increase in demand for water of 7.18 percent.

As already shown (Wagnick 1992) the six GCC countries have 49.7 percent of the world's desalination capacity and the Arab world as a whole has just over 56 percent. Although these proportions are bound to change, the Arab world will remain one of the world's foremost users of desalination for the foreseeable future. Technological advances may, and it is to be hoped will, reduce costs and increase efficiency in the industry, but it is up to the Arab world to ensure it is involved in future developments so that it may benefit from the advances financed by its patronage.

International Institutional Set-Up

Research carried out under the existing institutional arrangements in the major oil-producing countries has led to improvements in manufacturing standards, particularly with regard to protection against corrosion, optimum operating procedures, and the use of chemicals for water treatment and the protection of the desalination plants and their ancillaries. The knowledge of such innovations has, however, been confined mainly to dialogue between manufacturers and researchers who have used it to update technical specifications (Zubaidi 1987). There is a need for all the available experience to be brought together and streamlined so that improvements can be suggested in both the technology of desalination and the management of desalinated water.

This could be effected by an international entity with the following objectives:

■ The collection of data from all the world's desalination plants, particularly those in the GCC countries, placing special emphasis on the quality of feed and product waters, the pre-treatment of brine, energy consumption, performance, the main characteristics of operation and maintenance and the relative advantages and disadvantages of different procedures.

■ The identification and financing of promising fields of research such as construction materials and process chemicals, the protection of the environment, alternative energy sources, particularly solar energy and nuclear power, and the potential use of minerals recaptured from brine.

■ The provision of specialized advice from a roster of experts drawn up to review designs and alternative technologies.

■ The co-ordination of the above activities by contacts with governmental, academic, and manufacturing institutions that could try to make the following contributions to the furtherance of desalination technology:

Governmental Institutions
◆ To subsidize major research and to accelerate the development of promising technology.

◆ To assess the economics of using desalination as part of comprehensive water supply schemes that optimize plant operation, reduce "missing" water, increase storage capacity, use aquifer recharge and recovery, and maximize the re-use of sewage effluent.

◆ To study the optimum conjunctive use of all water, especially in arid zones, i.e., desalinated water, brackish water, and sewage effluent.

◆ To establish standards for the protection of the environment when desalination is used.

◆ To assess the prospects for privatization in terms of the type of management and procedures required, and the relationship between a water supply organization and the government concerned.

Academic Institutions
◆ To identify institutions interested in promoting research related to desalination.

◆ To host and subsidize research.

◆ To establish prizes for research in order to encourage international talent to participate more actively in the development of desalination.

◆ To propose desalination syllabi to be included in university courses.

Equipment Manufacturing Industry
◆ To establish standards for all materials and replacement parts.

◆ To publish manuals specifying the main components of desalination plants.

◆ To establish factories for manufacturing desalination plants and/ or their components in developing countries where industrial capabilities are available. Shipyards, some of which have been abandoned due to the prevailing economic recession, are particularly suited for the construction of MSF and/or ME desalination plants.

It is envisaged that the proposed international organization will rely basically on experience available in user countries and on the extensive experience in the design and manufacture of desalination plant in industrial countries. The management of such a set-up requires the backing of all concerned with the industry, and all elements with interests at stake should be represented on a board of directors with full autonomy to direct the institution toward its objectives.

Substantial financial backing would be needed to establish a unit of the caliber required to bring about major developments in desalination processes according to the broad outlines described above. The capital and running costs could be raised from sources that will benefit from the industry, either as users or producers. The following sources could be considered:

■ *Manufacturers:* These normally allocate budgets for research, which nonetheless are affected, sometimes drastically, by fluctuations of the market. Some firms' research duplicates that of their competitors, so it would help them to direct some of their research funds to support international research that could benefit all concerned.

- *User countries:* Countries using desalination have been adopting severe tendering conditions to protect themselves against all possible risks and introducing hard financial conditions and penalties that have resulted in an increase in the capital cost of construction work (El-Saie 1992). These conditions are partly due to lack of confidence in the industry. A well-established coordination unit would help to boost the confidence of the would-be owners of desalination plants and thus help to reduce the cost of desalination investments.

- *Funding Institutions:* These have a direct interest in the development of the desalination industry, particularly now that shortages of potable water are increasingly arising in remote areas of developing countries. The institutions' financial support will be very beneficial if low-cost and locally made desalination facilities can be developed.

The location of an international unit of this nature should be somewhere with easy access to sources of technology, on one hand, and to practical experience in operation and maintenance, on the other. These tend to be situated on opposite sides of the world, but a compromise should be one that benefits both user and producing countries, with the prestigious aspect of the location being of minor importance. The main considerations should be on ease of communication, availability of expertise and access to information and data.

Conclusion

The inception and development of desalination have taken place in unusual circumstances. While major technologies usually emanated from industrial countries to satisfy increasing demand in the same countries, demand for desalination arose in regions distant from the industrialized world. The oil boom in the Middle East created a rapidly growing desalination market, and in this situation, research was limited to in-house activities within the industry.

The choice among desalination options is not clear-cut but subject to many variables and it should not be left entirely to the influence of manufacturers. The requirements of industrial countries differ from those of developing countries where in the future there will be a greater demand for desalination. Developing countries, lacking in expertise and foreign currency, will be anxious to operate simple desalination plants that use spare parts either produceable locally or obtainable from a wide market not restricted to certain brands of products.

While the cost of desalination has been branded as high due to its high energy consumption, it is becoming evident that other options for providing water can be as costly, and have, in addition, restrictive disadvantages related to the need for long-term commitment and planning.

A reduction in the cost of desalinated water seems possible with efficient management of the full water cycle. The reduction of unaccounted-for water, the re-use of sewage effluent, and in the case of plants linked to electrical generation, the provision of balancing storage capacity to even out fluctuations in water supply and demand, are basic elements of the water cycle that have to be improved when desalination is considered.

The desalination industry is set to become an important major industry, probably with a positive environmental impact, that will be widely in demand in industrial and developing countries. Concerted efforts should therefore be made to make better use of the experience gained so far, particularly in oil-producing countries, combining it with the expertise available in industrial countries by establishing an international institutional set-up to promote the industry, increase the scope of public research work, and assist would-be owners of a desalination plant in its choice, installation and management.

Annex

Basis for Calculating Costs

The Peace Pipeline

With a proposed construction period of 10 years, the capital costs incurred in each year also incur a "carrying charge" depending on the rate of interest or discount rate used. With a total expenditure of $20 billion spread evenly over the 10 years, the true cost of the investment becomes $33.43 billion if a discount rate of 10 percent is adopted or $25.756 billion if the adopted rate is 5 percent.

With a 50-year lifespan, the "present worth factor" at 10 percent is 9.9148, i.e., the total investment cost divided by this factor set aside each year at 10 percent will reach the same amount at the end of 50 years as would have been reached in the same time if the whole cost of the original investment had been invested at the same rate. The annual capital cost at 10 percent is therefore $3,371,727,000, while at a 5 percent rate of discount this falls to $1,410,831,567.

A delivery to 15 million people of .35m³ per person per day implies a total annual delivery of 1,917,562,500m³. The annual capital costs alone to be supported by each m³ would be $1.758 using a 10 percent rate of discount and $0.735 using a rate of 5 percent.

The Growth of the Desalination Industry

A number of costings for specimen MSF and RO plants were reviewed, and the components for (a) plant and machinery directly relating to desalination in the capital costs, and (b) replacements, spare parts and specialized requirements for desalination in the operating costs, were isolated. The totals were averaged across the plants reviewed, and reduced to costs per m³ of daily capacity. In these calculations, all site works and buildings, all local labor and other local inputs, etc. were disregarded, as were all costs attributable to electricity production rather than water.

The demand on the industry proper was then calculated in the three parts:

■ The growth element, which represented the addition required in any year to total daily capacity. This was 7.18 percent, the figure required to double capacity in a decade.

■ The operational element. Spare parts, replacements, membranes, chemicals, and an arbitrary addition of 10 percent of operational and maintenance expenses to allow for miscellaneous requirements. This was calculated against total capacity in each year.

■ The replacement element. Given an average plant life of 20 years, in any one year the plants established 20 years previously will need replacing.

The total of these three elements was deemed to represent the demand on the industry in each year for which it was calculated. Of course, the calculations are based on a number of assumptions: (1) Constant prices based on 1989–90 figures, i.e., no inflation, and no technological change causing increases or decreases in real prices; and (2) The proportions of the various methods used will remain constant. Movement between MSF and RO will alter the capital inputs somewhat, but with some compensatory movement in annual operating costs. These two are probably the major assumptions required, but it is unlikely that their relaxation would alter the basic thrust of the conclusions drawn.

9

Global Climate Change and Its Consequences for Water Availability in the Arab World

F. A. Bazzaz

With population growth and improvements in standard of living, water scarcity is becoming an important issue in many regions of the world. The naturally dry Arab region faces this problem in a particularly intense form. Ongoing and possible future changes in climate, with both worldwide and regional dimensions, could greatly aggravate the problem of assuring adequate supplies of water and of water-based food to Arab populations. Increases in atmospheric temperature, and differences in the quantity and timing of precipitation are climate changes that would provoke cascades of adaptations and sub-effects. They would have profound effects on human life, especially in a region where temperatures are already high and water scarce. Such climate changes are potential results of ongoing increases in the atmospheric concentration of carbon dioxide and certain other "greenhouse" gases generated by combustion and other human activities. Although the analysis of climate change is at an early stage and specific predictions cannot be made for individual regions of the earth, this paper urges that water resource planning in the Arab world should actively recognize that familiar historical climate patterns cannot be reliably assumed to continue indefinitely into the future, but may be entering a period of change.

In the already hard-pressed Arab region, mostly desert or semi-desert, there are heightened concerns about the supply of water for domestic, industrial and agricultural uses to satisfy rapidly increasing demands in all these sectors. The region's growth of population is rapid, and there is strong evidence that human activities, especially overgrazing and over-gathering of plants for fuel, combined with shifting climatic conditions are contributing to desertification in parts of the region. Currently, much fossil water from underground sources is being used for agricultural production, and the water table is receding greatly in some areas, but the desire for self-sufficiency in food production, whether on a country or regional basis, continues to have a great impact on the use of water resources in the region. Nonetheless, some water uses, especially for crop production in certain areas, are uneconomic and there will have to be a shift in allocation of some water from agricultural to industrial and domestic uses. Certain reductions in water supply may be imminent as well: Turkey is now developing massive irrigation projects that could limit the water that historically has passed from its eastern region through to Syria and on to Iraq by as much as 60 percent if all 22 projected dams are completed (Vesilind 1993).

At the world level, the increase in human population and its increased per capita and total consumption of energy and other resources, have created much potential for greatly accelerated environmental change. While much change in climate has occurred in the past, the rate at which change is presently occurring worldwide is much faster than before. What has taken centuries in the past is taking only years now. It is predicted that the human environment will be very different in the next century from what it is now (Houghton et al. 1990; Lubchenco 1991). It is of utmost importance that the Arab region and the rest of the world be prepared for this eventuality.

Globally speaking, it is strongly suspected that several elements of the human environment are rapidly changing. For example, carbon dioxide concentrations in the atmosphere are rising, temperature may be rising, nitrogen and sulfur deposition are increasing, the stratospheric ozone layer is thinning. These events are the result of the emission into the atmosphere of large quantities of the so-called "greenhouse gases," such as carbon dioxide (CO_2), methane (CH_4), nitrous oxides (N_2O), and chlorofluoro-carbons (CFCs).

These gases absorb infrared radiation and therefore prevent heat, coming as visible radiation during the daytime, from escaping back into outer space at night. Because of this property, when these gases (and water vapor) are present in the atmosphere in low concentrations, they make life on earth possible. In the absence of greenhouse gases, especially water vapor and CO_2, the differences between daytime and nighttime temperatures would be so great that biological life could not be sustained. Currently, greenhouse gases ameliorate the temperature of the planet, making life possible. Living organisms on our planet have adapted through evolution to function within the current temperature boundaries.

Carbon dioxide, the most common of the greenhouse gases, is emitted in large quantities from combustion of fossil fuels such as oil, gas and coal, from deforestation and burning, and during volcanic eruptions. With the current rate of growth in energy use, a doubling of CO_2 in the atmosphere from the pre-industrial level of 280 ppm is expected to be reached sometime in the middle of the 21st century.

Several mathematical models (GCMs) based on general circulation principles have been developed to predict climate change. While differing somewhat in the details of their predictions (Schneider 1989; Houghton et al. 1990, 1992; Cess 1993) these models predict that with the increase in concentrations of greenhouse gases in the atmosphere there will be an increase of 3.5 to 5.2°C in the mean annual global temperature. Most scientists agree that this is a reasonable prediction, although a few scientists suspect that increased cloudiness will alter the effect, as discussed below. Furthermore, the general circulation models predict that there will be changes in the patterns of rain and snowfall. These two issues — temperature rise and change in rainfall — are directly relevant to water resources and their future management.

The amount of global warming is expected to depend on the concentrations of the greenhouse gases in the atmosphere. The amount of CO_2 in the atmosphere depends largely on the rate of fossil fuel consumption. The Intergovernmental Panel of Climate Change (IPCC) has developed predictions of the degree of warming based on several fuel consumption scenarios including 1) an actual decline in use, 2) business as usual and 3) an increase in consumption.

However, the direct linkage between variations in the presence of greenhouse gases and temperature increase is as yet unclear. For example, some scientists argue that the increase in greenhouse gases in the atmosphere will increase the cloud cover and this may prevent some of the solar radiation from reaching the earth. Thus, this increased reflectivity by the increased cloud cover may prevent the anticipated rise in global temperature. Furthermore the emission of sulfur dioxide (SO_2) into the atmosphere through human activities and from large volcanoes may reduce the radiative forcing in certain regions and act to reduce warming. These natural emissions can be very large. For example, it is estimated that up to 20 million tons of SO_2 were emitted due the eruption of Mt. Pinatubo in the Philippines in June 1991 and about 7 million tons were emitted from Mt. El Chichon in Mexico in 1982.

Despite these debates and the great uncertainties associated with these predictions, the consensus of the scientific community leans strongly toward increased global temperature with an increase of greenhouse gases in the atmosphere (Houghton et al. 1990, 1992). In fact, some scientists (e.g., Hansen et al. 1988; Oechel et al. 1993) claim that temperature rise is already detectable.

Global change scenarios all predict not only change in the global quantity of precipitation, but also changes in the patterns of its fall. This will in turn influence the patterns of discharge from rivers and may increase the variability and the magnitude of flooding and drought in affected locations.

Current global change simulations with General Circulation models (GCMs) suggest that there will be a global increase of up to about 15 percent in total global precipitation under the doubled CO_2 conditions of the future. However, geographic distribution, rather than the global mean, is more relevant to a particular region. The spatial resolution of the GCMs is still too tentative to make specific predictions for the Arab region. These models will continue to be improved and their predictions made more exact (Cess et al. 1993), and the actual effects of climate change in real time events in real places, will probably come in an elusively variable, almost random, fashion, a characteristic called stochasticity. But, even without precise knowledge of the exact patterns of hydrology in the new climate, prudent policy dictates careful consideration of the consequences of such a change.

Recent model simulations suggest that an increase in greenhouse gases in the atmosphere will bring an increase in night temperature. This increase is important to plant growth, as it can cause an elevation in plant respiration and therefore a decline in overall growth in certain regions where the temperatures are already high. Furthermore, elevation of air temperature may require crops to consume more water for cooling via transpiration. Of particular importance to the Arab region and other desert or semi-desert environments are any modifications in the formation of dew. Dew is the condensation of water on cooler surfaces during the night when the humidity of the air is higher than it is during the day. Because of radiation from plant surfaces during the night, they become cooler than the surrounding, relatively more humid air, and condensation (dew) occurs on the plants' surfaces. It is well established that desert plants receive much of their needed moisture to survive and grow from dew. Without adequate dew, many plants will die or their growth will be reduced. If this lessening of dew is widespread, the effects of climate change on desert vegetation will be severe, and may result in a great increase in desertification, a problem that already plagues many regions of the Arab world.

GCMs also predict that the temperature rise associated with greenhouse gas emissions will not be globally equal. Some regions will experience a more than average rise and others a less than average rise. All available models predict that northern and southern polar latitudes will experience a much larger temperature rise (up to 12°C) than the equatorial latitudes. If this is the case, there can be positive feedback on the temperature rise, because northern latitudes contain very large quantities of stored carbon in the form of organic material that is presently unavailable for the decomposing microbes, as most of it is presently contained in frozen soils (Oechel et al. 1993). In the predicted warmer climate of the future, these materials can quickly decompose, adding more and more carbon dioxide and/or methane to the atmosphere, which in turn can cause more temperature rise. Because the head waters of the major rivers in the Arab region are located both in the tropics and the temperate zones, the impact of climate change over a broad area of the earth is directly relevant to the region.

Besides the changes in temperature, changes in snow cover in eastern Turkey and in the frequency of storms in tropical Africa both could have potential impacts on the hydrology of the Arab region. Model simulations show a tendency for an increase in the frequency of convectional precipitation, usually meaning more intense local rains at the expense of gentle and persistent rains (Gordon et al. 1991; Pittock et al. 1991). There is also an indication that there has already been a modest decrease in snow cover associated with temperature rise in the Northern Hemisphere. The Canadian General Circulation Climate model (CCC) predicts a significant increase in the number of cyclonic storms. Melting of the polar ice can be caused by the large temperature rise and the positive feedback of soil warming in high latitudes. In the tropics, Haarsma et al. (1992) found that the number of simulated tropical storm disturbances increased with the doubling of the CO_2 in the atmosphere. All of these changes can alter the amount of water discharged into the Euphrates, the Nile, and the Tigris and the temporal patterns of the discharges.

The melting of ice in the Arctic and Antarctic oceans and the expansion of ocean water worldwide associated directly with the increase in temperature could cause a significant increase in sea levels. This would have major consequences for fresh water availability and agricultural production in the Arab region, especially low lying areas and population centers near bodies of water.

Even modest sea-level rise can greatly affect the following:

- Discharge from major rivers into the elevated seas.

- Intrusion of salty sea water into river deltas, especially the Nile and Shatt-al-Arab.

- Population centers located near sea shores.

- Desalination plants located close to shorelines.

The Nile delta and the regions near the mouth of Shatt-al-Arab are particularly vulnerable to sea rise. It is well known that there has to be a certain quantity of discharge from the Nile River to prevent the salt water of the Mediterranean from intruding into the Nile delta region. Should there be a decreased outflow of fresh water, much of the Nile delta will become unsuitable for agriculture, resulting in a disastrous situation for the Egyptian economy and a catastrophe for the region. This anticipated rise in the level of the Mediterranean under global change conditions, no matter how small it is, requires careful planning and coordination of response among the Nile riparians (Ethiopia, Sudan, Egypt). The area surrounding the mouth of Shatt-al-Arab and low lying areas in the Gulf may be subject to similar changes, and in that case would also require policies for adaptation.

There is still much debate about how much sea levels will rise if the global temperature increases, particularly if the temperature increase is greater in higher latitudes. It is anticipated that a sea rise will be less than one meter, but even at this level the implications for coastal zone erosion and effects on shell fish breeding grounds could be enormous.

The Direct Effects of Elevated CO_2

While there is some debate about the extent of temperature rise, there is no debate about the rising CO_2 concentration in the atmosphere, which has increased from 280 ppm at the start of the industrial revolution to its current level of 358 ppm. Currently, CO_2 is rising at about 1.0 ppm per year and the global increase is correlated with the rate of consumptions of fossil fuels.

Aside from being a greenhouse gas, carbon dioxide has the potential to increase plant growth. Assuming that plant nutrients, water, and biocides are available, it could increase agricultural productivity. It is estimated that with the doubling of CO_2, and with advanced technology, agricultural productivity in the advanced countries could increase by as much as 30 percent of its current quantity (Kimball et al. 1993). Additionally, by reducing the openings of pores found on leaf surfaces (stomata), CO_2 can act as an anti-transparent, therefore reducing the consumption of water needed by crops to produce a given amount of biomass and yield. These theoretically positive effects of rising CO_2 levels have been hailed by many agricultural scientists in the U.S. and Europe. They believe that doubling of CO_2 in the atmosphere will be of much benefit to agricultural yields and plant growth, a view that is advocated in a videotape, "The Greening of the Planet Earth" (Institute for Biospheric Research 1992), which has been widely distributed in the United States. However, such predictions must be considered with great caution. Model simulation suggests that while more atmospheric CO_2 could favor a significant increase in agricultural production in the developed countries, there may be a decrease in productivity in the developing countries. Rosensweig and Perry predict such a decrease from their model studies (unpublished data).

Moreover, assertions that CO_2 will increase plant growth usually ignore the "quality" of the crop produced, especially its protein content. It is established that plant tissues grown in a high CO_2 environment have lower nitrogen, and therefore protein, content than tissues grown in ambient CO_2 (e.g. Bazzaz 1990; Bazzaz and Fajer 1992).

It is important to keep in mind that such predictions are subject to modifications and therefore must be treated with caution. It is unlikely that agricultural production will increase much with the increase of atmospheric CO_2 except for crops grown under glass or under plastic tents where the other plant requirements can plentifully supplied. And, although CO_2 can enhance plant growth, there are several reasons to suspect that for natural vegetation, CO_2 may not necessarily be a good greenhouse gas (Fajer and Bazzaz 1992).

A recent survey of published literature shows that most plant species respond to increased CO_2 in a positive direction, but with major differences between different plants with respect to growth enhancement and water use (Porter 1993). While generally increased growth is likely to be true for agricultural crops grown under ideal conditions, studies with tree species suggest that growth enhancement may be of short duration in some species (Bazzaz et al. 1994; Norby 1992). More critically, however, plants that have the so-called C_4-photosynthetic pathway (e.g., many grasses of sub-tropical and tropical origin that are common in the Arab region) are much less responsive to increased CO_2. The biomass of short-statured Mediterranean grasslands from California also respond very little to increased CO_2 concentration (Williams et al. 1988; Ehleringer and Fields 1993).

Natural vegetation in the Arab region receives some rain and dew. As mentioned previously, with the increase in temperature during the night as predicted with global change, it is possible that water condensation may be critically reduced, depriving these plants of much needed water and greatly impairing their survival, growth, and reproduction. When vegetative cover is reduced because of the reduction in growth and survival of plants, the desert surface becomes exposed and wind erosion increases, spreading desertification. Increased grazing, caused by a reduction in nitrogen content of plants in a high CO_2 environment (Lincoln et al. 1993), can exacerbate wind erosion and expand desertification. Thus, under a plausible changing climate scenario of increasing night temperature and overgrazing, the process of desertification could be intensified and become more widespread in the Arab region.

Many of these scenarios and predictions are based on the mathematical models mentioned above as general circulation models (GCMs). At present, there are several models in use and more precise GCMs are being generated (Cess et al. 1993). Close examination demonstrates differences and disagreements among current models, but Schneider (1989) compared those available at the time of his study and concluded that, despite uncertainty and divergences in their predictions, there was sufficient agreement among them to urge that action to reduce the emission of greenhouse gases must be immediately taken. The GCMs now in development need to be produced at a finer scale to form the basis for specific subregion predictions, which are not possible at the present time. Focusing on the Arab region, in this part of the world even present models generally agree in predicting much change in mean annual temperature, precipitation patterns and soil moisture.

Taken together, factors of global climate change and modified hydrology can have significant effects on agriculture and natural vegetation and on wildlife. These possible impacts can be summarized as follows:

■ Increased winter precipitation leads to higher river discharge as well as increased severity and frequency of flooding.

■ Increased temperature in the summer will lead to longer and more frequent droughts, which like the floods mentioned just above can have disastrous consequences for agricultural production and human health.

■ Increase in the frequency of extreme weather events, especially in the upper watersheds of rivers such as the Euphrates and the Tigris, such as unseasonable freezing and thawing and very cold and very warm spells, can greatly influence discharge patterns.

■ Change in depth of water in rivers and reservoirs can influence water temperature, amounts of dissolved O_2, and the suitability of the water for human consumption and agricultural use.

■ Climate change can modify both the temperature and velocity of rivers; oxygen, which is required for a healthy and productive river, is inversely proportional to water temperature.

■ Dams to control floods and water discharge create sluggish water behind them (low O_2) and fast-moving water below them, which is well oxygenated but more turbulent.

Certain action responses to the prospect of climate change are called for:

■ Planning for the future, considering all elements of changed hydrology, economics, environment, and security. Short-term gains versus long-term costs must be critically evaluated.

■ Regional and international cooperation and coordination. For example, the International Committee for the Hydrology of the Rhine Basin is developing a water management model under different climate change scenarios. This model can be used as an example for future action by the riparians in the Arab region.

■ Adoption of a global view to regional issues.

■ Steps should be taken to reduce the emission into the atmosphere of CO_2 and other greenhouse gases. A great deal can be accomplished by a drastic increase in the efficiency of energy use. While this is currently not a global policy, it should become so in the near future.

■ Serious efforts should be made to find alternative sources of energy. The energy consumption picture may change very quickly and drastically if fusion technology suddenly becomes a possibility. A serious search for alternative use for their great natural resource, oil, for purposes other than fuel, must be a top priority for the Arab oil producers.

Scientific knowledge must be developed to refine these predictions and to increase their level of certainty. This issue is of great economic and social importance to the Arab region. It must never be evaded, as an ostrich buries his head in the sand, with assumptions that global change does not exist or that it will go away. Nor should the planners react to this like Chicken Little, running around shouting "the sky is falling, the sky is falling...." The sky is not falling, but the climate is changing, and there is a real need for critical assessment and prudent action. Environmental management and development are not alternatives to each other. To view them in this light as in opposition to each other is both naive and dangerous. Planners and policy makers must strive to reconcile the two.

There is currently much concern about "food security" and the need for countries to produce a large proportion of their basic food needs, or even to export food, no matter what the cost. The present status of geopolitical reality argues against adhering closely to this outlook, although it was understandable and valuable in the past. With the ending of the Cold War period and its super-power rivalry, it is expected that the sharing of resources among nations will be further emphasized and that a more global view of resource availability and supply patterns will be taken. Planners should consider in their deliberations and actions the changing meaning of food security in a world changing both climatically and geopolitically. Food production is intimately linked to water use, so much so that it is said that countries that export food grown with fossil water are really exporting the irreplaceable water, rather than just the food.

Despite the uncertainty in global climate change forecasts and the lively debates now going on in the scientific and policy-making communities, Arab planners should not ignore the warnings of climate change. It is easy to say that the region has other priorities at the moment, but global change impact, despite its great stochasticity, must be considered in long-term planning. Without proper consideration and a balanced outlook now, the region may suffer long-term consequences.

10

Conflict and Water Use in the Middle East[*]

Thomas Naff

Scarcity and Conflict

International and transboundary water use conflicts are never simple. The components of such disputes are in themselves invariably complex, none more so than the two that are most basic and omnipresent: the issues of scarcity and security.

Scarcity is the first strand — security being the other — of a double helix along whose intertwined curves lie the constituent elements of hydropolitical conflict. At the most basic level, actual scarcity may be said to exist when real demand (i.e., need) exceeds real supply. Although the maxims of supply and demand may determine actual shortages, the concept of water scarcity encompasses many discrete but interrelated factors that govern supply for any given demand: climate, perceived and real need, quality (complicated by a wide variety of standards for different functions of water in river basins across the globe), location and reliability of source, consumption, technical capacity, accessibility, demographic growth patterns, distribution of population and water resources, efficiency, organization and management, use of fertilizers, loss and waste, stocks of water (extant, available, and safe-yield), and policy decisions on the rate of consumption and distribution. There is also a kind of psychological scarcity, a scarcity in the eye of the beholder. This kind of scarcity exists when, for whatever reason, people perceive or believe a shortage exists — whether or not physical reality justifies the impression — and they behave

---

[*] Portions of this paper have been extracted from a forthcoming book to be published by SUNY Press. I am indebted to Professor J. Dellapenna (Villanova Law School) for his guidance and insights, which have strengthened this chapter; he is, course, in no way responsible for any of its shortcomings.

accordingly. Perceptions of the amount and quality and availability of water are usually a part of a people's attitude toward the environment (Cotgrove 1982; Whyte 1986).

While there are numerous reasons for water scarcity, they all tend to be variations on six basic causes, which, taken together, will delimit supply and demand: climate variations (principally drought); degradation of water quality by human activity at a rate faster than the source can be renewed; depletion of a source, such as an aquifer, at a rate faster than it can be replenished; out-of-basin diversion or storage of surface water; redistribution for other uses or to another place; and consumption. In the Middle East, these causes stem, in one way or another, from a single overriding, immutable determinant of scarcity that accounts for the region's aridity and, for that matter, of the aridity of other parts of the globe as well: the way in which the earth functions as a stupendous heat pump run on solar energy, which generates a constant process of intense evaporation within a broad zone that encompasses the Middle East region (Kolars 1990, 59).

The consequences of scarcity are as complex as its causes. For example, water and other associated environmental scarcities often set afoot large-scale migrations from the countryside to the cities, creating large, dislocated, underemployed or unemployed floating multitudes — particularly in third world countries where this phenomenon is more typical — that become serious drains on the economy, create political hazards, and generate a distortion in the national economic balance in favor of city over rural dwellers. Such conditions raise the possibility of conflict. In situations of high population growth, increasing strain on water resources, and behavior that depletes the resource at an unsustainable rate or even destroys it, resource scarcity then promotes social inequities, political tensions, state weakness, and authoritarian regimes; it thus becomes a determinant of both security and conflict (Gurr 1985, pp. 55–58; Ophuls 1977, pp. 8–9; Postel 1992).

In the Middle East, the composite effects of climate, poor supply, maldistribution and escalating populations are revealed in exponential discrepancies of water supply per person across the region, ranging from a per capita supply of 115 m³ in Libya to as much as 5000 m³ in Iraq in rainy years (the average is about 2000 m³). A disturbing related trend has emerged in recent decades: over the last 30 years, the average available supply of water for the entire Middle East has fallen rapidly from somewhat more than 2000 m³ per capita to less than 1500 m³ per capita (Kolars 1992, pp. 103–6, Tables 1 and 2; Tvedt 1992, pp. 14–33). Presently, more than half of all Middle Eastern countries are confronting serious water short-ages.

The region has very little margin of safety where water supply is concerned, especially given a population that is projected to double within the next quarter century. Unless this situation is reversed without further delay, several key actors in the major river basins — Jordan, Israel, the Occupied Territories, Egypt, Syria, and Iraq — face a series of destabilizing economic and political crises within the foreseeable future, the conse-quences of which will reverberate throughout the region and in much of the western world. Scarcity, especially mismanaged scarcity combined with maldistribution, contributes significantly to the creation of an environment of uncertainty and instability in the basic political, eco-nomic, and social institutions of society, most destructively in situations where the reciprocal factors of ecological marginality and rising poverty obtain — a condition that characterizes most Arab countries.

Security and Conflict

The idea of security and the causes of conflict have been historically and conceptually interlaced. The concept of security in the modern epoch has been explained in terms of perceived threats of violence in some form of organized mayhem, usually warfare, to national sovereignty or territorial integrity, by an outside force. This notion of security involving a threat to an established group's power or possibly to its existence, by an outside or "other" agent, has applied as well to civil wars with "loyalist" or "insider" factions contending against "rebels" or "outsiders." The concept of secu-rity is perhaps the most problematic of the notions associated with conflict because it is insufficiently descriptive; that is, as a general idea, it is ambiguous at its core. Its descriptive meaning varies with circumstance and time (Cranston 1967). To give any degree of precision to the meaning of security, one must always ask "security of, from, or against what?" Unless the problem is handled punctiliously it can be analytically misleading. To be analytically helpful, the term *security* needs to be given as clear a meaning and as specific an application as possible. The bedrock meaning of security comes down to being secure from harm or annihilation — being safe from something or someone that can damage, but from which there is possible recovery, or from being reduced to non-existence. The former can lead to the latter, so one seeks safety from both. This is the meaning of security that will be used in this study.

Contemporary ideas about what constitutes security remain closely linked to the nation-state, although in an increasingly international political environment the connection has been stretched somewhat to regional and international levels. However, the nation, encompassing religion and ethnicity and expressed ideologically as nationalism, remains the most virulent, widespread, emotional, and influential mode of political and cultural identity, especially when it is coterminous with ethnicity. Religious or quasi-religious ideology is a normal concomitant of almost all brands of nationalism, lending them mantles of morality and legitimacy. The abiding potency of the idea of nationhood in world affairs has received ample confirmation in the vicious internecine wars in the post-communist era. These conflicts give the appearance on all sides of deriving from an atavistic revival of tribalism, nourished by religio-ethnic myths, creating what have been aptly called "Tribes With Flags" (David 1988). Such powerful self-consciousness engenders an equally powerful sense of the "other" or "them," seen as opposed to "us," which promotes a very strong bent toward a self-absorbed cultural nationalism. This inherently aggressive outlook, with its built-in tendency to assume the role of an injured innocent, creates a very dangerous security situation prone to conflict.

Like water itself, the concept of security (and that of conflict) is complex and multifaceted (Romm 1992; Gleick 1990; Dougherty and Pfaltzgraff 1971). Those who subscribe to the doctrine of political realism in international affairs define security in variations on the following theme: the capacity of a state to secure its safety and perceived national interests from violence by means of such assets as military power (projectable or defensive), population (size and competence), economic strength, and vital resources, relative to other states who are seen as real or potential enemies or whose status or relationship is deemed important. Thus, a state will always attempt to maximize means to its security in direct ratio to felt threats.

It cannot be argued that military power, economic structure, and state interest — the heart of realist/rational choice theory and its variations — are not fundamental determinants of security and foreign policy in all nations. But it can be demonstrated that overweighting these elements and de-emphasizing others — such as environmental and water resource problems — that cannot be confined to a single country alone, leads to distortions of reality. Because the realist analytic approach is based on constricted assumptions, realists are apt to "squeeze environmental issues into a structure of concepts including 'state,' 'sovereignty,' 'territory,' 'national interest,' and 'balance of power.' The fit is bad, which may lead theorists to ignore, distort, and misunderstand important aspects of global environmental problems." (Homer-Dixon 1991, pp. 84–85)

While the traditional meaning of security, focused as it is on machines of war and economic arsenals together with strategies for employing them, is sound as far as it goes, it is obviously too cramped. A world approaching a new millennium with too many inhabitants, with its vital natural resources diminishing too rapidly, with the scale of its political and ecological problems growing quickly from local to global levels, a world that is also interdependent and technologically driven, clearly requires a new, meaningfully extended definition of security. A sample list of issues —most of which involve water and other resources—that could legitimately be included under an enlarged security rubric is not difficult to compile:

- Agriculture, which is militarily and economically important and represents food security, which is different from food sufficiency. Food security requires a guarantee of enough food to satisfy a population's minimal nutritional needs over a long period of time, a policy usually expressed as self-contained, domestically produced sufficiency; food sufficiency requires that there is on-going sufficiency of food for the needs and development of a society, attained chiefly by trade from whatever sources. Food security in the arid Middle East will always be a wasteful and ill-fated policy; the ultimate reality about food security is that it is absolutely dependent on water security. Food sufficiency, on the other hand, while more realistic does require an economy that generates enough exports to cover the cost of large food imports. Agriculture also falls under the rubrics of environmental and resource (i.e., water) security.

- Demographic pressures in combination with other factors, such as drought, tend to create serious social, economic, and political instability along with large-scale internal and external migrations, all of which can produce security threats to the regime.

- Resource scarcities, especially water, often have transboundary consequences and may be accompanied by the danger of environmental colonialism, a vulnerability widely felt among poor and weak nations. If a resource such as water or oil is a significant source of economic or political power, then it can be fitted into a realist or power analytical frame. Because nations cannot survive without enough water, water is intrinsically a security issue. There is a link between environmental degradation and security since environmental abuse limits water supplies.

■ Health issues, some of which — such as AIDS — can have a devastating impact on a nation's capacity to maintain basic economic and military security by wiping out much of its youth, especially in poorer countries whose medical establishments can be easily overwhelmed by a deadly epidemic.

■ Ideological and cultural differences involving human rights, national-ism, religious extremism, and authoritarianism, leading to security threats and possible conflict.

■ Nuclear or chemical accidents.

■ A variety of economic issues. For example, product dumping can lead to trade wars. Serious economic shocks in a single key nation such as the U.S., Japan, or Germany can have repercussions across the globe. Because of interdependence, large multinational corporations operat-ing on a global scale in basic enterprises could bring about a subtle loss of national sovereignty.

These are only issues created by human societies. There are plenty of naturally occurring calamities that have security implications: floods, drought and desertification, earthquakes, and contagions are a few typical examples (Kirk 1992; Choucri 1992, pp. 101–7).

Moreover there are serious methodological and analytical difficulties embedded in the concept of security. The causal relationship between a specific resource, environmental, or demographic problem and a security (or conflict) issue is neither plain nor linear. Such problems tend to have complex feedback interactions with other complex political, social, and economic issues, resulting in a non-quantifiable reciprocity that produces multiple effects. In situations of constant tension and hostility, such as exists, for example, in the Jordan and Euphrates basins, a resource issue like water scarcity is a constant underlying security factor that could act as a trigger for conflict; but precisely how and why it would trigger warfare rather than another reaction is not clearly known, as water could in the same circumstances act as a catalyst for negotiations. Thus many problems that may be hung on an environmental peg, especially water, must be examined as dependent variables of other factors such as population, culture, social relations, values, and political, military, and economic conditions (Kirk 1992, p. 58; Homer-Dixon 1991, pp. 5–6).

The multilayered linkages between environmental factors (broadly defined) and security and conflict are as yet poorly understood; consequently, sound generalizations are difficult to make. Historical parallels or comparisons can be misleading unless allowance is made for the evolution of the international system from past to present. Further insights into the cause and effect relationships between degradation of the natural environment and security issues, together with the mechanics involved, need to be gained before a new workable, theoretically solid definition of security can be put forward. In this regard, it would be well to give careful heed to a warning sounded about the dubious wisdom of binding environmental/ resource security concepts with those of realist national security:

> ...the nationalist and militarist mindsets closely associated with national security thinking directly conflict with the core of the environmentalist world view.... If the nation-state enjoys a more prominent status in world politics than its competence and accomplishments warrant, then it makes little sense to emphasize the links between it and the emerging problems of global habitability. Nationalist sentiment and the war system have a long-established character that are likely to defy any rhetorically conjured redirection toward benign ends. The movement to preserve the habitability of the planet for future generations must directly challenge the tribal power of nationalism and the chronic militarization of public discourse. (Deudeney 1991, p. 28)

Another useful approach to redefining security as a policy issue would be to eschew attempts at forging a single newly synthesized meaning and accept that there are in reality at least two distinct classifications of security belonging to the same social scientific genus, which are organically connected and share common attributes. In the first instance they could be differentiated as traditional and non-traditional. In this context, traditional notions of security emphasize the political, military, and economic protection of the nation while non-traditional concepts emphasize broadly conceived environmental safety, which applies both within national boundaries, and transcends them. While many factors, such as vital natural resources and population, straddle each kind of security, underpinning both definitions is a common policy design: to ensure survivability and sustainability, whether applied to regime, nation, region, tropical forest, transboundary water system, ocean, air, etc. Because of close interconnections, both goals — conventional and unconventional security — must ultimately be achieved to attain either (Lonergan 1992, pp. 8–9; Gleick 1990).

The Idea of Conflict

Conflict is as complicated a concept as security, and then some. It requires for its fuller comprehension a prior grasp of notions such as issues, situations, and its opposite, cooperation. Moreover, one must take into account such factors as values, ideologies, symbols, motivations, goals, and the origins and processes of conflict, while, at the same time, making necessary distinctions among all of these elements (Frey 1993, pp. 57–59; Coser 1956). There are many types of conflict that are generally recognized, and often given their own definitions, attesting to the elasticity of the term. Political, economic, ethnic, religious, racial, resource, international, trade, tribal, clan, and family dissensions are among the most common that may be indexed under a typology of conflict. The size and importance (which is what is usually meant by "scale" and "level" in these discussions) and the intensity of a disagreement must also be taken into account in rendering a definition of conflict.

Moving up from individual or small groups, a conflict (violent or nonviolent) may be acted out at a local, village, national, regional, interstate, multinational or global level, and the most widespread conflicts can involve state participants who do not share borders but rather are situated far from one another geographically — though because water is normally used within basin systems, hydro-conflicts commonly involve contiguous and other basin actors (Gleick 1992). It should be borne in mind that not all conflicts are violent; violence need not be involved for strife or friction to qualify as conflict. In fact conflict can exist in a latent state until animated by such events as scarcity or perceived frustration of need or desire. Disagreements can (and do) simmer along for very long periods of time without resolution, but not without damage.

As posited, all of the factors that enter into considerations of security — traditional and nontraditional — are integral to conflict as well. The intricacies of conflict have been multiplied in the latter half of this century by the rapid degradation of the environment on a global scale, in significant part as a function of rising demographic trends and concurrent economic development, resulting in very serious resource scarcities in many regions of the world. This circumstance has increased competition for resources, animated aggressive nationalist sentiments, and created many flash points of possible conflict, subjecting the international system to greater strain than ever before and making the resolution of conflicts exponentially more complex, therefore more difficult to achieve. This latter characteristic is especially peculiar to conflicts over water, since water is vital and pervasive, has so many essential uses, and does not respect national boundaries in the course of its flow. Moreover, there is complication from the sheer number of factors always present in water problems: atmospheric, hydrological, chemical, technological, managerial, political, socioeconomic, legal, and strategic to name a few of the more obvious ones. Not only must all such factors be taken into account in the quest for solutions, but the inherent complexities are compounded by a web of feed-back relationships among conflict factors, particularly when two or more national actors contend over the same supply of water in an international basin.

Attempts to understand the many-layered relationship between water and conflict can be greatly helped by good, "fine-grained" theory, whose function it is to explain. As in the case of security theory, useful conflict theory must also encompass and explain environmentally and ecologically caused strife on a scale and complexity heretofore unmet. Although there is a growing body of theoretical literature on the nature and causes of conflict — more than two dozen original and adapted theories, and a similar number of models have been offered in the last two decades — "unhappily, general conflict theories are not very well developed and, at best, furnish too coarse-grained a perspective to illuminate specific water issues." And as regards water conflicts "general conflict theory is, simultaneously, not really general (it omits important aspects of conflict phenomena) and too general (it does not bring out the key features of water resource conflict as distinct from any other type of conflict)." (Frey 1993, pp. 57, 59; Homer-Dixon 1991; Rogers 1991)

No single theory or model has as yet been developed that can deal with the layered political, socio-economic, legal, and strategic entanglements of fresh water that underlie hydro-conflicts. Consequently, a high level of uncertainty attaches to virtually all of them. If existing water conflict models are to work at all, they must be based on narrowly conceived, fairly simple assumptions and relatively small data sets too restricted to contain all the intricacies of water, thus running the risk of being overwhelmed by complexity or possibly producing very circumscribed, over-simplified results that could be either self-fulfilling or self-evident, or worse, erroneous.

Because water is so multifaceted, has so many applications, cross-cuts so many issues, and involves so many interrelationships, it tends to defy easy or comprehensive categorization in conflict typologies and theories. Any attempt to categorize water as a conflict issue must therefore employ a multi-dimensional typology or a combination of typologies. This makes accurate predictions of behavior in potential water-based conflicts elusive at best. Consequently, most predictive theories of conflict tend to break down in specific hydropolitical case studies.

Among predictive models, there is one, the Power Matrix model, that, though simple, does nevertheless capture enough of the key politically significant qualities of water — extreme salience, scarcity, maldistribution, and sharing — to work at an elementary level. It currently produces results that allow rough-hewn, reasonably accurate predictions for the conflict potential of water (Frey 1993, pp. 61–62; Frey and Naff 1985, pp. 77–79; Naff and Matson 1984, pp. 192–94). With a small variation, this model is employed in the discussion on water and conflict that follows.

Law, Water, and Conflict: Basic Principles

The cornerstone of international fresh water law is the assumption that the allocation of scarce resources requires legal means, rather than coercive force, if sharing is to be equitable and conflict is to be avoided. In principle, long-term cooperation among sovereign riparians, particularly where water is scarce, would be well nigh impossible outside the buttressing framework of law (Naff and Matson 1984, pp. 158–80; Teclaff 1967, 1985). But international riparian law can be effective only when riparians commit themselves to law as the first means for the delineation and regulation of rights and responsibilities, and the amelioration of grievance.

Historically, international riparian law has been underdeveloped, eluding the efforts of jurists to sort out its complexities and persuade nations to subject their competing claims to a standardized code of legal principles. Those complexities have sometimes made the process appear muddled. Although in the era of the United Nations some headway has been made, progress has been so slow and achievement so meager that some observers have concluded that no universal code of international riverine law is possible. Nevertheless, experience, scholarship, and jurisprudence (and, perhaps, not a little blind faith) have produced four basic legal principles that are generally invoked when riparians contend: absolute sovereignty, absolute or territorial integrity, community of co-riparian states, and limited territorial sovereignty (Caponera 1992, pp. 212–27; Garretson, Hayton and Olmstead 1967).

Absolute sovereignty (sometimes called the Harmon doctrine) decrees that a riparian may do what it will with water (or any resource) within its boundaries without constraints. It may use it up, pollute it, dam it, send it downstream in any quantity or condition. In contradistinction, the principle of territorial integrity requires that the river's natural flow be uninterrupted in its downstream course, that the lower riparians have a right to the full flow and quality of the water. The theory of co-riparian communalism stipulates that the entire river basin constitutes a single, geographic and economic unit that transcends national boundaries, whereby the basin's waters are either invested in the whole community or shared among the co-riparians by agreement, the underlying assumption being that optimum use of the basin's waters mandates a cooperative, integrated development of the entire drainage basin. The notion of limited territorial sovereignty supplants the opposed principles of absolute sovereignty and absolute integrity by according recognition to a riparian's jurisdiction over the transboundary waters that flow through its territory, but places limits on the exercise of its control over those waters in such ways as to insure the downstream states a reasonable share of that water in reasonable condition. Older principles such as first-in-use-first-in-right, historical utilization, beneficial (or optimal) use, good neighborliness, etc., are generally subsumable under these four principles. Whatever the legal principle, all of the rules devised for the sharing and apportionment of water are rooted in the notion that nations are obliged to cooperate in matters involving vital natural resources, especially when scarce.

Law: Equitable utilization and no appreciable harm

In modern times, a blending of the traditional notions of co-riparian community and limited territorial sovereignty has produced a hybrid legal principle that has gradually emerged as the preferred approach among juridical scholars, international law organizations, and state litigants. At the heart of this concept are the basic principles of equitable utilization and no appreciable harm. As will be seen, in this context equity does not connote equality.

In customary international law, every state is under an obligation not to cause harm to another, not only by direct action, but by allowing the use of its territory in ways that result in harm to the rights of other countries. No appreciable harm provides that while a state is entitled to use the waters of a river that traverses its territory, it may not do so in such a way as to cause appreciable harm to the river's other riparians. This proposition does not explicitly proscribe *any* harm whatsoever, and though "appreciable harm" has proven impossible to define precisely, it clearly means more than merely "perceptible" but not necessarily "substantial." That is, it must be harm of a certain gravity or significance beyond simple inconvenience. In its fortieth session the International Law Commission (ILC) of the UN adopted this definition, believing that the concept could be objectified and that compliance could be judged on factual bases and thus embody factual standards of behavior and liability (United Nations 1988). In its meetings of June–July 1993 the ILC, in response to UN members' proposals, substituted the word "significant" for "appreciable" but did not alter its definition of the principle, thereby allowing the terms to be used interchangeably.

Equitable utilization (or equitable apportionment) states that riparians of an international waterway are obliged to use, develop, and protect the watercourse in an equitable and reasonable manner and are duty-bound to do so cooperatively. Each riparian has a right of utilization — reasonable and beneficial — equal to that of every other co-riparian. It is the right of utilization, the right to a seat at the table, that is equal for riparian neighbors, and "equitability" in this context does not mean an equal share of the water. Rather, equitability implies the idea of proportionality, a share and usage proportional to a riparian's population and its social and economic needs, consistent with the rights of its co-riparians. Reasonable (or rational) usage may be explained as exploitation of water, or any other natural resource, in such a way as to conserve the resource "for the benefit of the present and future generations through careful planning and management." (Barberis 1986, p. 49)

It is worth noting that both the ILC and the Institut de Droit International, have publicly embraced the "no appreciable harm" concept as the paramount rule governing international fresh water issues, particularly water quality matters. However, that position is not unequivocal. Many members of those legal bodies, along with a sizable number of legal scholars believe that "equitable utilization" should be the cardinal prescription in practice. Clearly, the two rules are closely related and both are often invoked, whether primarily or secondarily, in the same instances because transboundary basin disputes always involve an upper and lower riparian (McCaffrey 1993; Bourne 1992). Some governments unfortunately insist on interpreting "no appreciable harm," in a literal, narrow, nationalistic way. When they argue that any action without prior agreement that causes any reduction of the flow or usability of water, however small, constitutes appreciable harm, they virtually negate "equitable utilization." If carried to its logical conclusion this construction of the "no appreciable harm" idea becomes self-nullifying. (Egypt, Israel, and Argentina are among those that have adopted this posture.)

The Nile River affords a good case in point. Egypt, for whom any sustained, significant reductions in the flow of the Nile could spell disaster, has taken a narrow view of the "no appreciable harm" proposition and argued that this principle should be the standard legal reference, rather than "equitable utilization." Supposing, hypothetically, Ethiopia, as part of its economic development and recovery program, were to build a dam substantially more than 15 meters in height on the Blue Nile, a major feeder of the main stem of the river, and use the captured water within Ethiopia. That would reduce the flow of the Nile to Egypt by a certain amount annually. Supposing further that the Egyptians decided to adjudicate the issue rather than to settle it by the superiority of their arms; they would certainly invoke the principle of "no appreciable harm," narrowly construed (along with the doctrine of absolute integrity), and reject Ethiopian arguments based on rights conferred by "equitable utilization" (and absolute sovereignty). If the principle of "no appreciable harm" prevailed, either by a court judgment or imposed by military force, "equitable utilization" would be negated, but at the same time, Ethiopia would be denied the legitimate right of economic development, thus causing it appreciable harm. Conversely, were the Ethiopian stance to prevail, Egypt would be appreciably harmed. In this circumstance, the result of a judgment either way would be a high social cost. When the successful invocation of the "no appreciable harm" principle produces substantial social costs and inflicts significant harm to the economic and legal rights of another party — as is clearly possible — the principle contradicts itself (Dellapenna 1992).

Law: Empowerment and constraint

This hypothetical case study was simplified to make a point. But it is important to understand generally how the law functions in this context. Basically, what the legal process does, for both domestic and international actors, is to enable (or empower) them by legitimating their claims, and, conversely, to constrain them by limiting the claims they are permitted to make. But to do so effectively, it must have the necessary legal institutions in place — courts, police forces, various government bodies that legislate and regulate by some codified legal system and represent the legitimate interests of the citizenry at individual and corporate levels.

This same institutional requirement applies in the international sphere as well, in the form of international courts and organizations that are supranational and are empowered to enforce judgments by recognized international legal means. In the international sphere, treaties are the key legal instruments, but to enable the judicial process to function effectively, treaties must include arrangements for settling conflicts by rules of law through appropriate legal institutions. These goals have been difficult to achieve in international law, particularly as regards transboundary and international rivers. Thus, riparian and other conflicts continue for the most part to be dealt with by specific treaty agreements or by power relationships or, sometimes by mediation in combination with the other two choices, but without necessary reference to or application of law: "In the absence of a neutral enforcement mechanism, international law has nothing better to offer for sanctioning violations than the law of the vendetta." (Dellapenna, undated, pp. 8–9; Bilder 1982, p. 23) In some highly local instances, cultural factors and customary practices are successfully employed to settle riparian disputes, but it is difficult to generalize such cases because they are so culturally and locally specific.

Law: The Helsinki Rules

The Helsinki Rules are essentially a compendium of useful guidelines, drawn up by the International Law Association (ILA) for the application of law instead of coercion in the settlement of riparian disputes. The Rules deal with the four prevailing principles and their offshoots, but in the end they come down in favor of equitable and beneficial utilization. As stipulated in the key articles (III and V), each basin actor is entitled within its territory to a "reasonable" and "equitable" share in the beneficial uses of the waters in an international drainage basin as long as it does not cause substantial injury to another basin state (ILA 1967). But, unlike the UN's ILC in its efforts to fashion a set of generally acceptable principles for the application of international riparian law, the ILA did not attempt to provide an objective or operational definition of such pivotal concepts as "equitable" and "reasonable." It is not only the complexity of the issues being defined that accounts for the generality or even vagueness that often characterizes these legal usages but also the recognized need for built-in pragmatic flexibility where international or transboundary water questions are concerned.

Because the International Law Association is a private, professional association without official status, the Helsinki Rules have no official standing. The Rules represent little more than the collective opinion of a group of experts in the relevant fields of international law. Such opinions carry only the weight of a secondary source of international law insofar as they provide evidence of customary international law. However, the essence of the Helsinki Rules has been encapsulated in the draft proposals of the ILC, which is an officially constituted international body of legal experts, an official organ created by the United Nations. Although at this time, its work is also merely secondary evidence of customary international law, the ILC's position is expected to become the basis of a multilateral treaty on the peaceful non-navigational uses of fresh water. Should it be ratified by the members of the UN, the treaty will then become a direct primary source of international law (Dellapenna, undated private communication; McCaffery 1993).

Law: Groundwater and the Bellagio Draft Treaty

Until recently, the rules governing surface water sharing were applied to groundwater as well, but that circumstance has been changing since the appearance of the 1988 Bellagio Draft Treaty Concerning Transboundary Groundwater and the 1991 ILC report. Groundwaters, since they are connected parts of surface water systems, constitute, legally and politically, international or transboundary watercourses. Like counterpart surface water, groundwater does not respect political boundaries, often traversing several as it flows seeking its own level or outlets. For example, the northeastern African aquifer moves under Libya, Egypt, Chad, and Sudan. Saudi Arabia, Bahrain, and the UAE overlie the same aquifers while the Qa Disi aquifer underlies both Saudi Arabia and Jordan. The most legally and politically controversial shared groundwater in the region, the West Bank mountain aquifer, also known as the Yarqon-Taninim, lies mainly under occupied Palestinian terrain. It percolates into Israel across the Palestinian-Israeli Green Line but is wholly controlled by Israel by virtue of the occupation.

The chief difficulty hampering jurists who aim to establish precise definitions and devise rules for the sharing of underground water is a serious paucity of data on most aquifer systems; many important aquifers are not even fully mapped yet. Consequently, international law and legal institutions for the peaceful and equitable management of transboundary groundwater resources barely exist, and those few laws and institutions that do are notoriously weak. The need for an effective model treaty has become urgent (Barberis 1986, 1–64; Tecklaff and Utton 1981). The Bellagio Draft Treaty, which aims to fill that need, is founded on the principle that underground water rights should be regulated by mutual respect, good neighborliness, reciprocity, and collective agreement, and it acknowledges that the fulfillment of these notions requires joint management of the resource. The fundamental goal of the 20-article draft treaty is to promote optimum use of available groundwaters, facilitated by strategies for conflict avoidance or resolution in the face of rising demands for very limited supplies (Hayton and Utton 1989).

Law: Islamic water law

There is another body of law, sharia, or Islamic law, which by its nature is religious law, whose rules regulated water issues in the Middle East for almost a millennium and a half. Although at the world level westernized codes of law have come to predominate in the last century and a half, sharia is still applied in many Islamic nations where, in some instances, the spirit of traditional Islamic water law has been incorporated into more recent secular legal codes that have been adopted. With the resurgence of religious fervor in the Muslim world there have come demands for the application of sharia in all aspects of life in Muslim societies. What Islamic law has to say about the hydrologic culture of the Middle East region, and the relevance of Islamic law to present water conditions must therefore be seriously considered. Indeed, this is a basic requisite since in Islam, a Muslim society is defined as one that adheres to sharia (al-Mawardi 1983; al-Rahbi 1973; al-Nabban 1970; ben-Adam 1967; Wilkinson 1990 and 1977). Moreover, Islamic water law compares very well with Western canons on water.

The significance of water in Islamic legal thought is disclosed in the double meaning that the word sharia carries. In the first instance, it reveals the moral path that Muslims must pursue to attain salvation, and at the same time, in a more technical (and older) sense, it denotes access to the source of pure drinking water that must be preserved for humans. Specific hard and fast rules of Islamic law are relatively few — general moral guidelines are more characteristic. Where water is concerned, sharia tends to be less rigid than in other areas of Muslim jurisprudence. It is more the spirit of the law than the letter that is applied in water matters; that is, there is more application of custom *('urf)* and of reasoning than of strict doctrine. By and large, because received customs represent the collective norms of the social group and contain rules of behavior considered essential to the well being of the community, subscribing societies tend to feel bound to observe them (Lloyd 1966, pp. 201–2; Maktari 1971, pp. 6–7).

Law: Customary law in Islam and the West

Although customary laws differ from one Muslim society to the next, and though there are differences between Muslim and Western customary laws, they do share certain common traits. Customary water law is of fundamental importance to Western legal systems and to sharia alike. Further common to both, customary law as a juridic model combines advantages with serious vexations. Rooted in communal experience,

custom offers societies living under both sharia and Western legal systems
the benefits of legitimacy, familiarity, adaptability, and flexibility, which
allow for positive, practical rulings. Given the wide-ranging diversity of
conditions and situations from river basin to river basin the world over, the
exploitation of these qualities is often essential to conflict avoidance.

Beyond the general characteristics of *'urf,* it is worth noting certain
other qualities of Islamic law that have a bearing on water issues: sharia is
not a national law in the sense that American or European or Japanese legal
systems are. Generally, Islamic law has been applied regionally. Because
there are four major schools *(madhahab)* of sharia that are employed
diversely in different parts of the Islamic world, there have always been
wide variations in the interpretation and application of Islamic law
according to the different schools and even within the same school as
practiced in different Muslim nations.

However, the significance of the extra-national or extra-territorial
nature of sharia is that, by this quality, it is constitutionally international
within the Muslim community. That is not to say it is formally or
institutionally codified as "international" in the way that there is a separate
body of law in the West that is designated as such, and to which individual
nations are asked to adhere. It is, rather, a generalized set of divinely
ordained moral guidelines that are organized into systems of positive law
based on evidence and precedents. These broad moral rules are incumbent
upon both the Muslim individual and the community, that is, nation.
Sharia, being the literal, perfect word of God, is considered to comprehend
all circumstances and exigencies of the human condition, universally,
without national or international distinction. Sharia recognizes and em-
bodies the concept of a law of nations, and since at least the nineteenth
century when Muslim nations began practicing reciprocal diplomacy
according to European rules, Western and Islamic understanding of that
notion have been in harmony (al-Shaybani 1958–60; Naff 1984). There
is, therefore, no innate reason why sharia is not adaptable to any of the
contemporary international principles of water law being proposed by
various international legal organizations.

Law: A profile of Islamic water law

As stated above, Islamic law per se offers few specific, hard-and-fast
rules governing the sharing and use of water. Water appears in the Quran
only about half a hundred times, without a clear legal character or link to
sanctions; rather, the emphasis is on water as the source of life: "Have not
the unbelievers then beheld that the heavens and the earth were a mass all

sewn up, and then We unstitched them and of water fashioned every living thing?" The traditions *(hadith)* of the Prophet Muhammad offer no more precise legal language than the Quran, as for example: "He who withholds water in order to deny the use of pasture, God withholds from him His mercy in the Day of Resurrection." (Arberry 1980, Sura xxi, 30; al-Mawardi 1983, p. 158; ben-Adam 1967, p. 76; Water Authority of Jordan 1993)

Sharian water law derives in principle and for purposes of taxation from juridical rules governing land. Muslim jurists have consistently treated water, land, and crops as indivisible, and water rights have generally been restricted to amounts considered to be adequate for a given crop area. This is based on one of the few stipulations the Prophet is said to have articulated in a *hadith* concerning water, that the sum of water to be drawn was not to exceed that which is needed to cover a cultivated plot to two ankles' depth (al-Mawardi 1983, p. 156; ben-Adam 1967, pp. 71–76). This provision hypothetically fixed the basic legal principle for allocating water in Islamic law. By and large, the relatively few *hadith* concerning water appertain to the rights of ownership to wells and springs, to rights of access to water, the obligation to share water, and prohibitions on selling water. Although for purposes of use, allocation, and adjudication water is segregated according to source (river, well, and spring water, and further into rain, snow, and hail) sharia in fact recognizes only two broad categories of water within which all others are comprehended: owned and not owned.

Most Muslim jurists consider water generally to be beyond the pale of private ownership, to be *mubah* or *res nullius*, that is, a substance that cannot be owned unless it is taken in full possession, such as water contained in a jar. If water is claimed by the state, the ruler is considered to hold it in trust for the community or nation because the Prophet is said to have declared in a *hadith* that "...mankind are co-owners in three things: water, fire, and pasture." No person or ruler may appropriate a river or sell, rent, or lease its water nor may he tax such a resource; only a product that results from its use may be subject to a levy by the state (al-Rahbi 1973, pp. 636–38, 646–48; al-Nabhan 1970, p. 247; Encyclopedia of Islam 1991, Vol. 6, pp. 860–61).

A profile of the legal personality of Muslim water law reveals it to be highly pragmatic, largely customary, and supple in its application of moral principles as guidelines. In summary thus: no persons may be denied water that is necessary for their survival or livelihood; while animals have clear legal rights to water, humans take precedence in use; drinking water for man and beast and for domestic uses take priority over agricultural needs; once all drinking and domestic requirements of the community are satisfied, those living upstream have antecedent rights based on the assumption that the natural course of canalization, and therefore of settlement, proceeds from the upper reaches of a watercourse onward downstream; on the principle of first-in-use, first-in-right, upstream riparians enjoy priority, because in Islamic law, in the absence of convincing proof otherwise, they are presumed to be the first settlers; when, however, new societies are settled upstream after the establishment of downstream communities, the usage rights of the new community are subject to adjudication and their withdrawals must not adversely affect historical prior rights; the hoarding of surplus water, after all of the needs of the community are met, is forbidden; water is considered to be an overriding community interest, and both Islamic law and the Prophet's traditions deem as immoral its treatment as a product for commerce or speculation. Finally, as an addendum to this summary, sharia rules governing the appropriation of water originate in those that regulate the appropriation of land, to wit, expropriation and use must derive from an input of labor, e.g., building an irrigation canal. Only the fruit of labor matters. It is the irrigation channel and the irrigated field and its crop that may be owned in inalienable right *(mulk)* by virtue of the labor that created them, not the water that flows through the one into the other. Water is the product of Allah's labor, not man's, and therefore can be used only transitorily in accordance with sharia and '*urf*. (Discussion based on the relevant Arabic and Western sources cited above.)

A word is in order about the apparent anomaly in the presumption that the sequence in which a watercourse is settled is from upstream to downstream, and about the first-in-right principle based on that assumption. One might easily conclude that perhaps desert dwellers with little experience of river basin settlement would develop such a conception, but the governing factor was probably the direction in which canalization of water proceeded. In point of fact, historically, settlement in most river basins, particularly those that involve heavy off-stream use of water, normally proceeds upward from the lower end of the basin because the lower portion tends to be more level, affording easier agricultural development and urbanization than more elevated upstream regions. Thus, priority in utilization as a principle of law has usually favored the downstream users.

Law: Law, treaties, and conflict

Although riparians who make contentious claims over shared rivers rarely resort to legal measures in international courts of law — in this respect, the Middle East is typical — they nevertheless always assert that particular legal theory that best justifies their demands, using it more as a bargaining ploy than as an objective, detached legal argument. International law as an instrument of regulation on transboundary fresh water issues is at present inconclusive and weak. This circumstance has allowed riparian issues to be manipulated as part of power relationships not only in the Middle East, but in other world regions as well. There is nothing inherently lacking in legal theory or in law itself, Islamic or Western, that has produced such a condition. The basic problem, and it is precisely at this point that politics and law come together where water is concerned, is the absence of prior formal political agreements — treaties — that govern the general and specific terms of shared waters, together with essential international or inter-riparian institutions that assure compliance among the users.

Worldwide, some 286 international fluvial and other fresh water treaties have been concluded. Of those, about two thirds concern North American and European river systems, the rest are scattered around the globe. In the Middle East region, with one notable exception, international treaties regulating the sharing, use, and quality control of water are virtually non-existent; it follows that there are no legal institutional arrangements either. The exception, the 1959 Egyptian-Sudanese apportionment agreement on the Nile, involves only two of the ten Nile riparians. Political and ideological rancor or outright hostilities have defeated sporadic efforts to fashion multilateral (or even bilateral) cooperative agreements for the use of the other major river basins in the area, the Jordan, Euphrates, and Tigris.

Such agreements are the essential first steps toward transforming legal theory into the institutional application of law. Only with the political agreements in place, whether they are multinational, such as the Law of the Seas, or simply basin-focused, such as the 1959 Egyptian-Sudanese treaty, which deals only with a part of a single river basin, can there be created an adequate array of effective legal instruments for solving disputes that arise over shared water resources. While law cannot provide all the needed answers and must await political settlements, law is nevertheless indispensable to finding and maintaining legitimate, sustainable solutions.

Water Use Conflicts in the Middle East

Water conflicts are notorious in the history and mythology of world civilizations. Gun fights over watering holes are a familiar feature of American "westerns." Other famous water-inspired conflicts, such as the rivalries over the wells of Beersheba between Abraham and the Philistines, and later between Isaac and Guar over a well called Esek (which itself means "quarrel"), come to us from biblical "easterns." (The latter disagreement was settled simply by the digging of a new well: Genesis: 19-22). Islamic jurists have traditionally gone to considerable lengths to interpret the law in ways to avoid conflict over water, but in the end, there are reliable *hadith* that justify the use of arms to gain access to water: "If I were not to find a passage for the water but on your belly I would use it!" said 'Umar b. al-Khattab, companion of the Prophet and second Caliph (al-Rahbi 1973, p. 651).

Conflict: Complex causes of conflict

Why does water cause so much conflict? Generally, because it is essential to life: "There is virtually no human artifact or commodity that is produced in the absence of water; agriculture is impossible without it and so are most manufacturing processes." But specifically, because water flows: "Its unregulated flows are likely to be erratic, and in an arid country, the consequences for any user unable to capture water the moment it is needed are likely to be dire. Also, the unpredictable character of stream flow can create a tense environment of uncertainty that is disruptive of social relations." (Maas and Anderson 1978, p. 2) In the Middle East, water exhibits all of these elements of conflict.

As a contemporary issue of security and international relations, water displays certain distinguishing characteristics:

- Water as an issue is pervasive, highly complex, and utterly vital (for many of its functions, it has no substitute for human and animal use).

- Because of its complexity, water is fragmented as a strategic and foreign affairs issue, tending to be dealt with piecemeal, problem by problem, rather than comprehensively, both domestically and internationally.

- Water is always a terrain security issue, especially when scarce, since all concerned parties feel compelled to control the ground on or under which water flows.

- The relationship between water dependency and security is perceived as vital and often absolute. Water issues are seen as zero-sum, especially where two or more mutually antagonistic actors compete for the same water source.

■ As a zero-sum security issue, water carries a constant potential for conflict.

■ International law as a means of settling and regulating fresh water issues remains rudimentary and relatively ineffectual without prior treaty arrangements in place.

In sum, the inescapable strategic reality of water is that under severe shortage, which is the prognosis for the Jordan River basin, water becomes "a highly symbolic, contagious, aggregated, intense, salient, complicated zero-sum power-and-prestige-packed crisis issue, highly prone to conflict and extremely difficult to resolve." (Frey and Naff 1985, p. 77; Naff and Matson 1984, p. 8)

But, if water is such a volatile strategic issue, why has it not led to overt conflict in the modern Middle East as it did in antiquity? The short answer is that it has, but usually as a contributing factor submerged within the context of other contentious issues among parties already embroiled. Moreover, people everywhere tend to have a condemnatory attitude toward conflicts that result in depriving someone of water, and in the latter half of this century, as the world has become more economically interdependent and politically complex, and as the number of sovereign nations has increased, the consequences of water wars tend to ramify more broadly, causing the international community to exert a greater restraining influence.

A fuller answer is much more complicated and, in part, must be extracted from the tangled relations among natural resources, the environment, population growth, and the state. Each of these factors is in itself a putative source of conflict, but it is rare that only a single one of them will produce strife. They always interact reciprocally with one another and with other causal factors, directly and indirectly; the combinations can become daunting. So it is not always easy to spy out water or some other environmental factor as being a root or secondary cause of a conflict. However, if a regime cannot deal effectively with the results of resource scarcity and environmental degradation by maintaining the delivery of essential social services, the consequences could be discontent, anger, and challenges to its authority, all of which could lead to serious conflict (Goldstone 1991, pp. 1–62, pp. 349–497).

Resource and environmental issues and situations can be related to conflict in at least three ways: as proximate causes, as the means of conflict, and as the rationalization of conflict. Historically, vital resources such as water have been used more as the means or rationalization of conflict than seen as its causes, and water has tended to play a multiple role in generating conflict. In the causal equation of conflict, renewable resources are more important than non-renewable ones as the roots and proximate causes of conflict, and will become progressively more so if current environmental trends continue. Technology can also play an indirect role in precipitating water and other resource conflicts by making possible exponentially greater extractions of the resource, by indirectly damaging or destroying the resource through the technologically-caused overuse of an interdependent resource (e.g., logging, mining), and by producing side-effects or by-products that severely damage or destroy the resource by, for example, polluting it. Thus, how a resource is used is as important as whether it is renewable or non-renewable, but in the end, "Humankind is more dependent on environmental conditions than on technology." (Homer-Dixon 1991, pp. 85–98)

Pivotal among all of the elements related to conflict is the human factor, specifically in its growth trends. Rapidly escalating population growth (natural increase plus immigration), is unsustainable in arid and semi-arid regions of the world such as the Middle East. The dynamics of population growth, in combination with economic growth and technological progress that raise popular expectations of higher standards of living, engender a spiraling demand for resources. Under such pressure, a resource deficit becomes increasingly serious, especially if the resource is already meager or maldistributed and has no substitute, as is the case for water in the Middle East. These scenarios are real and replicated around the globe; they are generally associated with resource and environmentally related conflict (Choucri 1991).

Conflict: The superordinate principle

In the modern era, another possible constraint on water-driven conflicts in the Arab region may have been the influence of "superordinate goals" (or interests) that presumably can function even in the absence of trust among actors. Simply put, when cooperation clearly benefits all concerned, particularly if the issue revolves around something as vital as water, otherwise hostile groups, acting on interests that are superordinate (overriding), tend to exhibit a willingness to seek an accommodation or to cooperate rather than to confront or fight one another. Furthermore, such cooperation may produce positive changes in how the actors perceive one another, making other issues of contention more tractable (Muzafer 1958, pp. 349–56; Frey and Naff 1985, p. 78).

Hypothetically, the anxiety that always attends water scarcity could, in some of the Middle East's river basins, animate a consciousness of common, overriding goals, but not without uncommon leadership and luck (Frey and Naff 1985, p. 78). Evidence in support of the superordinate principle can be adduced. For example, between 1953 and 1955, Eric Johnston, President Eisenhower's special envoy, mediated discussions among Israel, Jordan, and Syria over the apportionment of the Jordan River's waters. Although negotiations over the Johnston Plan failed and animosity and distrust continued, the principal users of the Jordan system adhered approximately and informally to the technical terms of the 1955 Johnston Plan until the 1967 Six Day War radically altered the situation. In 1959, Egypt and Sudan negotiated a treaty on sharing the Nile. This pact remains in force and contains a clause mandating arbitration, mediation, or referral of disagreements to the World Court. The influence of superordinate interests may be discerned in the hydropolitics of other regions of the world as well. Despite deep antagonism, India and Pakistan, with facilitation from the World Bank, not only agreed on an equitable apportionment of the Indus River, but in all of their clashes spared one another's water installations and did not allow hostilities to obstruct the work of a joint Indo-Pakistani water management administration (Biswas 1992, pp. 201–9; United Nations 1984).

Given the unrelenting state of hostility among most Middle Eastern riparians and the number of occasions when those animosities could have erupted into substantial warfare, there have been relatively few instances of sustained belligerency over solely water issues. However, in the Middle East, the superordinate principle has not been sufficiently strong to inspire a more positive disposition between Arabs and Israelis (or, for that matter, between Turks and Arabs, between Arabs and Iranians, and among some Arabs).

Conflict: The power factor

The superordinate principle, while valid in some circumstances, is nevertheless subject to limitations and variations where water is concerned. Chief among these is the matter of power relationships among a basin's actors. The factors of interest (or need) and riparian position are related to this pivotal issue. The key to whether the superordinate principle will function is the power symmetry or asymmetry in a given basin. If the relative power among a basin's users is approximately symmetrical, there is a greater chance that superordinate goals will influence policies. However, in some circumstances, for example where hostility is deep and intense and the actors are roughly equally powerful, then the chances for conflict increase. Such would be the case in the Jordan basin if those conditions obtained. If power is asymmetrical, that is, one actor holds such a predominant or hegemonic position in relation to the other users as to be able to determine whether and in what circumstances cooperation will occur, then superordinate interests will induce cooperation only if they are sufficiently compelling to the basin's most powerful state. Without the concurrence of the basin actor with overriding projectable power or sufficient defensive power, cooperation will not occur. There can be an exception to this if the issue is so unimportant that an indifferent or passive stance is adopted, but this is virtually never the case where real or potential water scarcity exists (Naff and Matson 1984, pp. 11, pp. 182–96).

■ Only when Egypt, the Nile's premier power, agreed, was the 1959 treaty with the Sudan signed.

■ Repeated efforts to establish a cooperative regime for the Euphrates among Turkey, Syria, and Iraq have been unproductive for a combination of reasons: intense ill-will between the regimes of Syria and Iraq; historical grievances and mistrust by the two lower riparians toward Turkey; and the lack of urgency felt by Turkey, an upper riparian with sufficient defensive power, to promote a cooperative regime energetically. Latterly, the imbalance has only increased in the radically altered situation in the basin since Iraq's defeat in the Gulf War. Turkey is now the basin actor with the most projectable power, in addition to its defensive capacity.

■ In the Jordan basin before 1967, Syria, Jordan, and Israel were respectively the upper, middle and lower riparians. If for a time superordinate interests induced them to cooperate tacitly in observing the Johnston Plan, that influence was too feeble by the early sixties to overcome the animosities generated by the diversion of some of the basin's waters for Israel's national water carrier on the one hand, and the Arab threat of a counter diversion on the other. Water later became a factor in the outbreak of the 1967 war. Since 1967, any hope for cooperative water sharing in the Jordan basin without a prior political settlement, has evaporated. Israel, the indisputably dominant power, has been able to satisfy its own needs unimpeded and has not as yet been compelled by an overriding superordinate motive to alter the situation. Neither Syria nor Jordan, singly or in some political/military coalition, have demonstrated an ability to improve their hydropolitical status vis-à-vis Israel, or to produce change through cooperation.

Conflict: A power matrix model

Asymmetrical power alone cannot definitively explain the relationship between water and conflict. Other causal agents such as interest/need and riparian position, which interact with relative power relationships, must also be factored in. Taken together they provide, at an elementary level, the matrices of a three-part power model of water conflict: interest or need, including the perceptions and motivations of the actors; riparian position; and projectable/defensive power. It may be useful to stress again in this context that need is a factor relevant to water per se in ways that do not apply to other conflict issues (Frey and Naff 1985, pp. 77–80; Lemarquand 1977).

The first factor, need for water, or felt interest in it, reflects the motivations and perceptions of riparian actors and directs them toward cooperation or strife. If interests are seen as being advanced or reinforced by other parties, the impulse will be toward collaboration; if needs are perceived as being frustrated by others, the pressure will be toward conflict. If the hindrance is sensed as deliberate, unnecessary, and illegitimate, and if it occurs near the point of need satisfaction, the likelihood of conflict is heightened. The impact of such factors is cumulative. Frey (Naff and Matson 1984) has pointed out that owing to certain power considerations, blockage does not necessarily result in hostilities. The actor perceiving itself as harmed may possess too little power to alter matters, or, even if it is potent enough, it may reckon the costs of exercising its power too high to warrant action.

The second power-related factor is riparian position, which accords special advantages to the upstream powers over downstream competitors. The upstream riparian may be in a position to determine the quantity and quality of water passing downstream by such tactics as consumption, diversion, contamination, and regulation of flow. Such control obtains only if a riparian is situated at the main source of the river. Being upstream but above the major flow of the river carries little advantage.

The third and most important factor is projectable power, though defensive and internal power can also be significant influences. A riparian's projectable power is its ability to impose its own will on its rivals at whatever distance necessary thereby enabling it to govern their behavior in water issues. Defensive power, if sufficient, allows an actor to use its impregnability as a deterrence to enforce change in the behavior of other actors. Moreover, by its defensive strategies such an actor can also shape the behavior of competing water users. For example, in a purely hypothetical situation, if Syria were powerful enough to deter aggression by its upper and lower riparian neighbors in the Euphrates basin, she could not be forced to cooperate in any Iraqi-Turkish agreement over apportionment, use, or flow of the water thereby preventing consummation of such an accord and forcing Turkey and Iraq either to drop the matter or find another means for dealing with it.

Internal power exercised to control water-related actions, particularly as regards water distribution and costs, may have a significant effect on external riparian policies. When in the late 1950s and early 1960s Israel built its national water carrier to distribute drinking and irrigation water more widely and open the Negev for settlement, it diverted Jordan River water out of the Jordan basin over the objections of its Arab neighbors. This added a hydropolitical dimension to Israel's Arab policies that was manifested in Israeli threats of military attacks against any attempts by the Arab riparians to undo Israel's actions by river diversions of their own, including the aborted Maqarin Dam on the Syrian-Jordanian border.

This paradigm of riparian power matrices yields the following assessment of conflict potential when applied in the following model, by way of illustration only, to three of the region's principal river systems, the Jordan, the Euphrates, and the Nile. The model is designed to be a simple and easily usable instrument for determining at a basic level relative riparian power relationships and for predicting the potential for conflict (see Table 1).

Each factor, i.e., interest/need, power, and riparian position, is assigned a weighted value for each of the basin nations; in this illustration a low-to-high scale of 1–5 is used. The totals are reached on the basis of available hydropolitical, military, and other data such as competence of leadership, quality of the armed forces, and strength of alliances, plus the experience and perceptions of the analyst. To mirror the power reality in the three basins more accurately, military power is given a weightier value for Israel, Turkey, and Egypt. The reasons for this in the case of Israel are that country's overwhelming conventional military strength coupled with its possession of a considerable nuclear arsenal and the means to deliver it anywhere in the Middle East, an overall capacity unmatched by Israel's potential enemies in the area. Turkey is rated heavily due to Syria's ongoing embroilment in Lebanon and Syria's need to secure its Israeli borders, and

T A B L E   1

Model for Determining Relative Power and Conflict Potential in Major Middle East River Basins

		Interest/ Need	Power	Riparian Position	Total
Jordan Basin	Israel	5	9	5	19
	Jordan	5	2	2	9
	Syria	3	3	2	8
	Lebanon	1	0.5	2	3.5
Euphrates Basin	Turkey	5	8	5	18
	Syria	5	3	3	11
	Iraq	4	2	1	7
Nile Basin	Egypt	5	7	1	13
	Sudan	4	1.5	4	9.5
	Ethiopia	3	0.5	4	7.5

Adapted from Frey,1993. The scale is 1-5, except for power, which is scaled 1-10 in order to reflect reality more accurately; Iraq's military power is based on conditions after the 1990–91 Gulf War.

because Turkey's relative military position was enhanced by the destruction of Iraq's armed forces in the Gulf War. Egypt is ranked high due to the very poor quality of the armed forces of Sudan and Ethiopia as compared to those of Egypt, taking into account their relative ability to sustain conflict for any length of time. In the region as a whole, the loss of Soviet assistance and arms is also reckoned. The overall power ranking of the riparians is determined by the sum of the three criteria. The higher the total, the greater the relative strength of the individual riparian. By calculating the power relationships of basin states that must share the same water supply, the analyst is then able to predict (roughly) the potential for water-based conflict under various conditions as reflected in the following propositions.

Conflict: Power, conflict, and water uses in the Middle East
There are three propositions underlying this analysis:

- The greatest potential for conflict exists when a lower riparian is a more powerful actor than the upper water-controlling riparian and perceives its needs to be deliberately frustrated. As indicated, such was the case in the Jordan basin when before 1967 Israel was in a disadvantageous lower position, possessed the most relative power, and had very high interest and need, which it saw threatened by its rivals; the prospects for strife were high and strife there was. As asserted, water as an issue was an important factor in the outbreak of the Six Day War.

The building of the Aswan High Dam in Egypt, which was in part intended to symbolize Egypt's place in the van of Arab politics and of Arab achievement in general, has had a significant impact on virtually all major sectors of Egypt's economy and has even altered the environment. The dam is essential to Egypt's continued economic growth, especially in light of the country's population explosion; therefore, a sufficient flow of the Nile to ensure that the dam functions at full potential is a matter of exquisite sensitivity to Egypt's rulers, who are never forgetful of their country's position as the basin's lowest but most powerful riparian. Thus, it is an axiomatic policy of every Egyptian regime that it will go to war, if necessary, to prevent either of its closest upper riparian neighbors, Sudan and Ethiopia, from reducing in any way the flow of the Nile.

■ When an uppermost riparian is the most powerful actor in an international basin, that disparity (or asymmetry) of power inhibits conflict potential. After 1967, Israel's seizure and annexation of the Golan Heights resulted in its control of headwaters of the Jordan; this circumstance combined with Israel's dominant military power precluded major basin conflict except on those occasions when Israel initiated invasions of Lebanon as in 1982. Turkey's advantage in controlling some 96 percent of the headwaters of the Euphrates River was augmented by the destruction of Iraq's projectable power in the 1990–91 Gulf War, giving Turkey greater-than-ever dominance in the basin, but in this case, Turkey's lower degree of need has so far restrained aggressive assertion of that ascendancy.

■ When relative power symmetry coexists in a basin with asymmetry in interest and position, there will be a moderate but consistent potential for conflict. Tacit or informal cooperation often characterizes such circumstances, especially with third party intervention (this was approximately the condition in the Jordan basin in the 1950s and early 1960s). However, this balance is always precarious, especially where tensions normally tend to be high and can quickly escalate to flash points. It will be threatened in times of prolonged critical scarcity, and when water becomes a significant element in a persistent and comprehensive inter-riparian political rivalry. The conflict potential of water then rises relative to the probability of politically motivated hostilities, and once water becomes directly involved, it may seriously enlarge a conflict.

The implications of the relationship between water and conflict in international river systems are the obvious ones: the key determinants of water-driven strife are scarcity and maldistribution, followed closely by perceived need and the relative power status (together with felt deprivation and frustration) of the international basin's riparians; these factors are reciprocal, and in combination reinforce one another thus tending to magnify the potential for conflict in times of crisis. Moreover, each of these hydropolitical elements is often part of or significantly influenced by larger aggregated antagonisms, basin-wide or regional, thereby broadening the potential for conflict.

These are precisely the conditions that pervade the Middle East, particularly as regards the Jordan basin. The time of critical scarcity has arrived and events show that the scope for tacit collaboration and tolerance has all but eroded. Water has already been a major rallying cry for the Palestinian Intifada in the Occupied Territories and for conservative parties in Israel. Once an issue becomes a slogan in the streets and media, it becomes very difficult to manage through normal channels of negotiation. The potential for water-inspired conflict in the Middle East is integral to the region's milieu of pervasive tension, and rises accordingly as tensions are heightened.

The bond between water-based problems and their social repercussions, such as destabilization and conflict, particularly where water is scarce or seriously degraded and is shared in a tense environment (exactly the case in the Jordan basin), is, like all other aspects of water, complex and not easily analyzed, and makes policy formulation difficult. Conflict-laden hydropolitical factors are often concealed in ideological, ethnic, and political strifes and can make them more complex and violent (Homer-Dixon 1990, pp. 4–5). Accordingly, in addition to hostilities between or among nations, water conflicts can take many forms. These forms include intersectoral rivalries, competition among interest groups, propaganda and ideological warfare, internal strife, and so on, usually in combination. All these, if serious or prolonged, contribute to the political destabilization of a given basin.

The Paramount Question and a Recommendation

One should not assume, in the absence of treaty agreements among the Arabs and Israelis (accords that would have to be underpinned by some kind of international guarantees), that more than marginal basin cooperation can be achieved. Treaty agreements, with the necessary legal obligations and structures to uphold them, are the key to successful hydropolitical cooperation in the Middle East, at the basin level or the regional level or both.

The peace-seeking negotiations between Israel and its various Arab neighbors are focused on the ultimate territorial disposition of the Jordan basin and south Lebanon. Hence, the paramount question arising out the hydropolitical phase of the negotiations is this: Given Israel's control of virtually all of the basin's major water resources, its overpowering military superiority, together with a deep reluctance to yield up significant portions, if any, of the territories it presently holds, and with them its hydrological advantages, what incentives or power can the Arabs use, singly or collectively, to persuade the Israelis to accept substantial territorial and political changes in the status quo?

In default of a clear answer at present, certain steps can nevertheless be taken in the direction of cooperation, even in the absence of sufficient trust or treaty arrangements, steps that would, at all events, be necessary for eventual full cooperation. At least two factors make these actions possible: 1) all the key actors, at least in the Jordan basin and very probably in the Euphrates and Nile basins as well, finally appear to be serious about preferring negotiation to conflict, and 2) the mounting scale of the basins' water crises, aggravated by rising populations and declining economies, has been a compelling motive. Like the main issues, interim actions must involve the principles of flexibility, equity, proportionality, data sharing, law, and a sense of fairness. And, since all of the negotiators in the current diplomatic exchanges, especially the Arabs, are practicing what I call "side-effect" diplomacy, and seek to use the tactic of "strategic discrepancies" in making trade-offs, outside assistance and/or mediation will be required even for preliminary measures.

Recently, I have elsewhere offered in print thirteen proposals that could lay the bases for eventual cooperation without absolutely requiring a prior political settlement. These proposals should even help promote a settlement. I want here to focus on only two reciprocal ideas that are salient because they involve real cooperation in areas where all parties agree common grounds exist:

■ Since it is unlikely that cooperation can be coerced or induced at the highest political levels, the most promising approach would be to encourage cooperation at a lower but still significant level, among scientists and technical experts. If scientists and technocrats in the area, together with the officials they advise, can communicate sufficiently to develop shared understanding of the water situation, of available and new technologies, and of potential solutions, they could constitute a community of informed specialists throughout the region, and become a strong force for cooperation by pressing for and guiding effective water policies.

■ For the creation of such a community of experts, it would be necessary to constitute a technical infrastructure for hydropolicy that addresses water problems at both basin and regional levels by establishing two types of water institutes (or clearing houses) with data bases: one for river basins and another for comprehensive regional hydrological issues. They would be situated (either within the region or, if necessary, outside it for an interim period) so as to reduce ideological barriers to participation. The work of these institutes would emphasize science, technology, and data. These institutes, comprising staff, fellows, train-ees, and other personnel from the region and from the world's other major basins, would perform several functions: conduct basic and applied research; provide the expertise, research, educational opportu-nities, and data necessary to develop the entrepreneurial, human, and technical resources presently lacking; generate data bases and hydro-logic, economic, and other social scientific analytical tools; act as conference settings; serve as centers for accurate record keeping and information dissemination; and foster interaction among basin and regional specialists.

Since 1960, the water issue in the Middle East has become increasingly militarized, while at the same time the region's water problems have grown more acute. Consequently, there have been more shooting incidents associated with water since 1967 than in all of the previous decades of the century. It would appear that the superordinate principle will, within the decade of the 1990s, be put to its severest test: whether under the combined pressures of demographic changes, security threats, and high levels of militarization, a severe water crisis will result in a basin-wide or regional conflict, or whether the competition for water — so vital to life — can serve as a catalyst for cooperation among the regions riparians as the price of mutual survival.

11

The Agenda for the Next Thirty Years

Peter Rogers

Introduction

Historical and current trends in water use in the Arab world are now receiving much attention, with emphasis on the tight constraints on water in many of the countries and sub-regions. Indeed, the situation is so constrained that one has the impression that many water professionals and government officials would rather not think about the future when water use will be even further constrained. The future use of water must be directly faced, however, if any sensible policies are to be adopted in the present to bring that future under control. In this chapter, the future is raised and addressed because we can only make sense of present choices by considering carefully the very contrasting futures to which different decisions — or non-decisions — of today will lead.

For today's choices, how much of the future do we need to consider? Five years, ten years, twenty years, or more? The nature of the investments in water infrastructure gives us some idea of the length of time before investments made today will need to be replaced. For large structures like dams and reservoirs, 50 years seems appropriate; for treatment plants and components of the urban distribution and wastewater systems, and for irrigation systems, 20 years may be more suitable. For the purposes of this paper we have chosen an intermediate time horizon of roughly thirty years, bringing us to about the year 2025. Thirty years is a very long time into the future and the reliability of forecasts consequently will not be very high. Looking in the other direction, 1963 was about thirty years ago. Consider the major convulsions the world has undergone since then: the war in Vietnam, wars in the Middle East, the oil price hikes, rapid economic growth in Europe and in the Asian rim countries and Japan, the discovery of human impacts upon the ozone layer and of the global "greenhouse" heat balance, unprecedented growth of the world's population, and the breakup of the Soviet Union, to name just a few. All this within the relatively short time period that thirty years seems in retrospect. Some developments in this period have had major impacts upon water and the aquatic environment, for example, the enormous spread of washing machines and flush toilets in residences and the rapid development of water-consuming service industries, such as food processing. Will the next 30 years see major political and economic changes? Almost certainly so.

For instance, the rise of oil prices during the 1970s changed the terms of trade between the developed countries and the oil-exporting countries, particularly those in the Arab world. This led to greater national wealth and general improvement in the standard of living in many Arab countries, and huge increases in the demands for middle-class amenities. An important part of this was large increases in the amount of water demanded, first for domestic use, and later for agricultural and industrial purposes as well.

There are 20 countries in the Arab world, stretching from Mauritania on the Atlantic to Oman on the Indian Ocean. Apart from language, culture, and religion, they share an important physical characteristic: they all are very arid. Within the region there are, of course, wide divergences between the individual countries, but nevertheless absolute water availability is restricted in each. Morocco, Lebanon, and parts of Algeria are relatively well watered and Libya, Egypt, and Saudi Arabia are very poorly served. Figure 1 of the Summary to this book shows the location of the countries, the location of international rivers, the Nubian sandstone aquifer, and the major surface and groundwater recharge zones. Figure 2, also in the Summary, is a rainfall map of the region showing that almost 90 percent of the region receives an average annual rainfall of less than 250 mm. Below this level of rainfall, irrigation is needed to ensure most agricultural production. Another feature of this arid region is a substantial number of international river basins and aquifers shared by at least two countries. Several river basins have 3 riparians, and the Nile is shared by 9 countries. Table 1 (based upon Gischler 1979) lists 25 international rivers in the region.

Of the 20 countries in the region 7 are quite small in land area and 4 are quite large (Figure 1). Algeria and the Sudan top the list for land area, followed by Saudi Arabia and Libya. Mauritania and Egypt are substantially smaller and they are followed by Somalia, Yemen, Morocco, and Iraq. When arable land is considered (Figure 2), Sudan leads, followed by

TABLE 1

List of International Rivers in the Arab World*

(based on Gischler 1979)

No.	International River	Countries in Downstream Direction	Area Drainage Basin in km²
1	Senegal	Guinea, Mali, Senegal, Mauritania	450,000
2	Atui	Mauritania	
3	Oued Draa	Morocco, Algeria	15,100
4	Oued Daoura	Morocco, Algeria	
5	Oued Guir	Morocco, Algeria	
6	Tafna	Morocco, Algeria	6,900
7	Medjerda	Algeria, Tunisia	23,000
8	Nile	Tanzania, Kenya, Burundi, Rwanda, Uganda, Sudan, Egypt, Zaire, Ethiopia, Congo Central African Republic	2,800,00
9	White Nile	Sudan	353,550
10	Sobat	Ethiopia, Sudan	244,900
11	Blue Nile	Ethiopia, Sudan	325,000
12	Atbara	Ethiopia, Sudan	220,700
13	Gash	Ethiopia, Sudan	21,000
14	Baraka	Ethiopia, Sudan	41,694
15	Lagh Bor	Ethiopia, Kenya, Somalia	
16	Juba	Ethiopia, Somalia	200,000
17	Shebeli	Ethiopia, Somalia	260,000
18	Yarmouk	Syria, Jordan	7,252
19	Nahr el Assi (Orontes)	Lebanon, Syria	16,900
20	Euphrates	Turkey, Syria, Iraq	350,000
21	Khabur	Turkey, Syria	31,800
22	Tigris	Turkey, Syria, Iraq	296,500
23	Great Zab	Turkey, Iraq	26,473
24	Karun	Iran, Iraq	
25	Jordan	Syria, Lebanon, Israel, Palestine	18,300

* Locations of rivers, identified by numbers above, are shown in Figure 1 of this volume's Summary.

FIGURE 1

Arab States — Areas (in million Ha)

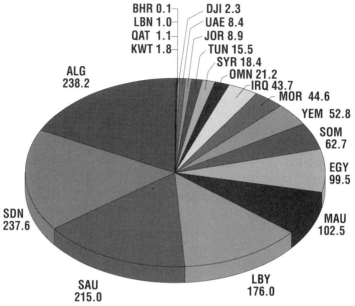

Source: Human Development Report 1992

FIGURE 2

Arable Area (in million Ha)

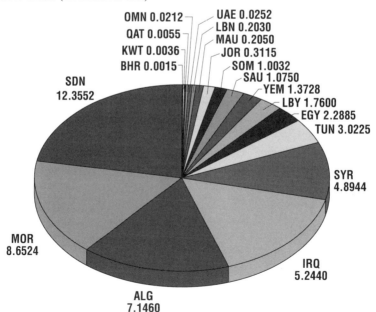

Source: Human Development Report 1992

Morocco, Algeria, and Iraq. Syria and Tunisia both have more arable land than Egypt.

A determining feature for any consideration of future resource use in the Arab world is the rapid population growth that is being experienced in all of its countries. From 1965 to 1990 the total population of the 20 countries more than doubled from 108 million to more than 220 million. Figure 3 shows the distribution of the population and its growth from 1965 to 1989 for each of the 20 countries, among which the largest populations are those of Egypt, Morocco, Algeria, and the Sudan.

In analyzing water scarcity, population by itself is not necessarily a determining factor, but rather the density of population on the resource base. In terms of raw population density Bahrain (Figure 4) is by far the most densely settled, with Lebanon and Kuwait well behind. However, in terms of population density on arable land (Figure 5) Egypt places second behind Bahrain with a density of almost 20 persons per cultivable hectare. Compare this to 1.3 persons per arable hectare for the United States and one can immediately understand the potential food crises of a country like Egypt, particularly when faced with continued rapid population growth.

Population growth leads to increased total water use, of course, but does not by itself affect per capita rates of water use. Rising incomes do lift per capita water demand, because they increase consumption possibilities in general, and they also influence tastes and preferences. So this economic factor must be taken into account, in addition to population growth, to understand the rate of growth of water use. Along with population size and income level, the degree of urbanization is an important indicator of increasing water use. Figure 6 shows that urbanization has occurred rapidly in all the Arab countries between 1960 and 1991.

Figure 7 indicates a fairly high level of access to safe drinking water for Arab urban populations. The rural populations are, however, much worse off, with less than half of the people having access to potable water in many countries. Figure 8 shows water withdrawn per capita for all uses. Iraq is by a large measure the heaviest water user followed by Egypt, the Sudan, and Yemen. Figure 9 gives a breakdown of the water withdrawals by sector. In all of the large water-using countries, agriculture uses by far the largest amounts of water; well over 90 percent goes to agriculture in Egypt, Iraq, the Sudan, and Morocco.

FIGURE 3
Total Population, Arab States: Time Series

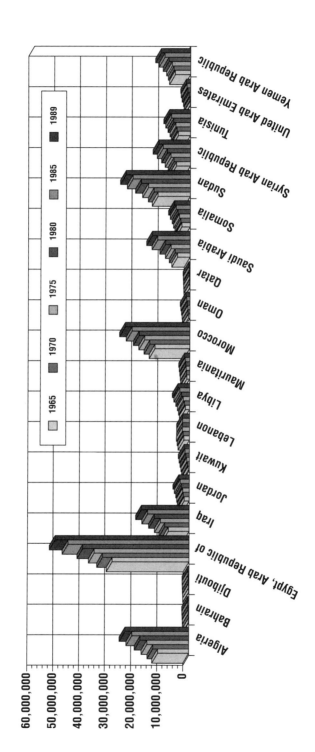

Source: United Nations 1990

FIGURE 4
Population Density — 1987

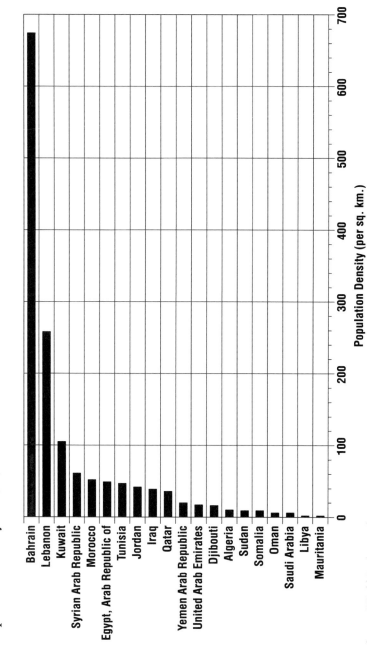

Source: World Bank 1990d

FIGURE 5
Population Density on Arable Land

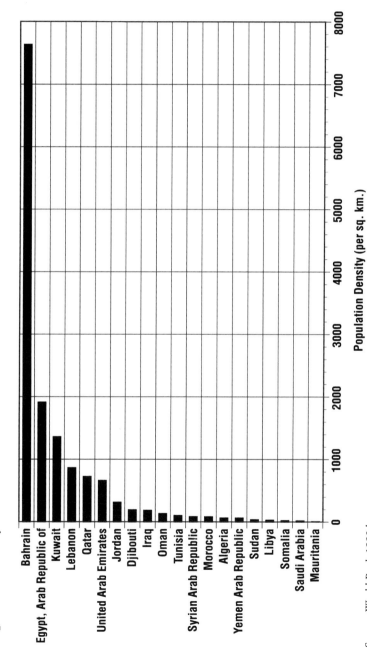

Source: World Bank 1990d

F I G U R E   6

Arab States — % Urbanization

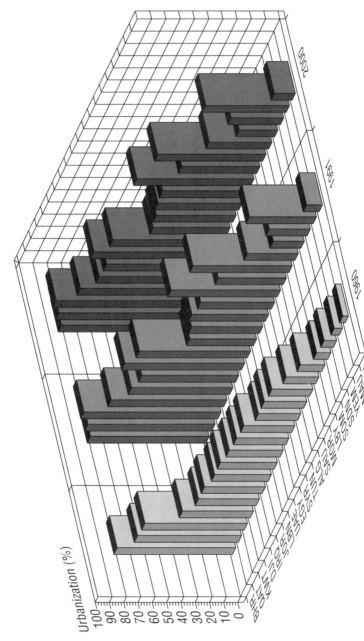

Source: World Resources Institute 1992

F I G U R E 7
Access to Safe Water

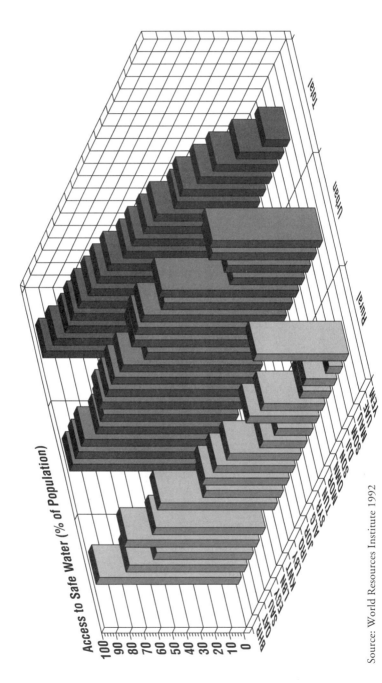

Source: World Resources Institute 1992

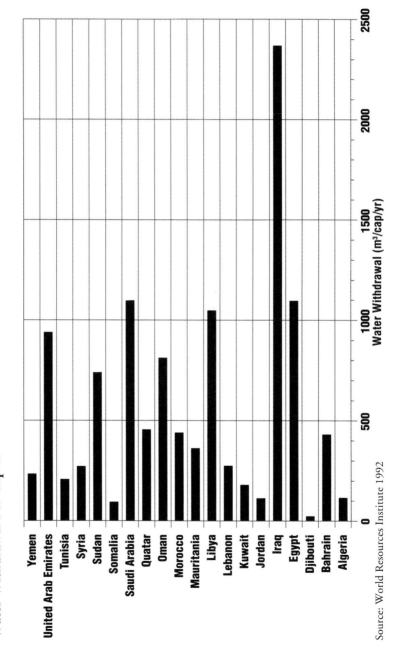

FIGURE 8

Water Withdrawal Per Capita

Source: World Resources Institute 1992

FIGURE 9
Water Consumption by Sector

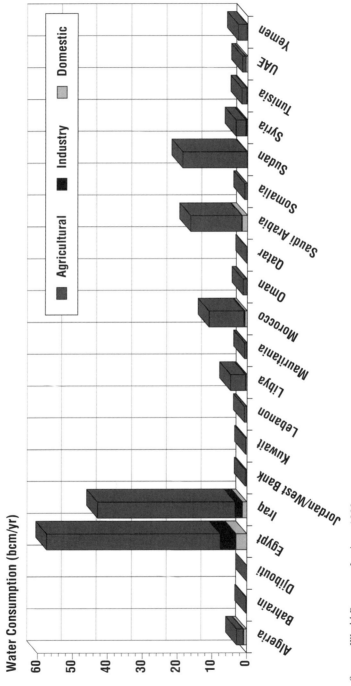

Source: World Resources Institute 1992

Supply Options

In the Arab world, as elsewhere, there are two major ways of meeting the demand for water: conventional supply options, and conserving water in one use to free it for another application. Under conventional supply options we typically consider direct use of rainwater and its capture in cisterns and tanks, surface water diversions from rivers and lakes, abstraction of groundwater from shallow and deep aquifers, and desalination of brackish water or seawater.

In addition to these sources, water may also be saved by various conservation methods. This is not new water; however, conserved water is just as useful for meeting demands as water from the more conventional sources. The conservation sources of water typically encompass the recycling of municipal and industrial wastewater, the recycling of agricultural drainage water, increasing the yields of watersheds or aquifers by management of the watersheds, reduction of urban water system leakages, demand reduction by pricing or other rationing methods, use of water-saving devices, process change within industry, and change to water-saving crops in irrigated agriculture. Not all of these options are available in every case, but many studies of water supply focus exclusively upon the so-called "conventional" supply options, and the perspective should definitely be enlarged to take into account these latter possibilities.

In considering supply possibilities, engineering practice, economic analysis, and plain common sense suggest that the least expensive options be exercised first, followed by the next least expensive, and so forth, until the planned demand can be met. Consider a hypothetical example of a municipality planning to expand its water supply capacity by 4 mcm per day. While the following data are hypothetical they are typical of the magnitudes that water planners would be faced with in Arab countries expecting large increases in urban populations. An amount of 1.0 mcm/day of surface water could be developed for about $0.10 per m³, an amount of 0.5 mcm/day of shallow groundwater for $0.17 per m³, an amount of 1.0 mcm/day of deep groundwater for about $0.50 per m³, up to 1.0 mcm/day of water could be imported by tanker for about $0.60 per m³ and the same amount by pipeline for $0.80 per m³, an amount of 1.0 mcm/day of brackish water could be desalinated by electrodialysis or reverse osmosis for about $1.0 per m³, and an unlimited amount of water could be supplied by desalination of sea water by multi-stage flash distillation or multiple effect distillation for about $2.0 per m³. It is now possible to derive from this information a supply schedule for meeting the future demand. Table

TABLE   2 A

Traditional Water Supply Options

	Amount mcm/day	Cost $/m³	Cumulative mcm/day
Surface Water	1	0.1	1
Shallow Groundwater	0.5	0.17	1.5
Deep Groundwater	1	0.5	2.5
Water Importation: Tanker	1	0.6	3.5
Water Importation: Pipeline	1	0.8	4.5
Desalination: Brackish Water	1	1	5.5
Desalination: Sea Water	infinite	2	unlimited

2a presents these hypothetical data and Figure 10 shows the implied supply schedule.

The same logic holds for the case of non-conventional supply options based upon all forms of conservation. Imagine that it is possible to reduce non-beneficial evaporation and seepage losses in agriculture by an amount of 1.0 mcm/day at a cost of $0.05 per m³, to recycle 0.5 mcm/day of water in industry at a cost of $0.13 per m³, to eliminate 0.2 mcm/day of urban unaccounted-for water at a cost of $0.15 per m³, to recycle 2 mcm/day of urban wastewater at a cost of $0.70 per m³, and to reduce urban demands by 0.1 mcm/day by demand management techniques at a cost of $0.90 per m³. Table 2b gives the possible ranges of conservation options and their costs per unit. If these are incorporated with the conventional supply options then the new supply schedule for the potential investment would look like Figure 11.

Figures 10 and 11 demonstrate vividly how the supply options are radically expanded simply by thinking about the problem in a wider context that includes conservation measures. The water planner now has a much wider range of alternatives available for consideration, and following a rational strategy based upon developing the least expensive water first, would now be able to meet a 4 million m³/day expansion of the

TABLE   2 B

Unconventional Water Supply Options

	Amount mcm/day	Cost $/m³	Cumulative mcm/day
Reduction Agricultural Losses	1	0.05	1
Industrial Recycling	0.5	0.13	1.5
Reduction of Urban Leakage	0.2	0.05	1.7
Wastewater Recycling	2	0.7	3.7
Demand Management	0.1	0.9	3.8

FIGURE 10
Conventional Supply Curve – Hypothetical

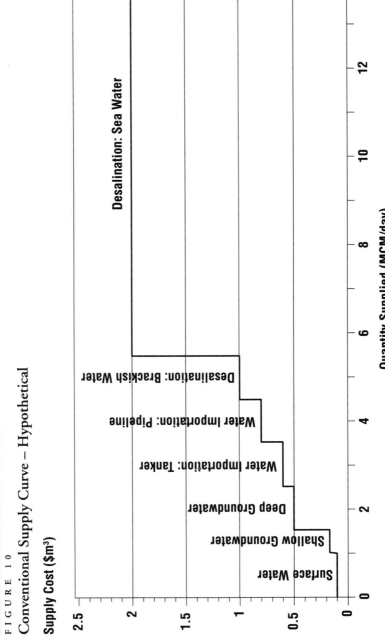

Supply Cost ($m³)

FIGURE 11
Unconventional Supply Curve – Hypothetical

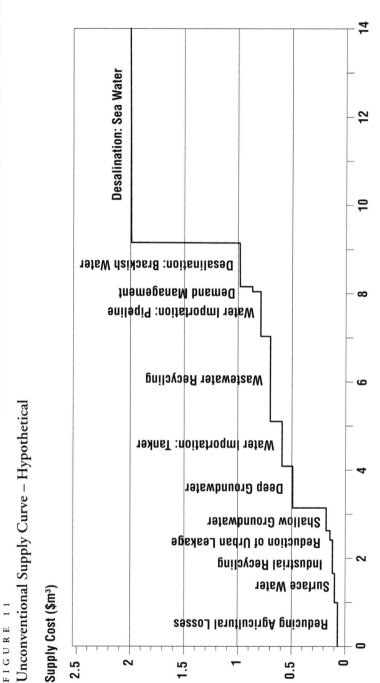

system by using a combination of traditional supplies and conservation. In this case expensive importation of water by both tanker and pipeline can be avoided. If the water planner had a demand curve for water demanded by urban consumers it would be possible to overlay it on the supply curve and identify where the curves intersected; this intersection point (quantity and price) would determine the economically efficient level at which to supply water. Unfortunately, it is very difficult to make reliable econometric estimates of demand curves for urban water supply in situations where water prices have been administered and where they have not varied much over time or space. Few such curves have been estimated for the United States (Gomez 1987 reports several for Latin American cities), and the general results show price elasticities in the range of minus 0.3 to minus 0.7, with average prices of about $0.40 per m^3.

Even though demand curves are not available for the Arab countries, Al-Alawi and Abdulrazzak (in this volume) and Jellali and Jebali (in this volume) report typical tariffs in various countries. Table 3 shows how these tariffs compare with typical United States prices. It is interesting to note how they cluster around typical U.S. prices for urban water supplies. The table is particularly instructive in showing how far away the typical tariffs are from the real costs of desalination at $1.5 to $2.0 per m^3. Moreover, the table shows clearly the problems of irrigating crops in these arid regions. At $.40 per m^3 the equivalent irrigation cost is $4,000 per hectare-meter. It is hard to imagine being able to afford the cost of this input for most field crops when U.S. farmers are finding it hard to survive with irrigation water costing less than $300 per hectare-meter. The cost of providing an additional, incremental water supply, which by the logic examined above is always more expensive than existing water supplies, makes it difficult to imagine any large expansion of irrigation water supplies in this region. This is the basis of the restrictive assumptions made about diversions of water to agriculture in the next section.

Future Water Demands — An Analytic Model

The demand for water is not something set by arbitrary notions of "need." Rather, it is reached by a complex interaction of economic and social forces. Several writers (e.g., Falkenmark 1989; Postel 1992) see "water crises" of varying magnitudes around the world, and particularly within the Arab world. They take a water resource base of one thousand cubic meters available per capita per year to represent a kind of "water barrier" below which nations and regions will become increasingly susceptible to various economic and social ills. The economic data from the 20

T ·A B L E   3
Water Costs and Water Tariffs

Price $/m^3$	Price $/1000gal$	Price $/Acre-ft$	Price $/ha-m$	Typical Water Tariffs $/m^3$
0	0	0	0	
0.01	0.04	12.29	100	
0.02	0.08	24.58	200	
Typical U.S. rates for irrigation water				
0.03	0.11	36.88	300	
0.04	0.15	49.17	400	
0.05	0.19	61.46	500	
0.06	0.23	73.75	600	
0.07	0.26	86.04	700	
0.08	0.30	98.34	800	
0.09	0.34	110.63	900	
0.1	0.38	122.92	1,000	0.14 Algeria (JJ)
0.2	0.76	245.84	2,000	0.15 Kuwait irrigation (AA)
0.3	1.13	368.76	3,000	0.18 Kuwait industry (AA)
				0.36 Morocco (JJ)
Typical U.S. rates for municipal supply				
0.4	1.51	491.68	4,000	
0.5	1.89	614.60	5,000	
0.6	2.27	737.52	6,000	0.56 Tunisia (JJ)
0.7	2.65	860.44	7,000	0.58 Kuwait domestic (AA)
0.8	3.03	983.36	8,000	
0.9	3.40	1,106.28	9,000	0.9 UAE (AA)
1	3.78	1,229.20	10,000	
1.1	4.16	1,352.12	11,000	
1.2	4.54	1,475.04	12,000	1.21 Qatar (AR)
1.3	4.92	1,597.96	13,000	1.35 Mauritania (JJ)
1.4	5.30	1,720.88	14,000	
Typical costs for brackish desalination				
1.5	5.67	1,843.80	15,000	
1.6	6.05	1,966.72	16,000	
1.7	6.43	2,089.64	17,000	
1.8	6.81	2,212.56	18,000	
1.9	7.19	2,335.48	19,000	
Typical costs for seawater desalination				
2	7.56	2,458.40	20,000	
2.1	7.94	2,581.32	21,000	
2.2	8.32	2,704.24	22,000	
2.3	8.70	2,827.16	23,000	
2.4	9.08	2,950.08	24,000	
2.5	9.46	3,073.00	25,000	

Notes:

AA refers to Jamil Al-Awadi and Mohammed Abdulrazzak, "Water Problems in the Arabian Peninsular" (this volume).

JJ refers to Mohammed Jellali and Ali Jebali, "Water Resources Management in the Maghreb Countries" (this volume).

countries in the Arab world seems to contradict this dismal outlook. In Figure 12 the data for GNP per capita are plotted against the total amount of potential water resource per capita in each of the 20 countries. This is a surprising graph. First, only Iraq, Mauritania, Lebanon, Syria, Somalia, Morocco, and Egypt have potential water resources in excess of the "water barrier," the remainder of the countries lying more or less below 500 cubic meters per capita per year. The 13 countries below the "water barrier," however, lie along an extremely wide range of per capita incomes ranging from less than $500 (Sudan) to more than $18,000 per capita (United Arab Emirates). Clearly, in the case of oil-exporting countries, another liquid mineral, petroleum, accounts much more for these figures than does water. But even when the oil-rich states are set aside, as in Figure 13, water availability seems to account for little of the differences between per capita income in the different countries. Note that the three countries that have the most water on a per capita basis (Iraq, Egypt, and the Sudan) all lie within the low-income end of the spectrum.

To predict comparative water demands for the region, ideally one would have reliable forecasts of population size and composition, and economic growth parameters for each country. In the absence of such information it may still be possible to arrive at rough estimates of the probable course of demand over the next three decades if simplifying assumptions are made. We make these assumptions below for the sake of identifying potential conflicts in water use, and to stimulate a discussion of this approach to establishing country-wide and regional water policies.

For the purposes of this exercise, the following assumptions are made:

■ From 1990 onwards the future growth of GNP in each country returns to the rates obtaining before 1973. (In many of the countries the GNP has actually been declining in the past few years.) The situation in 1990 is projected forward at these rates until 2025.

■ The population in each country is predicted to increase at the rates given in the United Nations (1990) predictions.

■ Two cases are considered for irrigated agriculture. In the first, and probably the most realistic, no additional water will be made available for agriculture. All increases in production will have to come from improved efficiency within the agricultural system itself. In the second case water supply to agriculture is assumed to grow at a rate of 3 percent per annum.

FIGURE 12

GNP and Water Resources

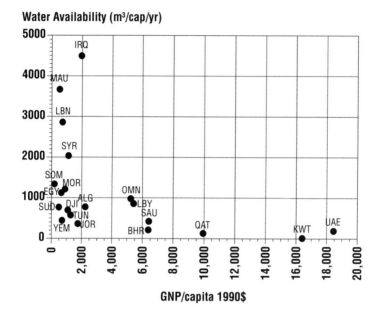

FIGURE 13

GNP and Water Resources, Excluding Major Oil Producers

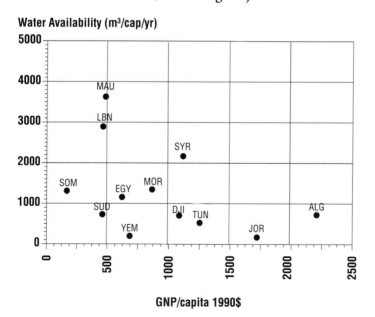

■ After 1990 the rural populations in all the countries are assumed to remain at the 1990 levels with all increases in population living in cities. This fits well with the first case in assumption 3 above when irrigation changes are restricted to efficiency improvements, but given the large amounts of underemployment in the rural areas it also fits the second case, in which more water is committed, as well.

■ In projecting water demands for the urban domestic sector two cases were considered. First, assume that domestic water demand grows from 1990 at the rate of urbanization. Second, assume that it grows from 1990 as a function of both the population and the income growth; or in other words, at the rate of urbanization plus the growth rate of per capita income times an income elasticity of 0.5.

■ Industrial sector water demand is assumed to grow at the rate of growth of GNP.

■ Assume no major efforts to ration industrial and urban demand by various demand management techniques. (These could be factored in later if it turns out that the demands exceed the resource base by the end of the period.)

Table 4 summarizes the results under three variants. The first, Case I, as suggested above assumes that the urban demand is not responsive to changes in income. The second, Case II, assumes an income elasticity of demand of 0.5, and the third, Case III, assumes the same income effect as case II and further assumes that irrigation water supply is increased at a rate of 3 percent per annum. Table 4 indicates that out of a total water resource availability of 282 bcm per year for the entire Arab world, by 1990, 168 bcm per year was already being diverted for use. This refers to withdrawals not consumption, which means that some of this water can be used repeatedly within the year. However, since the bulk of the water is withdrawn for irrigated agriculture, a large percentage of this (as high as 70 percent) will be lost to evaporation. So already in 1990, 60 percent of the total water available is being withdrawn in any year. This compares with a figure of 19 percent for the United States.

A surprising conclusion that can be drawn from the demand forecasting model is that, for Cases I and II, which prohibit any additional water to agriculture, the annual overall demand will rise by only about 45 bcm in total to between 200 and 215 bcm per year by the year 2025; leading to about a 75 percent use of the total available resource. If, however, irrigation water supply is allowed to grow at 3 percent per annum the total demand

TABLE 4 – PART ONE

Water Demand Forecasts for 2025 by Country and by Sector

Country	Resource Base (bcm)	1990 Withdrawal (bcm/yr)				Case I: 2025 Withdrawal (bcm/yr) Without Income Effect				Case II: 2025 Withdrawal (bcm/yr) With Income Effect				Case III: 2025 Withdrawal (bcm/yr) With Income Effect & 3% Incr. Agr			
		Dom	Ind	Agr	Total	Dom	Ind	Agr	Total	Dom	Ind	Agr	Total	Dom	Ind	Agr	Total
Algeria	19.00	0.85	0.15	1.85	2.85	2.60	0.44	1.85	4.88	4.37	0.44	1.85	6.65	4.37	0.44	5.21	10.01
Bahrain	0.10	0.09	0.02	0.11	0.22	0.19	0.02	0.11	0.32	0.19	0.02	0.11	0.32	0.19	0.02	0.32	0.53
Djibouti	0.30	0.00	0.00	0.01	0.01	0.01	0.01	0.01	0.02	0.02	0.01	0.01	0.03	0.02	0.01	0.01	0.04
Egypt	58.10	3.10	4.60	49.70	57.40	8.27	12.93	49.70	70.90	13.87	12.93	49.70	76.50	13.87	12.93	139.85	166.65
Iraq	81.00	1.28	2.14	39.38	42.80	4.10	2.14	39.38	45.62	4.10	2.14	39.38	45.62	4.10	2.14	110.80	117.04
Jordan/West Bank	1.40	0.13	0.03	0.29	0.45	0.46	0.05	0.29	0.80	0.63	0.05	0.29	0.97	0.63	0.05	0.82	1.50
Kuwait	0.00	0.30	0.01	0.08	0.38	0.39	0.02	0.08	0.48	0.53	0.02	0.08	0.62	0.53	0.02	0.23	0.77
Lebanon	7.80	0.08	0.03	0.64	0.75	0.15	0.06	0.64	0.84	0.20	0.06	0.64	0.90	0.20	0.06	1.79	2.05
Libya	3.83	0.29	0.19	4.28	4.76	1.05	0.41	4.28	5.73	1.52	0.41	4.28	6.20	1.52	0.41	12.03	13.95
Mauritania	7.40	0.09	0.03	0.61	0.73	0.36	0.06	0.61	1.04	0.52	0.06	0.61	1.20	0.52	0.06	1.73	2.31
Morocco	30.00	0.67	0.33	10.00	11.00	1.40	1.07	10.00	12.48	2.52	1.07	10.00	13.59	2.52	1.07	28.14	31.73
Oman	1.47	0.08	0.01	1.15	1.24	1.62	0.01	1.15	2.77	1.58	0.01	1.15	2.74	1.58	0.01	3.24	4.82
Qatar	0.05	0.08	0.01	0.11	0.19	0.14	0.18	0.11	0.42	0.61	0.18	0.11	0.89	0.61	0.18	0.31	1.09
Saudi Arabia	6.08	1.51	0.19	14.60	16.30	4.87	0.40	14.60	19.88	7.07	0.40	14.60	22.07	7.07	0.40	41.08	48.56
Somalia	11.50	0.02	0.00	0.79	0.81	0.14	0.00	0.79	0.93	0.23	0.00	0.79	1.01	0.23	0.00	2.21	2.44
Sudan	19.00	0.19	0.00	18.41	18.60	1.37	0.00	18.41	19.79	1.86	0.00	18.41	20.27	1.86	0.00	51.81	53.67
Syria	25.00	0.23	0.33	2.77	3.34	1.10	1.01	2.77	4.89	1.92	1.01	2.77	5.70	1.92	1.01	7.80	10.73
Tunisia	4.50	0.18	0.10	1.38	1.66	0.40	0.41	1.38	2.20	0.82	0.41	1.38	2.62	0.82	0.41	3.89	5.13
UAE	0.28	0.51	0.03	0.95	1.49	1.01	0.06	0.95	2.02	1.48	0.06	0.95	2.49	1.48	0.06	2.67	4.22
Yemen	5.00	0.14	0.07	2.50	2.72	1.10	0.27	2.50	3.87	2.12	0.27	2.50	4.88	2.12	0.27	7.04	9.42
TOTAL	281.81	9.82	8.27	149.61	167.69	30.73	19.53	149.61	199.86	46.15	19.53	149.61	215.28	46.15	19.53	420.98	486.65

TABLE 4 – PART TWO
Water Demand Forecasts for 2025 by Country and by Sector (continued)

Country	Resource base (bcm)	Total Projected Withdrawal 2025 (bcm)			Percent of Resource Base				Amount of Shortfall (bcm)				Population (in '000s)		Total Withdrawal per Capita (m³/yr/capita)			
		Case I	Case II	Case III	1990	2025 Case I	2025 Case II	2025 Case III	1990	2025 Case I	2025 Case II	2025 Case III	1990	2025	1990	2025 Case I	2025 Case II	2025 Case III
Algeria	19.00	4.9	6.7	10.0	15%	26%	35%	53%	0	0	0	0	24,960	51,830	114	94	128	193
Bahrain	0.10	0.3	0.3	0.5	216%	321%	321%	526%	0.1	0.2	0.2	0.4	503	1,014	429	317	317	519
Djibouti	0.30	0.0	0.0	0.0	3%	7%	9%	12%	0	0	0	0	440	1,159	23	18	24	32
Egypt	58.10	70.9	76.5	166.6	99%	122%	132%	287%	0	12.8	18.4	108.5	52,426	93,536	1095	758	818	1782
Iraq	81.00	45.6	45.6	117.0	53%	56%	56%	144%	0	0	0	36	18,080	46,260	2367	986	986	2530
Jordan/West Bank	1.40	0.8	1.0	1.5	32%	57%	69%	107%	0	0	0	0.1	4,009	10,807	112	74	90	139
Kuwait	0.00	0.5	0.6	0.8	*	*	*	*	0.4	0.5	0.6	0.8	2,143	2,789	179	173	223	275
Lebanon	7.80	0.8	0.9	2.1	10%	11%	12%	26%	0	0	0	0	2,740	4,476	274	188	201	459
Libya	3.83	5.7	6.2	14.0	124%	150%	162%	364%	0.9	1.9	2.4	10.1	4,545	12,873	1047	445	481	1084
Mauritania	7.40	1.0	1.2	2.3	10%	14%	16%	31%	0	0	0	0	2,024	4,993	361	207	240	463
Morocco	30.00	12.5	13.6	31.7	37%	42%	45%	106%	0	0	0	1.7	25,061	47,477	439	263	286	668
Oman	1.47	2.8	2.7	4.8	84%	189%	186%	328%	0	1.3	1.3	3.4	1,524	4,705	811	589	582	1025
Qatar	0.05	0.4	0.9	1.1	388%	847%	1788%	2184%	0.1	0.4	0.8	1	427	731	454	579	1223	1494
Saudi Arabia	6.08	19.9	22.1	48.6	268%	327%	363%	799%	10.2	13.8	16	42.5	14,870	40,426	1096	492	546	1201
Somalia	11.50	0.9	1.0	2.4	7%	8%	9%	21%	0	0	0	0	8,677	23,401	93	40	43	104
Sudan	19.00	19.8	20.3	53.7	98%	104%	107%	282%	0	0.8	1.3	34.7	25,203	60,602	738	327	335	886
Syria	25.00	4.9	5.7	10.7	13%	20%	23%	43%	0	0	0	0	12,355	35,250	270	139	162	304
Tunisia	4.50	2.2	2.6	5.1	37%	49%	58%	114%	0	0	0	0.6	8,057	13,425	206	164	195	382
UAE	0.28	2.0	2.5	4.2	532%	721%	890%	1505%	1.2	1.7	2.2	3.9	1,589	2,792	938	723	892	1510
Yemen	5.00	3.9	4.9	9.4	54%	77%	98%	188%	0	0	0	4.4	11,684	34,237	233	113	143	275
TOTAL	281.81	199.9	215.3	486.7					13.0	33.4	43.2	248.3	221,317	492,783				

* Resource base close to zero.

jumps to 486 bcm per year by 2025, almost twice the available resource base. This seems to be out of the question given the resource base and the costs of alternative supply options. However, many countries in the region are now planning to *increase* the amount of water going to irrigation in the future. Analyses provided by forecasting models like this one should cause them to reconsider a commitment to provide new irrigation. To an outside observer, the outcome of Case III says clearly that plans to provide more and more water to agriculture cannot be fulfilled, indeed, they can only fail very painfully.

The implications become more specific when individual countries are considered. Even though the total resource base is not exceeded region-wide in the first two forecast cases, some countries already exceed their own water resource bases. Table 4 shows that several countries are already using considerably more than their current annually renewable water supply (Bahrain, Kuwait, Libya, Qatar, Saudi Arabia, and the UAE) and had a combined deficit of 13 bcm per year in 1990. This is met by desalination, and by over-exploiting groundwater. These countries will be even more severely affected in the future, and more countries will experience new deficits in the future.

Even without additional water to agriculture our model predicts that Egypt, Oman, and the Sudan will join the deficit nations by 2025. When these individual shortfalls are added together, the total national shortfalls for the region increase from 13 bcm per year in 1990 to between 33 and 43 bcm per year by 2025, although there would be a small and uncomfortable surplus for the region as a whole in 2025. Unfortunately, these national shortfalls cannot be easily solved by transporting water between countries; some such transfers are potentially possible, but most are not feasible because of the long distances between the water surplus and water short areas. The shortfalls are also unlikely to be solved by reallocating water use in international river basins. The deficits will have to be met essentially as the six existing deficits are met by a combination of mining groundwater, desalination, and water conservation. One important form of conservation may indeed be the reallocation of water away from agriculture, perhaps actually reducing the quantities of water for irrigation below the 1990 levels.

A simple test of the reasonableness of the demand forecasting model is to examine the projections for urban water use. Table 5 gives the forecasts of the urban water use in 2025 under the scenarios of Cases I and II given above. Careful examination of Table 5 raises several questions. First, looking at the 1990 per capita urban water use for Oman (1321 liters per

T A B L E 5
Urban Water Use in 2025

Country	No Income Effect m3/ca/yr	l/c/d	g/ca/day	Income Effect m3/ca/yr	l/c/d	g/ca/day	Urban Population 1990	2025
Algeria	65.2	178.5	47.1	109.6	300.0	79.3	12,979,200	39,849,200
Bahrain	206.0	564.0	149.0	206.0	564.0	149.0	417,490	928,490
Djibouti	7.9	21.5	5.7	14.3	39.1	10.3	356,400	1,075,400
Egypt	125.8	344.5	91.0	210.9	577.5	152.6	24,640,220	65,750,220
Iraq	100.0	273.9	72.4	100.0	273.9	72.4	12,836,800	41,016,800
Jordan/West Bank	47.9	131.1	34.6	65.7	179.8	47.5	2,726,120	9,524,120
Kuwait	143.4	392.6	103.7	194.9	533.7	141.0	2,057,280	2,703,280
Lebanon	35.8	98.1	25.9	50.0	136.9	36.2	2,301,600	4,037,600
Libya	90.8	248.7	65.7	131.7	360.6	95.3	3,181,500	11,509,500
Mauritania	92.1	252.1	66.6	133.7	366.0	96.7	951,280	3,920,280
Morocco	32.9	90.0	23.8	59.0	161.5	42.7	20,299,410	42,715,410
Oman	483.2	1322.9	349.5	472.7	1294.2	341.9	167,640	3,348,640
Qatar	200.0	547.5	144.7	888.2	2431.8	642.5	380,030	684,030
Saudi Arabia	131.7	360.6	95.3	191.0	523.0	138.2	11,449,900	37,005,900
Somalia	7.8	21.3	5.6	12.7	34.9	9.2	3,123,720	17,847,720
Sudan	33.5	91.8	24.3	45.4	124.2	32.8	5,544,660	40,943,660
Syria	37.8	103.6	27.4	65.9	180.5	47.7	6,177,500	29,072,500
Tunisia	41.1	112.6	29.8	84.7	232.0	61.3	4,350,780	9,718,780
UAE	413.9	1133.2	299.4	607.3	1662.7	439.3	1,239,420	2,442,420
Yemen	42.5	116.4	30.7	81.7	223.5	59.1	3,388,360	25,941,360
MEAN	117.0	320.2	84.6	186.3	510.0	134.7		

capita per day) and UAE (1131 lpcd) the question of the reliability of the data arises. The 1990 per capita use in California is about 757 lpcd and we believe it unlikely that the these two countries would be so much larger than that figure. When these figures are projected, along with those of Qatar, unreasonably large demands are forecast. Nevertheless, these are all micro-states and the predicted total quantities are small in comparison to the other demands in the table and can essentially be ignored. By and large, the urban per capita demands predicted by our model seem plausible with the exception of Djibouti and Somalia, both of which seem to have much too small amounts of per capita urban demand. This may reflect suppressed demand conditioned by the present supply. We believe that the urban and industrial demands will lie somewhere between the results of Cases I and II. Case III does not look like a reasonable scenario. It does serve the purpose, however, of bracketing the potential upper limits of the demands.

Despite the crudeness of the present model and the temporal and spatial uncertainties inherent in the raw data and any projections based upon them, it should be plausible enough to serve as a useful point of departure for a discussion of policy implications.

Implications for Policy

Is there a water crisis in the Arab world?

To many observers it may seem strange to pose this question. To them the answer is resoundingly affirmative. The data marshaled in this paper, and in other papers in the symposium, however, raise serious questions about what is meant by a "water crisis" in the Arab world. A water crisis would imply a severe shortage of water for all reasonable uses leading to serious economic and social disruptions. A comparison with the "oil crisis" of the 1970s and early 1980s might help explain some of the important concepts involved. Even though oil is a non-renewable resource and water is generally a renewable resource, the similarities of the two so-called crises are instructive.

- First, as water is essential to life, so is energy essential to life in many cold parts of the world (for example, in Massachusetts one would simply freeze to death without adequate heating).

- Second, as with water now in the Arab world, before 1973 many people and industries were committed to using a particular form of energy and its attendant technologies for their livelihood.

- Third, the price of the resource to the consumers was kept artificially low by governments through price controls in the consuming countries.

- Fourth, it was deemed morally reprehensible and politically impossible to ration the resource by raising the price for such a fundamental life-sustaining resource.

- Fifth, there was as much talk about "energy independence" then, as there is of "food self-sufficiency" now in the Arab world.

The major objection often voiced to this comparison is that there is no substitute for water for sustaining life. While this is true, there is, however, an almost infinite supply of seawater which could be desalinated to produce fresh water, at a price, and provided there was enough energy available. Hence, the limit may not be fresh water, but energy, or capital to build the energy and desalinating systems. The resolution of the "energy crisis" has many lessons for the water policy-makers in the Arab world.

The single most important factor in the resolution of the "energy crisis" was the willingness of governments in the consuming countries to let the price of oil rise fairly rapidly. As a result of these price increases there were major adjustments by the consumers in the form of improved energy efficiency and substitution of other fuels. No Western governments fell because of rapid rise in energy prices, the concept of "energy independence" was fairly quickly abandoned by all except the United States with its strategic oil reserve, and the transition from a low-cost energy to a high-cost energy regime was relatively painless. On the supply side the existence of higher prices made oil exploration and the development of more expensive energy reserves economically attractive. This in turn expanded the resource base. Finally, as a result of the lessons learned dealing with the oil price shock the world is now better prepared to deal with similar rapid adjustments in other commodity markets.

The analogy to water in the Arab world should be fairly clear. Table 3 holds one of the keys. If water charges in the different Arab countries reflected both the real costs of providing the water and the opportunity costs associated with potential alternative uses of that water, then in many situations irrigating field crops such as rice and wheat would no longer be profitable. Large quantities of water would become available for other uses. New irrigation would be restricted to higher value crops such as fruit and vegetables. Table 4 demonstrates that this does not imply that the countries in this water situation would move out of agriculture altogether or even partially. Typically there is an order of magnitude in the differences of the quantities of water demanded by cities and industries and agriculture. In most countries, a reduction of 10 percent of the water currently going to agriculture would meet the increasing demands of cities and industries through the year 2025. In the worst case, Egypt, maybe as much as 25 percent would have to be diverted from current agricultural water supplies if there were no other unused sources of water.

Higher prices will lead to an expansion of the resource base by making more expensive supply options economically usable. Hence, adequate water can be supplied to all economically appropriate users. This is the opposite of a "crisis" situation, tempting one to say that the "water crisis," like the oil crisis, is a passing problem of misallocation rather than absolute shortage. This does not mean that the transitions involved will be painless. They will, however, be less painful than carrying forward into the next century massive sectoral misallocations of water. The recent economic collapse of the eastern, formerly communist, group of nations reveals the shortsightedness of such policies. It is also important to note that with careful planning the worst impacts of the reallocation of water upon economically disadvantaged groups can be mitigated. There is no reason for basic "lifeline services" for the poor to be incompatible with a major rationalization of the water sector.

Has a shortage of water been a serious drawback to development in the region?

It is often asserted that scarcity of water places a limit on economic development, but the data from the Arab world plotted in Figures 12 and 13 provide no corroboration for this, as was discussed above. The Arab countries are doing better economically than the much better watered sub-Saharan African countries, about as well as Pakistan, and less well than Turkey.

A more important question to ask is "Will a shortage of water be a drawback to the future development of the Arab countries?" The answer to this is less equivocal; under the pattern of economic development current in the Third World now the Arab countries will be left behind if they do not improve their infrastructure for industrial development. One important part of this infrastructure is the provision of adequate supplies of high quality water. Unless the policies regarding water allocation are changed to give industry the water it needs, then indeed the Arab countries are likely to be severely affected by water shortages in the future.

How important is resolving disputes over international rivers and aquifers?

As mentioned earlier there are at least 25 international rivers (see Table 1) and several important shared aquifers in the Arab region. While Figure 1 of the Summary indicates the international rivers and major aquifers, it does not include the many smaller aquifers underlying the border between Saudi Arabia and Jordan, or the borders of Syria, Iraq, Iran, and Turkey, and between Jordan and Israel. All internationally shared rivers have a potential for water disputes and those in the Arab world are no exceptions to this general rule. Conflicts over the use of the Jordan and the Tigris-Euphrates waters have tended to obscure real and developing conflicts in other basins such as the Nile and the Juba, and the potential competition among countries using groundwater from the Nubian aquifer. We know that resolving international disputes over water is not only a complex problem in law and diplomacy, but that applying developing legal principles and common economic and engineering approaches to these problems has major pitfalls (Rogers 1993).

Settling these international river issues would certainly be a wise and useful achievement. Nonetheless, if we consider together the nature of the conflicts and the magnitudes of the water deficits in this region, we see that even the most expeditious and equitable resolution of the conflicts would make little dent in reducing the scale of current deficits, nor the future deficits predicted by our model. It is these looming deficits that are, of course, the real problem. For example, resolution of international disputes in the Nile basin can hardly increase the water available to Egypt and the Sudan, and might actually reduce its amount. Similarly, resolving the Tigris-Euphrates disputes would affect Iraq and Syria, neither of which is predicted to be a severely affected country by Cases I and II in our demand forecasting model. All of the other water disputes, while they are significant to the parties, do not involve amounts of water large enough to resolve the deficits.

Agenda for the Next Thirty Years

The water agenda for the countries in the Arab world for the next thirty years should include the following areas of concern.

Studies and investigations

Agriculture

Farming is by far the largest water use in the Arab world and therefore merits the first and closest examination. Given the asymmetry in the quantities of water consumed, small percentage savings in agricultural water could make very large amounts of water available to the other sectors. Two areas of work that are most widely suggested are the improvement of irrigation efficiency and a shift to less water consuming crops. Large savings are usually promised for activities in these two areas, but Seckler (1993) and Abu-Zeid and Seckler (1992) raise some questions as to whether these savings are actually realizable. Using Egyptian data, Seckler argues that for systems in which there is irrigation drainage-water recapture, either by surface drains or by tubewells, the actual level of irrigation efficiency must consider all of the water that is recycled by the drainage capture works. He claims that for the Egyptian system the realized irrigation potential is substantially higher than what is recorded when only the field efficiencies are considered. Secondly, he marshals evidence showing that little can be saved by crop switching. Most of the water saved by the supposedly water-saving crops depends either upon growing them in a cooler season, or on the shorter time that the new crops are in the field in comparison with traditional, so-called high water-using crops.

Hence, there is a great need to carry out detailed agronomic and agroeconomic studies to elucidate for the individual countries in the region just how much water might be saved by these approaches. The issue is very important, and the need is to look at real field situations and not rely upon conceptual or desk studies.

Municipal and industrial water use

The issue of urban and industrial water use is also very important since the bulk of the water saved from agriculture will go to urban supply. Detailed studies need to be made to compare and contrast conventional and unconventional supply methods as shown in Figures 10 and 11. The issue of desalination needs to be handled carefully. Table 3 indicates that desalination as it is currently envisaged is probably too expensive to be considered as an alternative to most other types of water supply over the next few decades in most locations. Nevertheless, additional work needs to be done on site-specific cases to elucidate the real costs of desalination as an option. Conservation of water for urban and industrial uses appears to be much more economically attractive than desalination in most situations. The issue of concern here is to establish precisely how much water could be supplied by various levels of conservation, both within the urban system and also in the local watershed and aquifer environments.

Hydrological and geophysical explorations

Detailed geophysical investigations need to be carried out to establish the sources and amounts of recharge to fractured rock zones. This will entail substantial work on the ground and with geographical information systems involving remote sensing and ground truthing. There is a wealth of geographical data from the petroleum industry that needs to reviewed from a geo-hydrological point of view.

Economic evaluation of the benefits of cooperation on international rivers and aquifers

The detailed analysis of international river basins with an eye to identifying Pareto-efficient solutions is also strongly recommended. As mentioned above there is great merit in arriving at good political and economically efficient solutions for their own sake. These are very good confidence builders for further economic and social interactions.

Institutional development

Comprehensive water agencies

Institutions dealing with water need to be strengthened in each country. In most countries the control of water is fragmented between various agencies. Rarely does one ministry have responsibility for all aspects of ground and surface water, water supply and wastewater disposal. A first prerequisite for truly multi-disciplinary studies is the existence of institutions that will help foster and support such activities. Therefore, it is of paramount importance to establish genuine full-sector water agencies.

Agricultural water user institutions

Farmers' user-groups or irrigation cooperatives are commonly neglected. These should be in the non-governmental sector so that they can provide a counterweight in each country to the government department's own technical agencies. They should be flexible enough to arrange for water sharing within and between years, and able to manage water transfers between different user groups.

Regional and bilateral institutions

A series of multilateral or bilateral regional water agencies in the Arab world could be very useful. The existence of such agencies will legitimize the need for broad-scale thinking by water professionals. They would also provide the forums for the exchange of technical and economic approaches to solving the problems of the region. In addition they provide the venues in which water problems would receive some political prominence.

Developing private sector institutions

Finally, the private sector should be much freer and active than now. Several countries are now experimenting with various forms of private participation in water planning, management, and control. Much flexibility and improved economic efficiency can be achieved with the appropriate mix of public and private agencies. This may require changes in the legal, political, and banking systems.

Specific actions

Water pricing

Each country in the region should make a commitment to calculate the real costs of water to each user or user group. The subsidies should be clearly identified and the analysis be used over a period of time to develop tariff structures that more nearly reflect the true resource and opportunity costs of the water.

Ecosystem maintenance

Much of the literature, including this paper, discusses water abstraction from the environment and water use as though it were acceptable, and even desirable, to divert up to 100 percent of the water potential of a country or region for human use. This does not make sense from the point of view of the environment itself. If all its water is diverted, an aquatic ecosystem will be severely compromised. Significant quantities of water need to be left in flowing streams, lakes, reservoirs, and wetlands to maintain fauna and flora. Countries should join the Ramsar Convention to protect wetlands, and should work with their neighbors to maintain fish populations in the inland and coastal fisheries.

Biographical Notes

Mohamed J. Abdulrazzak, after completing his studies at the University of Arizona and Colorado State University, worked as a field hydrologist in the Ministry of Agriculture and Water in Riyadh, Saudi Arabia, before commencing a teaching career at King Abdulaziz University in Jeddah. He is now an Associate Professor, teaching a variety of classes in hydrology and water resource management. He has served as Vice Dean of Research and Postgraduate Studies, and is currently Chairman of the Department of Hydrology and Water Resources Management. Dr. Abdulrazzak has also served as a member of the Commission on Hydrology, and has been involved, as both principal and co-investigator, in several major research projects on desertification and various aspects of his country's water resources, including weather modification and environmental hydrology.

Jamil S.K. Al Alawi is Undersecretary for Electricity and Water in the Ministry of Works, Power and Water of the State of Bahrain. He studied mechanical engineering in the United Kingdom. He is Chairman of the Gulf States Interconnection Committee, President of the Water Science Technology Association for the Gulf States, and also President of the International Desalination Association.

Abdulaziz Naser Al-Saqabi received his engineering degree from the University of Colorado in 1980, and served as a site engineer for the municipality of Kuwait. He worked for a time in the private sector until joining the Kuwait Fund for Arab Economic Development, where he is Assistant Engineering Advisor of the Water and Sanitation sector.

J. A. Allan is Professor of Geography at the School of Oriental and African Studies of the University of London. He has studied and taught in various capacities in the Middle East, North Africa, India, and Europe, including work at the International Institute of Aerospace and Environmental Science in the Netherlands, and has advised the World Bank, the European Commission, FAO, and ODA. His research, which includes linking satellite imagery with geographical and hydrological information systems, has focused on the water resources of the Middle East and North and Northeast Africa, with stress on the shaping of water policy to take long-term economic and environmental imperatives into account.

Yahia Bakour, after studies in Alexandria, Egypt, received his Ph. D. in agricultural economics at the University of Meissen in Germany in 1973. He served both in the Ministry of Agriculture of the Arab Republic of Syria and the College of Agriculture of the University of Damascus, and has been twice elected President of the Syrian Agriculturist Union. He is Secretary General of the Arab Agriculturalist Union, being first elected in 1980, and is a trustee of the International Food Policy Research Institute (IFPRI) in Washington, D.C. Dr. Bakour's publications include two books on agricultural cooperatives, and many papers, including treatments of agricultural education and food security in the Arab region. Dr. Bakour is now Director General of the Arab Organization for Agricultural Development, headquartered in Khartoum.

Shawki M. Barghouti is Division Chief for Agricultural Operations at the World Bank in Washington. After taking his bachelor's degree from Cairo University, he received his doctorate in agriculture from the University of Wisconsin. Dr. Barghouti worked with the government of Jordan before joining the United Nations Food and Agriculture Organization in East Africa. He also worked with the Ford Foundation and the International Center for Agricultural Research in the Dry Areas (ICARDA) before joining the World Bank in 1980. At the Bank, Dr. Barghouti has participated in organizing the June 1991 international workshop on water resources, and in the writing and editing of several World Bank books on irrigation and water management.

Fakhri A. Bazzaz is the H. H. Timken Professor of Science at Harvard University. After studies at the University of Baghdad and taking his doctorate at the University of Illinois, Dr. Bazzaz became head of the Department of Plant Biology and Acting Director of the School of Life Sciences at the University of Illinois. He came to Harvard in 1984. He has published about two hundred scientific papers. A life member of Clare Hall of Cambridge University, England, and a Guggenheim Fellow in 1989, he became a fellow of the American Academy of Arts and Sciences in 1990, and currently chairs the Academy's Biology and Medicine Election Committee.

Taysir Dabbagh is Senior Engineering Advisor to the Kuwait Fund for Economic Development. After taking his engineering degree from Hammersmith College, London, he did post-graduate work at Dundee University, Scotland. Before joining the Kuwait Fund in 1978, Mr. Dabbagh worked for British Consulting Engineering, principally on water supply and sanitation projects.

Ali Jebali is a Director General in the Ministry of Agriculture of Tunisia. He completed civil engineering studies in Paris in 1972, and served as an engineer for the Tunisian Ministry of Agriculture. He directed Agricultural Water Management in the Rural Engineering department, before being appointed General Manager of Hydraulic Studies and Water Networks in the same ministry.

Mohammed Jellali is Director of the Hydraulics Administration of Morocco. After earning a degree in geology at the Ecole Nationale Supérieure de Géologie Appliquée at Nancy, France, he qualified as an engineer-hydrologist through studies at Orstom in Paris. Before taking over the direction of Morocco's Hydraulics Adminstration, he was Director of Water Research and Planning at the Ministry of Public Works, Vocational Training, and Managerial and Staff Training, in Rabat.

John Kolars is Professor Emeritus of Geography and Near Eastern Studies at the University of Michigan. He did his doctoral studies at the University of Chicago, and is the author of *Tradition, Season and Change in a Turkish Village*, *The Euphrates River and the Southeast Anatolia Development Project* (with William Mitchell), *The Litani River, Interrupted Development* (forthcoming), and numerous articles on the uses of international rivers in the Middle East.

Peter Lydon did his bachelor's and master's studies at the University of Toronto and Yale, and did further work in political science at MIT. After service with the U.S. State Department, he lives in Berkeley, California, where he is a visiting scholar at the Institute of Governmental Studies of the University of California. He is a co-author, with Peter Rogers and David Seckler, of the 1989 Eastern Waters Study on the water resources of the Ganges-Brahmaputra basin, and since then has worked on water issues in Bangladesh, and on California city planning subjects.

Yahia Abdel Mageed graduated from the Engineering University of Khartoum in 1951, and went on to studies in Hydrology at Imperial College, London. He is a former Minister of Irrigation of Sudan, Chairman of the Joint Technical Commission on Nile Water, and Secretary General of the United Nations Water Conference. Mr. Mageed, whose experience in irrigation and water resources development and management spans more than 50 years, has served on a number of expert groups in the UN system, including currently the Scientific and Technical Advisory panel of the Global Environmental Facility. He has written on water resources in the arid and semi-arid zones of Africa and on integrated river basin development and management, particularly of the Nile basin. He is now Chairman of Yahia Abdel Mageed and Partners, Consultants and Technologists, Khartoum.

Thomas Naff is Professor of Asian and Middle Eastern Studies at the University of Pennsylvania. After receiving advanced degrees at the School of Oriental and African Studies of the University of London and at the University of California at Berkeley, Dr. Naff taught at the American University at Cairo, where he was the co-founder of the Center for Arabic Studies Abroad (CASA). He then became the Director of the Middle East Center at the University of Pennsylvania, and later founder and director of the Penn Middle East Research Institute. He has taught, published, and lectured on a wide range of Middle Eastern subjects, covering the period from the advent of Islam to current projects on water issues in the Middle East.

Peter Rogers is the Gordon McKay Professor of Environmental Engineering and City Planning at Harvard University. After studies at the University of Liverpool and Northwestern, he took his doctorate at Harvard in 1966. He is interested in environmental and resource planning. He has developed formal models to treat resource issues that cross national boundaries, to incorporate the effects of energy and water technologies into macro-economic judgments, and to analyze the financing and planning of urban infrastructure. He has advised on Bangladesh's flood vulnerability and the crisis of the Aral Sea, and is the author of *America's Water*, an examination of U.S. federal water policy.

Mohamed Sadeqi is an Engineering Advisor at the Kuwait Fund for Arab Economic Development. After studies at Oakland University and Wayne State University, he took his doctorate at the University of Colorado. He has worked as a power plant engineer in Kuwait, and as a scientist at the Kuwait Institute for Scientific Research. His research experience is in energy generation and management, water production, and environmental protection.

Abdul Karim Sadik is Advisor to Executive Director of the World Bank in Washington. After studies at the American University of Beirut, he did his doctorate in Economics and Business Administration at the University of Florida. After teaching at the University of Florida, Dr. Sadik served from 1974 to 1991 as Economic Advisor to the Kuwait Fund for Arab Economic Development. He was also Executive Director for Kuwait of the International Fund for Agricultural Development (IFAD) from 1983 to 1991.

Peter Gething Sadler is Professor of Economics at the Cardiff Business School of the University of Wales. His special interests are regional economics and cost benefit analysis, and the economics of the Middle East, especially Kuwait. He has served as the Director of the Institute for the Study of Sparsely Populated Areas at the University of Aberdeen and of the Institute of Economic Research at the University College of Wales, and has consulted with UNCTAD and other organizations on Middle East economic affairs, water transportation, and environmental protection.

John Waterbury is Professor of Politics and International Affairs at Princeton University, and Director of the Center of International Studies there. Since completing his doctoral work at Columbia University, Dr. Waterbury has written *Hydropolitics of the Nile Valley*, and *The Egypt of Nasser and Sadat: the Political Economy of Two Regimes*. He is the co-author, with Alan Richards, of *A Political Economy of the Middle East: State Class and Economic Development*, and *Exposed to Innumerable Delusions: Public Enterprise and State Power in Egypt, India, Mexico and Turkey*.

References

Abate, Zewdie. 1992. *Observed context of environment and development in Ethiopia: An approach that needs development.* Occasional Paper No. 7. London: School of African and Oriental Studies, Centre of Middle Eastern Studies.

Abate, Zewdie. 1993. *Water resources of Ethiopia.* Reading, England: Garnet Press.

Abd al-Hay, 'Abd al-Tawab. 1988. *The Nile and the future.* Cairo: Ahram Center for Distribution and Translation. (In Arabic.)

Abdelrahman, H. 1992. "Irrigation systems in Oman." Proceedings of First Gulf Water Conference. Oct. 10–14, Dubai.

Abdusalaam, Abdelrahim S. 1978. *The Libyan oasis of Jalo: A geographical perspective.* Master's thesis in Geography, University of Michigan, Ann Arbor, Michigan.

Aboukhaled, A., A. Arar, A. Balba, B.G. Bishay, L.T. Khadry, P.E. Rijtema, and A. Taher. 1975. *Research on crop water use, salt-affected soils and drainage in Egypt.* Cairo: FAO Near East Regional Office.

Abu Rizaiza, Omar S., and Mohamed N. Allam. 1989. "Water requirements versus water availability in Saudi Arabia," *Journal of Water Resources Planning and Management,* American Society of Civil Engineers (ASCE), Vol. 115, No. 1 (January), pp. 64–74.

Abu Taleb, Maher F., Jonathan P. Deason, Elias Salameh, and B. Kefaya. 1992. "Water resources planning and development in Jordan: Problems, future scenarios, and recommendations." In Le Moigne, G., S. Barghouti, G. Feder, L. Garbus, and M. Xie, eds. *Country Experiences with Water Resources Management: Economic, Technical, and Environmental Issues.* World Bank Technical Paper No. 175. Washington, D.C.: The World Bank.

Abu Zeid, Mahmoud A., and M.A. Rady. 1992. "Water resources management and policies in Egypt." In Le Moigne, G., S. Barghouti, G. Feder, L. Garbus, and M. Xie, eds. *Country Experiences with Water Resources Management: Economic, Technical, and Environmental Issues.* World Bank Technical Paper No. 175. Washington, D.C.: The World Bank.

Abu Zeid, Mahmoud A., and Safwat 'Abdel-Dayem. 1990. "The Nile, the Aswan High Dam and the 1979–1988 drought." International Commission on Irrigation and Drainage. 14th Congress, Rio de Janeiro.

Abu-Zeid, Mahmoud, and David Seckler. 1992. "Round table on Egyptian water policy." Proceedings of a Seminar on Egyptian Water Policy, April 11–13. Cairo: Water Research Center, Ministry of Public Works and Water Resources.

Adams, Robert McC. 1965. *Land behind Baghdad—A history of settlement on the Diyala plains*. Chicago: The University of Chicago Press.

Ahmad, M.U. 1975. "The development of water resources of the Libyan Sahara." Proceedings of the Second World Congress of the International Water Resources Association, New Delhi. Vol. 3, pp. 1–10. Urbana, Ill.: International Water Resources Association.

Akkad, A. 1990. "Conservation in the Arabian Gulf countries." *Journal of Management Operations.* American Water Works Association. Vol. 82, pp. 40–50.

Al-Asam, S.M. 1992. "Modern irrigation systems and their role in water conservation in the United Arab Emirates." Proceedings of First Gulf Water Conference. Oct. 10–14, Dubai.

Al-Fusail, A.M., B.M. Al-Selwi, G.A. Said, and A. Badr. 1991. "Water resources and population distribution in the Republic of Yemen." Proceedings of First National Population Policy Conference. Oct. 20–29, Sanaa, Yemen.

Al-Gazi, A. 1990. "Migration movement between rural and urban centers in the Arab World." *Journal of Security.* Vol. 6, pp. 87–120. (In Arabic.)

Al-Ibrahim, A.A. 1990. "Water use in Saudi Arabia, problems and policy implementation." *Journal of Water Resource Planning and Management.* ASCE, Vol. 116, No. 3, pp. 375–388.

Al-Kenson, D.J. and Aykliston. 1986. "Water resources and water utilization in Qatar." In Proceedings of the Symposium on Water Resources and Utilization in the Arab World. February 17–20, Kuwait. (In Arabic.)

Al-Khairu, 'Iz al-Din 'Ali. 1975. *The Euphrates in light of public international law principles.* Cairo: Dar al-Kutub. (In Arabic.)

Al-Mahmoud, M.A. 1992. "Water resources development in Qatar." In Proceedings of the First Gulf Water Conference. Oct. 10–14, Dubai.

Al-Mansour, K. and M. Al-Arady. 1986. "Water resources and utilization in Bahrain." In Proceedings of the Symposium on Water Resources and Utilization in the Arab World. February 17–20, Kuwait. (In Arabic.)

Al-Mawardi, A.b H. 1983. *Al-akham al-sultaniyya.* Cairo: Dar Ashababli Tibaa.

Al-Mugran, A. 1992. "Strategy of water resources development in the GCC Countries." Proceedings of the First Gulf Water Conference. Oct. 10–14, Dubai.

Al-Nabban, M. F. 1970. *Al-itijah al-jama'i fi'l tashn'i al-iqtisadi al-Islami.* Cairo: Dar al-Fikr. (In Arabic.)

Al-Rahbi, 'A. 1973. *Fiqh al-muluk wa mitlah al-ritaj al-mursad.* Baghdad: Al-Irshad Press.

Al-Shaybani, M. 1958–1960. *Sharh kitab al-siyar al-kabir*, Salah al-Din al-Munajjid, ed., 3 vols. Cairo: Matba'a Masriya.

Al-Sufy, A.M. 1992. "Development of water resources suitable for consumption." Proceedings of the First Gulf Water Conference. Oct. 10–14, Dubai.

Al-Tokhais, A.S. 1992. *Groundwater management strategies for Saudi Arabia.* Ph.D. Dissertation, Colorado State University, Ft. Collins, Colorado.

Al-Weshah, R. 1992. "Jordan's water resources: Technical perspective." *Water International,* Vol. 17, pp. 124–132.

Alheritière, Dominique. 1987. "Settlement of public international disputes on shared resources: Elements of a comparative study of international instruments." In Utton, Albert and Ludwik Teclaff, eds. *Transboundary Resource Law,* pp. 139–49. Boulder, Colo.: Westview Press.

Allan, J.A. 1981. *Libya: the experience of oil.* London: Croom Helm.

Allan, J.A. 1983. "Natural resources as national fantasies." *Geoforum,* pp. 243–247.

Allan, J.A. 1988. "Libya's Great Water Carrier: Progress and prospects of the Great Man-Made River." *Libyan Studies,* Vol. 19, pp. 141–146.

Allan, J.A. 1989a. "Water resources evaluation and development in Libya, 1969–1989." *Libyan Studies,* Vol. 20, pp. 235–242.

Allan, J.A. 1989b. "Natural resources: not so natural for ease of development." In Allan, J.A., K.S. McLachlan, M.M. Buru. *State and Region in Libya,* London: School of Oriental and African Studies, Centre of Near and Middle Eastern Studies.

Allan, J.A. 1993. "Fortunately there are substitutes for water; otherwise our hydropolitical futures would be impossible." In *Priorities for Water Resources Allocation and Management.* London: Office of Development Assistance.

American Society of International Law. 1977. *Legal questions arising out of the construction of a dam at Maqarin on the Yarmuk River.* (Prepared for USAID.) Washington, D.C.: United States Agency for International Development.

Anderson, E. 1993. Personal communication to J.A. Allan.

Arab Fund for Economic and Social Development, Arab Monetary Fund, and Organization for Arab Petroleum Exporting Countries. 1992. *Unified Arab Economic Report.* Kuwait: Arab Fund for Economic and Social Development. (In Arabic.)

Arab League of States. 1982. *Economic Development.* Annual Statistical Book No. 6. Cairo: Arab League of States.

Arab League of States. 1993. *Report.* Cairo: Arab League of States.

Arab Organization for Agricultural Development (AOAD). 1986 and 1993. *Agriculture and Development,* Periodical. Khartoum: AOAD.

Arab Organization For Agricultural Development (AOAD). 1986. *Arab Food Security Programme, Vol. II — Natural Resources.* (In Arabic.) Khartoum: AOAD.

Arberry, A.J. 1980. *The Koran Interpreted.* Vol. 2. London: Allen and Unwin.

Asaad, Shawki, Nabil Rufaeel. 1986. "Rational water use in the Arab region." (ACSAD) In Proceedings of the Water Resources Seminar. February 17–20, Kuwait.

Auda, 'Abd al-Malik. 1993. *Egypt, Ethiopia and the Nile.* p. 82. Cairo: al-Ahram al-Iqtisadi, 5/31.

Authman, M.N. 1983. *Water and development processes in Saudi Arabia.* Jeddah, Saudi Arabia: Tihama Press.

BAAC (British Arabian Advisory Company). 1980. *Water resources of Saudi Arabia, Vol. 1*. Prepared for the Ministry of Agriculture and Water of Saudi Arabia. Riyadh.

Bakour, Yahia. 1991. *Planning and management of water resources in Syria*. Damascus: The League of Arab States, Arab Organization for Agricultural Development (AOAD), Regional Office in Damascus.

Bakour, Yahia. 1992. "Planning and Management of Water Resources in Syria." In Le Moigne, G., S. Barghouti, G. Feder, L. Garbus, and M. Xie, eds. *Country Experiences with Water Resources Management: Economic, Technical, and Environmental Issues*. World Bank Technical Paper No. 175. Washington, D.C.: The World Bank.

Barberis, J. 1986. *International Groundwater Resources Law*. Rome: Food and Agriculture Organization.

Barghouti, S. 1990. "Agricultural diversification: Implications for crop production, the case in the Middle East and North Africa region." *Horticultural Science*, Vol. 25, No. 12.

Barghouti, S. 1992. "Agricultural diversification and the role of irrigation." In Le Moigne, G., S. Barghouti, and L. Garbus, eds. *Developing and Improving Irrigation and Drainage Systems*. World Bank Technical Paper No. 178. Washington, D.C.: The World Bank.

Bari, Zohurul. 1977. "Syrian-Iraqi dispute over the Euphrates waters." *International Studies*, Vol. 16, No. 2, pp. 227–44.

Bazzaz, F.A. 1990. "Plant-plant interaction in successional environments." In J.B. Grace and G.D. Tilman, eds., *Perspectives on Plant Competition*. San Diego, Calif.: Academic Press.

Bazzaz, F.A. and E.D. Fajer. 1992. "Plant life in a CO_2-rich world." *Scientific American*, Vol. 266, pp. 68–74.

Bazzaz, F.A., S. Miao, and P.M. Wayne. 1994. "Growth enhancement of tree species grown in elevated CO_2 atmospheres decline at different rates." *Oecologia*.

Beaumont, Peter, Gerald H. Blake, and J. Malcolm Wagstaff. 1988. *The Middle East — A Geographical Study*, 2d ed. New York: Halsted Press.

Beaumont, Peter. 1989. *Environmental Management and Development in Drylands*. London and New York: Routledge.

Ben-Adam, Y. 1967. *Kitab al-kharaj*. Ed. and tr., A. Ben Shemesh. Leiden: E.J. Brill.

Berkoff, Jeremy. 1994. "Jordan: Issues in water pricing." Draft paper for the World Bank. Washington, D.C., The World Bank.

Beschorner, Natasha. 1992. "Le rôle de l'eau dans la politique régionale de la Turquie." In Monde Arabe: Maghreb-Machrek, La Question de l'eau au Moyen Orient, *Discours et réalités*, No. 138, Oct–Dec.

Beschorner, Natasha. 1992. *Water and instability in the Middle East*. ADELPHI Paper No. 273, London: Brassey's.

Bilder, R. 1982. "Some limitations of adjudication as an international dispute settlement technique." *Virginia Journal of International Law*, Vol. 1.

Biswas, Asit K. 1982. "Shared natural resources: Source of conflict or springs of peace?" *Development Forum*, Vol. 13.

Biswas, Asit K. 1992. "Indus water treaty: The negotiating process." *Water International*, Vol. 17, No. 4.

Biswas, Asit K., A.K. Samaha, M.H. Amer, and M. Abu Zeid. 1980. *Water management for arid lands in developing countries.*

Bos, E., My T. Vu, A. Levin, and R.A. Bulatao. 1992. *World Population Projections 1992–93 Edition.* A World Bank Publication. Baltimore: The Johns Hopkins University Press.

Bourne, C.B. 1992. "The International Law Commission's draft articles on the law of international watercourses: Principles and planned measures." *Colorado Journal of International Environmental Law and Policy*, Vol. 3, No. 65.

Boutayeb, N. 1985. "Planification et Gestion de l'Eau au Maroc". November 21–24, Conference Proceedings. Agadir (Morocco).

Braidwood, Robert J. and Gordon R. Wiley. 1962. *Courses Toward Urban Life.* Chicago: Aldine Publishing Company.

Bromley, David and Michael Cernea. 1989. *The management of common property natural resources: some conceptual and operational fallacies.* World Bank Discussion Paper No. 57, Washington, D.C.: The World Bank.

Brown, L.R. 1976. "Man, food, and environmental inter-relationships." In *Nutrition and Development*, ed. N.S. Scrimshaw and M. Behar.

Burdon, D.J. 1973. "Hydrogeological conditions in the Middle East." *Journal of Engineering Geology.* Vol 15, pp. 71–82.

Bushnak, A.A. 1990. "Water supply challenge in the Gulf Region." *Journal of Desalination*, Vol. 78, pp. 133–145.

Butzer, Karl W. 1964. *Environment and Archaeology.* Chicago: Aldine Publishing Company.

Caponera, Dante A. 1954. *Water laws in Moslem countries.* FAO Development Paper No. 43, Agriculture. Rome: FAO.

Caponera, Dante A. 1992. *Principles of water law and administration, national and international.* Rotterdam: A.A. Balkami.

Cess, R.D. et al. 1993. "Uncertainties in carbon dioxide radiative forcing in atmospheric general circulation models." *Science.* Vol. 262, pp. 1252–1254.

Chitale, M.A. 1992. "Water resources management in India: Achievements and perspectives." In Le Moigne, G., S. Barghouti, G. Feder, L. Garbus, and M. Xie, eds. *Country Experiences with Water Resources Management: Economic, Institutional, Technological and Environmental Issues.* World Bank Technical Paper No. 175. Washington, D.C: The World Bank.

Choucri, Nazli. 1991. "Resource scarcity and national security in the Middle East." In *New Perspectives for a Changing World Order*, ed. Eric H. Arnett. American Association for the Advancement of Science (AAAS) Publication no. 91-40s. Washington, D.C.: American Association for the Advancement of Science.

Collins, Robert. 1990. *The waters of the Nile, hydropolitics and the Jonglei Canal 1900–1988.* Oxford and New York: Oxford University Press.

Conway, D. 1993. *Climate change and water resources in the Nile Basin.* Workshop Paper at the School of Oriental and African Studies, University of London.

Coser, L.A. 1956. *The Functions of Social Conflict.* Glencoe, Ill.: Free Press.

Cotgrove, S. 1982. *Catastrophe or cornucopia: The environment, politics, and the future.* New York: Wiley.

Country Report. 1986. "Water resources and utilization in the United Arab Emirates." In Proceedings of the Symposium on Water Resources and Utilization in the Arab World. February 17–20, Kuwait. (In Arabic.)

Country Report. 1986. "Water resources and utilization in Kuwait." In Proceedings of the Symposium on Water Resources and Utilization in the Arab World. February 17–20, Kuwait.

Cranston, M. 1967. *Freedom.* 3rd ed. London: Longmans.

Dabbagh, Taysir A., and A. Al-Saqabi. 1989. "The Increasing Demand for Desalination." Proceedings of the Fourth Congress on Desalination and Water Re-use. Vol. I., pp. 3–26. The International Desalination Association. Amsterdam: Elsevier Science Publishers.

Dabbagh, Taysir A., and A. Al-Saqabi. 1991. "The exorbitant cost of water in Africa: Contributory factors." Proceedings of the All Africa Rural Water Supply and Sanitation Workshop (of the IBRD/World Bank). Washington D.C.: The World Bank.

Danish, S., A. Khater, and M. Al-Ansari. 1992. "Options in water reuse in Bahrain." Proceedings of the First Gulf Water Conference. Oct. 10–14, Dubai.

Darghouth, M. 1992. "Institutional aspects of irrigation development in North Africa: Experience with Morocco." In Le Moigne, G., S. Barghouti, and L. Garbus, eds. *Developing and Improving Irrigation and Drainage Systems.* World Bank Technical Paper No. 178. Washington, D.C.: The World Bank.

David, Peter, "Tribes with flags." *The Economist.* Survey of the Arab East. Feb 6, 1988. London: The Economist.

Davies, J. 1992. Personal communication to J.A. Allan.

De Jong, R. 1989. "Water resources of the Gulf Cooperation Council (GCC) countries — International aspects," American Society of Civil Engineers (ASCE) *Journal of Water Resources Planning and Management.* Vol. 115, No. 4, pp. 503–510.

De Mare, Lennart. 1976. *Resources-needs-problems: An assessment of the world water situation by 2000.* Report No. 2. Lund, Sweden: Department of Water Resources Engineering, Lund Institute of Technology, University of Lund.

Dellapenna, J. 1993. "Building international water management institutions: The role of treaties and other legal arrangements." Unpublished paper.

Deudeny, D. "Environment and security: Muddled thinking." *Bulletin of the Atomic Scientists.* Vol. 47, No. 3. April 1991.

Dougherty, J.E., and R.L. Pfaltzgraff. 1971. *Contending Theories of International Relations.* New York: Lippincott.

Dracup J.A. and J. Glater. 1991. "Desalination: The need for academic research." Submission to the U.S. House of Representatives Committee on Science, Space, and Technology, Subcommittee on Science. July 17, 1991. Washington, D.C.: U.S. Congress.

Edgell, H.J. 1987. "Geological framework of Saudi Arabia — groundwater resources." *KAU Journal of Earth Science.* Vol. 3, pp. 267–285.

Edmunds, W. M., and E. P. Wright. 1979. "Groundwater recharge and palaeoclimate in the Sirte and Kufra Basins, Libya." *Journal of Hydrology,* Vol. 40, pp. 215–241.

Egypt, Government of. 1959. "Agreement on full utilization of Nile waters between Egypt and Sudan, 8 November 1959." *Revue Egyptien De Droit International,* Text 15, pp. 321–329. Cairo.

Ehleringer, J.R. and C.B. Fields, eds. 1993. *Scaling Physiological Processes.* New York: Academic Press.

El-Zawahry, A.E., and A.A. Ibrahim. 1992. "Management of irrigation water in Oman." Proceedings of the First Gulf Water Conference. Oct. 10–14, Dubai.

Elahi, A. 1992. "Irrigation development in Sub-Saharan Africa: Future perspectives." In Le Moigne, G., S. Barghouti, and L. Garbus, eds. *Developing and Improving Irrigation and Drainage Systems.* World Bank Technical Paper No. 178. Washington, D.C.: The World Bank.

Encyclopedia of Islam, New Edition. 1991. C.E. Bosworth et al, eds. Vol. 6. "Ma'." Leiden, Holland: E.J. Brill.

Fadda, Nasrat R., and Eugene Perrier. 1989. *Water, land use and development.* Aleppo: ICARDA (International Center for Agricultural Research in the Dry Areas).

Fajer, E.D. and F.A. Bazzaz. 1992. "Is carbon dioxide a 'good' greenhouse gas? Effects of increasing carbon dioxide on ecological systems." *Global Environmental Change: Human and Policy Dimensions.* Vol. 2, pp. 301–310.

Falkenmark, Malin. 1989. *Natural resource limits to population growth: the water perspective.* Gland, Switzerland: International Union for the Conservation of Nature.

Falkenmark, Malin. 1989b. "The massive water scarcity now facing Africa — Why isn't it being addressed?" *Ambio,* Vol. 18, No. 2.

FAO. 1979. "Shared water resources in the Gulf States." *United Nations,* Vol. 1, pp. 310 ff. New York: United Nations, and Rome: United Nations Food and Agriculture Organization (FAO).

FAO. 1980–1992. *Production Yearbooks.* Rome: United Nations Food and Agriculture Organization (FAO).

Fox, I.K. and D.G. Le Marquand. 1978. "International river basin cooperation — The lessons from experience." *Water Supply and Management,* Vol. 2, No. 5.

Frederick, K. 1993. *Balancing Water Demands with Supplies: The Role of Management in a World of Increasing Scarcity*. World Bank Technical Paper No. 189. Washington, D.C: The World Bank.

Frederiksen, H.D. 1993. *Water Resources Institutions: Some Principles and Practices*. World Bank Technical Paper No. 191. Washington, D.C.: The World Bank.

Frey, Frederick, and Thomas Naff. 1985. "Water: An emerging issue in the Middle East?" *Annals*. American Association of Political Science. Vol. 431. Special issue, November 1985.

Frey, Frederick. 1992 "The Political context of conflict and cooperation over international river basins." Conference on Middle East Water Crisis. May 7–9, Waterloo, Ontario.

Frey, Frederick. 1993 "The Political Context of Conflict and Cooperation over International River Basins." *Water International*, Vol. 18, No. 1.

GAP. 1990. *GAP: the South-East Anatolian Project Master Plan Study, Final Master Plan Report*. Ankara: Yüksei Proje AS.

Garretson, Albert Henry, R.D. Hayton, and C.J. Olmstead, eds. 1967. *The Law of International Drainage Basins*. Published for the Institute of International Law, New York University School of Law. Dobbs Ferry, New York: Oceana Publishers.

Gischler, Christiaan E. 1979. *Water Resources in the Arab Middle East and North Africa*. Cambridge, England: Middle Eastern and North African Studies Press Ltd.

Gleick, Peter. 1990. "Environment, resources, and international security and politics." In *Science and International Security: Responding to a Changing World*, ed. E.H. Arnett. Washington, D.C.: American Association for the Advancement of Science.

Gleick, Peter. 1992 "Water and Conflict." *Occasional Paper Series of the Project on Environmental Change and Acute Conflict*, No.1, September, pp. 3–28. Joint Project of the University of Toronto and the American Academy of Arts and Sciences (AAAS). Cambridge, Mass.: AAAS.

Godana, Bonaya Adhi. 1985. *Africa's shared water resources: Legal and institutional aspects of the Nile, Niger and Senegal river systems*. Boulder, Colo.: Lynne Rienner Pub.

Goldstone, J. 1991. *Revolution and rebellion in the early modern world*. Berkeley and Los Angeles: University of California Press.

Gomez, C. 1987. "Experiences in predicting willingness-to-pay on water projects in Latin America." In F.W. Montinari et al., eds. *Resource Mobilization for Drinking Water and Sanitation in Developing Countries*. New York: American Society of Civil Engineers (ASCE).

Gordon, H.B., P. Whetton, A.B. Pittock, A. Fowler, M. Haylock, and K. Hennessy. 1991. *Simulated changes in daily rainfall intensity due to the enhanced greenhouse effect: Implications for extreme rainfall events*. Unpublished manuscript.

Gruen, George E. 1993. "Recent negotiations over the waters of the Euphrates and the Tigris." In Proceedings of the International Symposium on Water Resources in the Middle East: Policy and Institutional Aspects, A Pre-Symposium Proceedings. Urbana, Ill.: International Water Association and the University of Illinois.

Guariso, Giorgio and Dale Whittington. 1987. "Implications of Ethiopian water development for Egypt and the Sudan." *Water Resources Development.* Vol. 3, No. 2, pp. 105–14.

Gurr, Ted R. 1985. "On the Political Consequences of Scarcity and Economic Decline." *International Studies Quarterly.* Vol. 29, No. 1, pp. 51–75.

Haarsma, R.J., J.F.B. Mitchell, C.A. Senior. 1992. "Tropical Disturbances in a Global Circulation Model." *Climate Dynamics.*

Hannoyer, Jean. 1985. "Grands projets hydrauliques en Syrie." *Maghreb-Machrek,* No. 109, pp. 24–42.

Hansen, J., I. Fung, A. Lacis, D. Rind, S. Lebedeff, R. Ruedy, and G. Russell. 1988. "Global climate changes as forecast by GISS's three-dimensional model." *Journal of Geophys. Res.* Vol. 93, pp. 9341–9364.

Hardin, Russell. 1982. *Collective Action.* Baltimore: Johns Hopkins University Press.

Hayton, R. D., and A.E. Utton. 1989. "Transboundary groundwaters: The Bellagio Draft Treaty, revised and augmented by R. D. Hayton and A. E. Utton." *Natural Resources Journal.* Summer 1989, pp. 663–722.

Hillel, D.I. 1982. *The Negev: Land, water and life in a desert environment.* London: Praeger.

Homer-Dixon, T., 1990. *Environmental change and violent conflict: Understanding the causal links.* Emerging Issues, Occasional Papers Series. Washington, D.C.: American Academy of Arts and Sciences, Committee on International Security Studies.

Homer-Dixon, T., 1991. "On the threshold: Environmental changes as causes of acute conflict." *International Security,* Vol. 16, No. 2.

Houghton, J.T., B.A. Callander, and S.K. Varney, eds. 1992. *Climate change: The supplementary report to the IPCC scientific assessment.* Cambridge, England: Cambridge University Press.

Houghton, J.T., G.J. Jenkins, and J.J. Ephraums, eds. 1990. *Climate change: The IPCC scientific assessment.* Cambridge, England: Cambridge University Press.

Howell, P.P., M. Lock, and S. Cobb. 1988. *The Jonglei Canal: Impact and opportunity,* Cambridge, England: Cambridge University Press.

Human Development Report 1992. Published for the United Nations Development Programme (UNDP) by the Oxford University Press, Oxford, England,: Oxford University Press.

Institute for Biospheric Research. 1992. Videotape: *The Greening of the Planet Earth.* Phoenix, Arizona: Institute for Biospheric Research. Copyright held by Western Fuels Association, Washington, D.C.

International Atomic Energy Agency (IAEA). 1992. *Report: Technical and economic evaluation of potable water production through desalination of sea water.* Vienna: IAEA.

International Law Association (ILA). 1967. *Report of the Fifty-Second Conference — Helsinki.* London: International Law Association.

International Law Commission (ILC). 1979. *First Report on the Law of the Non-Navigational Uses of International Water Courses.* Submitted to the UN General Assembly, GE 79-61671. New York: United Nations.

International Law Commission (ILC). 1991. *Report of the ILC on the Work of its Forty Third Session.* New York: UN General Assembly, Supplement No. 10 (A/46/10).

Kally, Elisha and Abraham Tal. 1989. "A Middle East water plan under peace." In Meir Mehav, ed.: *Economic Cooperation and Middle East Peace.* London: Weidenfeld & Nicolson.

Keenan, J.D. 1992. "Technological Aspects of Water Resources Management: Euphrates and Jordan." In Le Moigne, G., S. Barghouti, G. Feder, L. Garbus, and M. Xie, eds. *Country Experiences with Water Resources Management: Economic, Institutional, Technological, and Environmental Issues.* World Bank Technical Paper No. 175. Washington, D.C.: The World Bank.

Khadam, M., N. Shammas, and Y. Al-Feraiheedi. 1991. "Water losses from municipal utilities and their impacts." *Water International.* Vol. 16, pp. 254–261.

Khazen, A., and A. Abid. 1991. "Gestion des ressources en eau de surface pendant la sécheresse des années 1987–88–89." Tunis: Ministry of Agriculture of Tunisia.

Khouri, J., W.R. Agha, and A. Al-Deroubi. 1986. "Water resources in the Arab world and future perspectives." In Proceedings of a Symposium on Water Resources and Uses in the Arab World. February 17–20, Kuwait. (In Arabic.)

Khouri, Nadim. 1992. "Wastewater reuse implementation in selected countries of the Middle-East and North-Africa." In *Sustainable Water Resources Management in Arid Countries.* Special Issue. Urbana, Ill.: International Water Resources Association (IWRA), with Canadian Journal of Development Studies.

Kimball, B.A. et al. 1993. "Effects of elevated CO_2 and climate variables on plants." *Journal of Soil and Water Conservation.* Vol 48. pp. 9–15.

Kirk, E.J. 1991. "The greening of security: Environmental dimensions of national, international, and global security after the Cold War." In *New Perspectives for a Changing World Order*, ed. E.H. Arnett. AAAS Publication No. 91-40s. Washington, D. C.: American Association for the Advancement of Science.

Kolars, John. 1990. "The course of water in the Arab Middle East." *Arab-American Affairs.* No. 33.

Kolars, John, and William A. Mitchell. 1991. *The Euphrates River and the Southeast Anatolia Development Project.* Carbondale, Ill.: Southern Illinois University Press.

Kolars, John. 1992. "Water Resources of the Middle East." *Canadian Journal of Development Studies, Special Issue: Sustainable Water Resources Management in Arid Countries.*

Kolars, John. 1992b. "Les ressources en eau du Liban: le Litani dans son cadre regional." Monde Arabe — Maghreb-Machrek, La question de l'eau au Moyen-Orient. *Discours et réalités.* No. 138. (Oct.–Dec. 1992).

Kolars, John. 1994. "Problems of international river management: The case of the Euphrates." In Biswas, Asit, ed. *International waters of the Middle East, from Euphrates-Tigris to Nile.* Bombay, India: Oxford University Press.

Kubursi, A.A. and H.A. Amery. 1992. "The Litani and the rebirth of Lebanon: the elixir of economic development and political stability." Conference on Middle East Water Crisis. May 7–9, Waterloo, Ontario.

Le Moigne, Guy. 1992. "Comprehensive water resources management." In Le Moigne, G., S. Barghouti, and L. Garbus, eds. *Developing and Improving Irrigation and Drainage Systems.* World Bank Technical Paper No. 178. Washington, D.C.: The World Bank

Leitner. G.F. 1991. "Total water costs on a standard basis for three large operating SWRO plants." In Report of the International Desalination Association (IDA) and World Bank Conference, Yokohama. Volume 2. Issue 1. Brighton, England: Ewbank Preece Ltd.

Lemarquand, D. G. 1977. *International Rivers: The Politics of Cooperation.* Vancouver: Westwater Research Centre, University of British Columbia.

Lincoln, D.E., E.D. Fajer, and R.H.Johnson. 1993. "Effects of enriched carbon dioxide environment on plant-insect-herbivore interaction." *Trends in Ecology and Evolution.* Vol. 8, pp. 64–68.

Lipschutz, R., and Holdren, J. 1990. "Crossing borders: Resource flows, the global environment, and international security." *Bulletin of Peace Proposals.* Vol. 21, No. 2.

Lloyd, Dennis. 1966. *The Idea of Law.* London: MacGibbon and Key; Baltimore: Penguin Books. 1964.

Lloyd, J.W. 1990. "Groundwater resources development in the Eastern Sahara." *Journal of Hydrology.* Vol. 119, pp. 71–87.

Lloyd, J.W. and R.H. Rim. 1990. "The Hydrogeology of groundwater resources development of the Cambio-Ordovician sandstone aquifers in Saudi Arabia and Jordan." *Journal of Hydrology.* Vol. 121. pp. 1–20.

Lonergan, Stephen C. 1992. "Redefining security." *Delta, Newsletter of the Canadian Global Change Program,* Vol. 3, No. 2. Ottawa: The Royal Society of Canada.

Lonergan, Stephen C. and David B. Brooks. 1993. *The economic, ecological and geopolitical dimensions of water in Israel.* Victoria, B.C.: Centre for Sustainable Regional Development, University of Victoria.

Lowi, Miriam. 1992. "West Bank water resources and the resolution of conflict in the Middle East." *Occasional Paper Series of the Project on Environmental Change and Acute Conflict.* No. 1, pp. 29–60.

Lowi, Miriam. 1993. *Water and power: the politics of a scarce resource in the Jordan River basin.* Cambridge, England: Cambridge University Press.

Lubchenco, J. 1991. "The sustainable biosphere initiative: An ecological research agenda." *Ecology.* Vol.72, pp. 371–412.

Lutz, E. and M. Munasinghe. 1991. "Accounting for the environment." *Finance and Development.* pp. 19–21. Washington D.C.: The World Bank.

Maas, A., and R.L. Anderson. 1978. *...And the Desert Shall Rejoice: Conflict, Growth, and Justice in Arid Environments.* Cambridge, Mass.: MIT Press.

MacNeil, Jim, ed. 1991. *Beyond Interdependence.* New York: Oxford University Press.

Mageed Yahia Abdel. 1985. "Conservation projects of the Nile — the Jonglie Canal." In Golubev and A. Biswas, eds., *Large Scale Water Transfer.* Oxford, England: Tycooly.

Mageed, Yahia Abdel. 1993. "Environmentally sound management of water resources." *Water Resources Development,* Vol. X, No. 2.

Maktari, A.M.A. 1971. *Water Rights and Irrigation Practices in Lahjh.* Cambridge, England: Cambridge University Press.

Mallat, Chibli. 1991. *Law and the Nile River: Emerging International Rules and the Shari'a.* London: School of Oriental and African Studies.

Mallat, Chibli. 1993. *The Renewal of Islamic Law.* Cambridge, England: Cambridge University Press.

Mandeel, Mohamed Amin. 1992. *Desalination and the Environment.* Kuwait: Department of Chemical Engineering, University of Kuwait. (In Arabic.)

Margat J. 1988. *L'eau dans le bassin méditerranéen: Prospective des besoins et des ressources.* Sophia Antipolis: Centre d'activités regionales du Plan Bleu pour la Mediterranée.

MAW: *See* Ministry of Agriculture and Water.

McCaffrey, S. 1993. "Water, politics, and international law." In *Water in Crisis,* ed. Peter Gleick. New York: Oxford University Press.

Meybeck, Michel, Deborah Chapman and Richard Helmer. 1990. "Global freshwater quality — A first assessment." Published on behalf of the World Health Organization and the United Nations Environment Programme. Cambridge, Mass.: Basil Blackwell.

Michael, Michael. 1974. "The allocation of waters of international rivers." *Natural Resources Lawyer.* Vol. 8, No. 1, pp. 45–66.

Mikhail, Wakil. 1992. *"Analysis of future water needs for different sectors in Syria."* Conference on Middle East Water Crisis. May 7–9, Waterloo, Ontario.

Ministère de l'Agriculture, Tunisie. 1990. *Stratégie hydraulique de la Tunisie.* Tunis: Ministère de l'Agriculture.

Ministry of Agriculture and Water (MAW), Saudi Arabia. 1984. *Water Atlas of Saudi Arabia,* Riyadh: MAW, Saudi Arabia.

Ministry of Planning (MOP), Saudi Arabia. 1984. *Fourth Development Plan, 1985–1990 for Saudi Arabia.* Riyadh: MOP, Saudi Arabia.

Ministry of Planning (MOP). 1990. *Fifth Development Plan 1990–1995 for Saudi Arabia.* Riyadh: MOP, Saudi Arabia.

Mohamed, A.K. 1986. "Water resources and development in the Republic of Yemen." Proceedings of Symposium on Water Resources and Utilization in the Arab World. February 17–20, Kuwait.

MOP: *See* Ministry of Planning

Murphy, Irene and Sabadell, Lenora. 1986. "International river basins: A policy model for conflict." *Natural Resources.* Vol. 12 (June), pp. 133–44.

Muzafer, S. 1958. "Superordinate goals in the reduction of intergroup conflict." *American Journal of Sociology.* Vol. 63, No. 4 (Jan).

Naff, Thomas, and Matson, Ruth, eds. 1984. *Water in the Middle East: Conflict or Cooperation?* Boulder, Colo.: Westview Press.

Naff, Thomas. 1984. "The linkage of history and reform in Islam: An Ottoman model." In *In Quest of an Islamic Humanism,* ed. A.H. Green. New York: Columbia University Press.

Naff, Thomas. 1992. "The Jordan Basin: political, economic, and institutional issues." In Le Moigne, G., S. Barghouti, G. Feder, L. Garbus, and M. Xie, eds. *Country Experiences with Water Resources Management: Economic, Technical, and Environmental Issues.* World Bank Technical Paper No. 175. Washington, D.C.: The World Bank.

Norby, R.J., C.A. Gunderson, S.D. Wullschleger, E.G. O'Neill, and M.K. McCracken. 1992. "Productivity and compensatory responses of yellow-poplar trees in elevated CO_2." *Nature.* Vol. 357, pp. 322–324.

North, A. 1993. "Saddam's water war." *Geographical Magazine,* Vol. LXV, No. 7, July 1993, pp 10–14.

Oechel, W.C., S.J. Hastings, G. Vourlitis, M. Jenkins, G. Riechers, and N. Grulke. 1993. "Recent change of Arctic tundra ecosystems from a net carbon dioxide sink to a source." *Nature.* Vol. 361, pp. 520–523.

Okidi, Charles Odidi. 1979. "International law and the Lake Victoria and Nile basins" in Charles Odidi Okidi, ed. *Natural Resources and the Development of the Lake Victoria Basin of Kenya.* Nairobi: Institute for Development Studies.

Ophuls, William. 1977. *Ecology and the Politics of Scarcity.* San Francisco: Freeman.

Pallas, P.P. 1980. "Water resources of Libya." In Salem, M.J. and Busrewil, M.T., eds. *The Geology of Libya,* Volume II. London: Academic Press.

Parnall, Theodore, and Albert Utton. 1976. "The Senegal Valley Authority: A unique experiment in international river basin planning." *Indiana Law Journal.* Vol. 51, No. 2.

Pearce, D., A. Markhandya and E.W.B Barbier. 1989. *Blueprint for a Green Economy.* London: Earthscan Publications.

Pearce, F.D., and K.T. Turner. 1990. *Economics of Natural Resources and the Environment.* New York: Harvester Wheatsheaf and Hemel Hempstead.

Pittock, A.B., A.M. Fowler, and P.H. Whetton. 1991. "Probable changes in rainfall regimes due to the enhanced greenhouse effect." *Proceedings of International Hydrology and Water Resources Symposium.* October 1991, Perth, Australia.

Ploss, Irwin and Jonathan Rubenstein. 1992. "Water for Peace." *The New Republic.* Sept. 7 and 14, pp. 20–22.

Pollard, R. 1992. *UNCED Governmental Earth Summit Agreements, including the Biodiversity Treaty.* Ampthill, Beds, UK: International Synergy Institute.

Ponting, Clive. 1991. *A Green History of the World.* London: Sinclair Stevenson.

Porter, H. 1993. "Interspecific variation in the growth response of plants to an elevated ambient CO_2 concentration." *Vegetatio.* Vols.104/105, pp. 77–97.

Postel, Sandra. 1992. *The Last Oasis: Facing Water Scarcity.* New York: Norton.

Pyne, R.D.G. 1988. "Aquifer storage recovery: A new water supply and ground water recharge alternative." Proceedings of American Society of Civil Engineers International Symposium on Artificial Recharge of Ground Water. Anaheim, Calif.: ASCE.

Rajagopalan, V. 1992. "The global problem of land and water constraints." In Le Moigne, G., S. Barghouti, and L. Garbus, eds. *Developing and Improving Irrigation Drainage Systems.* World Bank Technical Paper No. 178. Washington, D.C.: The World Bank.

Reed, E.C. 1980. "Report on water losses." *Aqua.* London: International Water Supply Association.

Rogers, Peter P. 1991. "International river basins: Pervasive undirectional externalities." Paper presented at *Conference on The Economics of Transnational Commons,* April 25–27. University of Siena, Italy.

Rogers, Peter P. 1993. "The value of cooperation in resolving international river basin disputes." *Natural Resources Forum,* May, 1993.

Romm, J.J. 1992. *Defining National Security.* Council of Foreign Relations Occasional Paper. New York: Council on Foreign Relations.

Sadukham et al., 1991. "Role of evaporative and membrane technology in solving drinking water problems in India." In Report of the International Desalination Association (IDA) and World Bank Conference at Yokohama. Vol. 2, Issue 1. Brighton, England: Ewbank Preece Ltd.

Salameh, Elias, and Andrae Garber. 1990. "Jordan's water resources and their future potential." In Proceedings of the Symposium of 27th and 28th October 1991. Amman, Jordan: Frederich Ebert Stiftung.

Salameh, Elias, and Helen Bannayan. 1993. *Water Resources of Jordan, Present Status and Future Potentials.* Amman, Jordan: Friedrich Ebert Stiftung and the Royal Society for the Conservation of Nature.

Salem, M. H. 1993. *Agriculture and Development in the Arab World.* Volume 2. Khartoum: Arab Organization for Agricultural Development (AOAD).

Salim, Omar M. and Sadek A. Kedri. 1992. "Water Resources of Libya, 1990–2000." Fifth Meeting of the Permanent Arab Committee for the International Hydrologic Program.

Savage, C. 1993. *Water bag transportation: A new option for water resources development.* A working paper at the Water Issues Workshop of the School of African and Oriental Studies, University of London, December 1993.

Schneider, S.H. 1989. "The changing climate." *Scientific American.* Vol. 261, pp. 70–79.

Seckler, David. 1993. *Designing Water Resource Strategies for the Twenty-First Century.* Center for Economic Policy Studies, Discussion Paper No. 16. Arlington, Virginia and Morrilton, Arkansas: Winrock International Institute for Agricultural Development.

Séminaire regional sur les Stratégies de gestion des eaux dans les Pays Méditerranéens, Synthèse des rapports nationaux. 1990. May 20–30, Algiers.

Sen, Amartya. 1981. *Poverty and Famines: an Essay on Entitlement and Deprivation.* Oxford, England: Oxford University Press.

Shahin, M. 1989. "Review and assessment of water resources in the Arab region." *Water International.* Vol 14, pp. 206–219.

Shapland, Greg. 1992. "Policy Options for Downstream States." London: School of Oriental and African Studies.

Starr, Joyce. 1992. *Water Security: The Missing Link in our Mideast Strategy.* Informal publication, pp. 35–48.

Svendsen, Mark. 1986. "Meeting irrigation system recurrent cost obligations." *ODI/ IIMI Irrigation Management Network Papers.* London: ODI.

Teclaff, Ludwik A. 1967. *The River Basin in History and Law.* The Hague: Martinus Nijhoff.

Teclaff, Ludwik A. 1977. *Legal and Institutional Responses to Growing Water Demand.* FAO Legislative Study No. 14. Rome: FAO.

Teclaff, Ludwik A. 1985. *Water Law in Historical Perspective.* Buffalo, N.Y.: Hein.

Teclaff, Ludwik A. and Albert E. Utton. 1981. *International Groundwater Law.* New York: Oceana Publications.

Teerink, J.R., and M. Nakashima. 1993. *Water allocation, rights, and pricing: Examples from Japan and the United States.* World Bank Technical Paper No. 198. Washington, D.C.: The World Bank.

Thornthwaite, C.W. and J.R. Mather. 1955. "The Water Balance." Third printing, *Publications in Climatology.* Vol. VIII, No. 1. Centerton, N.J.: Drexel Institute of Technology, Laboratory of Climatology.

Thornthwaite, C.W., J.R. Mather and D.B. Carter. 1958. "Three Water Balance Maps of Southwest Asia." Second Printing. *Publications in Climatology.* Vol. XI, Number 1. Centerton, N.J.: Drexel Institute of Technology, Laboratory of Climatology.

Tuijl, W.V. 1989. "Irrigation developments and issues in EMENA (Europe, Middle East, North Africa) Countries." In Le Moigne, G., S. Barghouti, and H. Plusquellec, eds. *Technological and Institutional Innovation in Irrigation.* World Bank Technical Paper No. 94. Washington, D.C.: The World Bank.

Tuijl, W.V. 1993. *Improving water use in agriculture: Experiences in the Middle East and North Africa.* World Bank Draft Technical Paper. Washington, D.C.; The World Bank.

Tunisia, Ministry of Agriculture. 1994. *Water Economy 2000 Study.* Tunis: Ministry of Agriculture, Direction Générale des Etudes et des Travaux Hydrauliques.

Tvedt, Terje. 1992. "The struggle for water in the Middle East." *Canadian Journal of Development Studies, Special Issue, Sustainable Water Resources Management in Arid Countries.*

Ukayli, M.A. and T. Husain. 1988. "Evaluation of surface water availability, wastewater reuse and desalination in Saudi Arabia." *Water International.* Vol. 13, pp. 218–225.

UNCED. 1992. Agenda 21. Final Report of the UN Conference on Environment and Development (UNCED). Rio de Janeiro: UNCED.

United Nations Development Program (UNDP). 1989. "Nile Basin Integrated Development." (Report of the Fact Finding Mission — AF/86/003-RAB/86/014 1989). New York: UNDP.

United Nations Educational, Scientific and Cultural Organization (UNESCO). 1972. *Etude des resources en eau du Sahara septentrional, Rapport final.* Paris: UNESCO.

United Nations. 1973. *Groundwater in Africa.* New York: United Nations Department of Economic and Social Affairs.

United Nations. 1982. *Ground Water in the Eastern Mediterranean and Western Asia.* Natural Resources/Water Series. No. 9. New York: UN Department of Technical Cooperation for Development.

United Nations. 1984. *Treaties Concerning the Utilization of International Water Courses for Other Purposes than Navigation.* Natural Resources/Water Series. No. 13. New York: UN Department of Technical Cooperation for Development.

United Nations. 1988. *Yearbook of the International Water Commission.* (A/CN.4/Ser.A/1988/Add.1.) New York: United Nations.

United Nations. 1990. *World Population Prospects.* New York: UN Department of International Economic and Social Affairs.

United States Office of Technology Assessment. 1988. *Using Desalination Technologies for Water Treatment.* Background Paper. Washington D.C.: Office of Technology Assessment.

Uqba, A.K. 1992. "The need for national and regional integrated water resources management in the Gulf Cooperation Council Region." Proceedings of First Gulf Water Conference. October 10–14, Dubai.

Utton, Albert E. and Ludwik Teclaff, eds. 1987. *Transboundary Resource Law.* Boulder, Colo.: Westview Press.

Vesilind, P.J. 1993. Water, the Middle East's Critical Resource. *National Geographic.* Vol. 183, pp. 38–71.

Wade, Neil. 1991. "The effects of the recent energy cost increase on the relative water costs from reverse osmosis and distillation plants." In Report of International Desalination Association (IDA) and World Bank Conference at Yokohama. Vol. 2, Issue 1. Brighton, England: Ewbank Preece Ltd.

Wagnick K. 1992. *Worldwide Desalting Plants Inventory. Report 12.* (Wagnick Consulting, Gnarrenburg.) Washington, D.C.: The World Bank, IDA.

Water Authority of Jordan. 1993. "The Qur'an and the Water Environment." Royal Scientific Society of Jordan. Cited in Brooks, D.B. "Adjusting the flow: Two comments on the Middle East Water Crisis." *Water International.* Vol. 18, No. 1.

Waterbury, John. 1979. *Hydropolitics of the Nile Valley.* Syracuse, N.Y.: Syracuse University Press.

Waterbury, John. 1982. *Riverains and Lacustrines: Toward International Cooperation in the Nile Basin.* Discussion paper No. 107. Research Program in Development Studies, Princeton University.

Waterbury, John. 1987. "Legal and institutional arrangements for managing water resources in the Nile Basin." *Water Resources Development.* Vol. 3, No. 2, pp. 82–104.

Waterbury, John. 1990. "Dynamics of basin-wide cooperation in the utilization of the Euphrates." Conference on the Economic Development of Syria; Problems, Progress, Prospects. Jan. 6–7, Damascus.

Waterbury, John. 1992. "Three rivers in search of a regime: the Jordan, the Euphrates and the Nile." 17th Annual Symposium, Center for Contemporary Arab Studies (CCAS), Georgetown University, April 9–10.

Whittington, Dale, and Elizabeth McClelland. 1992. "Opportunities for regional and international cooperation in the Nile basin." In Le Moigne, G., S. Barghouti, G. Feder, L. Garbus, and M. Xie, eds. *Country Experiences with Water Resources Management: Economic, Technical, and Environmental Issues.* World Bank Technical Paper No. 175. Washington, D.C.: The World Bank.

Whyte, A.V.T. 1986. "From hazard perception to human ecology." In Kates, R.W., and I. Burton, eds. *Geography, Resources, and Environment.* Vol. II. Chicago: University of Chicago Press.

Wiet, G., Yackubi, 1937. *Les Pays.* Textes et traductions d'auteurs orientaux. Cairo: Publications de l'Institut Francais d'Archéologie Orientale.

Wild, J. 1994. Personal communication to J.A. Allan concerning villages in the Taroudant region.

Wilkinson, J.C. 1977. *Water and Tribal Settlement in Southeast Arabia.* Oxford: Oxford University Press.

Wilkinson, J.C. 1990. "Muslim land and water law." *Journal of Islamic Studies.* Vol. 1.

Williams, W., K. Garbutt, and F.A. Bazzaz. 1988. "The response of plants to elevated CO_2 — V. Performance of an assemblage of serpentine grassland herbs." *Environmental Experimental Botany.* Vol. 28, pp. 123–30.

Wittfogel, Karl A. 1965. "The Hydraulic Civilization." In *Man's Role in Changing the Face of the Earth,* ed. W.L. Thomas, Jr. Chicago: University of Chicago Press.

Wolf, Aaron. 1992. "Towards an interdisciplinary approach to the resolution of international water disputes: the Jordan River watershed as a case study." Conference on Water Quantity/Quality Disputes and their Resolution. May 2–3, Washington, D.C.

World Bank. 1988. *Jordan Water Resource Sector Study.* Washington, D.C.: The World Bank.

World Bank. 1990a. *Arab Republic of Egypt, Land Reclamation Subsector Review.* Washington, D.C.: The World Bank.

World Bank. 1990b. *Jordan: Toward an Agriculture Sector Strategy.* Washington, D.C.: The World Bank.

World Bank. 1990c. *Operational Directive: Projects on International Waterways.* (OD 7 50, April) Washington, D.C.: The World Bank.

World Bank. 1990d. *Social indicators of development.* STARS. (Socio-economic Time Series Access and Retrieval System.) Version 1.1. Washington. D.C.: The World Bank.

World Bank. 1991. *Staff Appraisal Report, Uganda Third Power Project.* (Report No. 9153-UG, 5/29). Washington, D.C.: The World Bank.

World Bank. 1992a. *Egypt Agriculture Sector Study.* Washington, D.C.: The World Bank.

World Bank. 1992b. *World Development Report 1992: Development and the Environment.* New York: Oxford University Press.

World Bank. 1992c. *World Development Indicators.* Washington, D.C.: The World Bank.

World Bank. 1993. *Water Resources Management: A Policy Paper.* Washington, D.C.: The World Bank.

World Resources Institute. 1990. *World Resources 1990–91.* A report by the World Resources Institute in collaboration with the United Nations Environmental Programme and the United Nations Development Programme. New York: Oxford University Press.

World Resources Institute. 1992. *World Resources 1992–1993.* A report by the World Resources Institute in collaboration with the United Nations Environmental Programme and the United Nations Development Programme. New York: Oxford University Press.

World Water/WHO. 1987. *The International Drinking Water Supply and Sanitation Directory.* Edition Three. London: Thomas Telford.

Wright, E.P. and W.M. Edmunds. 1971. "Hydrogeological studies in central Cyrenaica, Libya." In C. Gray, ed. *International Symposium on the Geology of Libya.* Faculty of Science, University of Libya, pp 459–481.

Xie, M., U. Kuffner, and G. Le Moigne. 1993. *Using Water Efficiently: Technological Options.* World Bank Technical Paper No. 205. Washington, D.C.: The World Bank.

Zaki, E. 1992. "Irrigation water management in Sudan." In Le Moigne, G., S. Barghouti, G. Feder, L. Garbus, and M. Xie, eds. *Country Experiences with Water Resources Management: Economic, Technical, and Environmental Issues.* World Bank Technical Paper No. 175. Washington, D.C.: The World Bank.

Index

AAAID. *See* Arab Authority for Agricultural Investment and Development

Abate, Zewdie, 52, 73

Abdel-Dayem, Safwat, 51

Abdelrahman, H., 188

Abdulrazzak, Mohamed J., 178, 301

Abdusalaam, Abdelrahim S., 123

Abib, A., 161

Aboukhaled, A., 79

Absolute sovereignty, 262

Absolute/territorial integrity, 262-63

Abu Dhabi, 78, 191

Abu Rizaiza, Omar S., 132, 185

Abu Taleb, F. Maher, 23-24, 26, 31, 60

Abu Zeid, Mahmoud A., 14, 20, 25, 29-30, 31, 51, 112, 113-14, 313

Access rights, 270

Acquired rights, 40, 41, 56, 57, 139

Adams, Robert, 124, 126

Adhaim River, 127-28

Afrin River, 130, 133

Agenda, future. *See* Future agenda

Agriculture/irrigation:
  and allocation/reallocation, 1, 2, 3-4, 5-12, 83, 98, 270-71;
  challenges facing, 10-12;
  and climate, 244, 249, 250, 286;
  commercialization of, 13;
  and conflict, 257;
  and conservation, 28-29, 36, 143;
  and costs/cost recovery, 14, 25, 26, 27, 28, 301;
  and crops, 9-10, 29-30, 36, 297, 313;
  daily management of, 26;
  and demand management, 97;
  and desalination, 301;
  development of, 125-27;
  diversification of, 13;
  and economic issues, 2, 10, 13, 21;
  and efficiency, 20-21, 28-29, 89, 142, 313;
  and the environment, 11-12;
  expansion of, 2, 5-8;
  and farmers's income, 10;
  and the future agenda, 286, 289, 297, 301, 303, 305, 308, 311, 313-14, 315;
  and government policies, 9-10, 12-13, 25, 26, 161;
  and institutional arrangements, 315;
  and intensification of farming, 6, 11-12;
  and Islamic law, 270-71;
  modernization of, 20-21, 36;
  and new water, 77;
  overview of, 67, 89;
  and pricing policies, 11, 36;
  and the private sector, 36;
  and productivity, 6, 9, 29;
  and saline water, 11;
  and security, 257;
  and studies/investigations, 313-14;
  and supply and demand, 4, 286, 289, 297, 301, 303, 305, 308, 311, 313-14, 315;
  and supply management, 19-20;
  and technology, 10-11, 13, 14, 28;
  and transboundary water, 41, 45-46;
  and transfers, 315;
  types of, 12;
  and wastewater/reuse, 16, 18, 78-79, 95, 231;
  and water quality, 142.
  See also Crops; Food imports; Food self-sufficiency; *specific country, or region*

Ahmad, M. U., 82

Akkad, A., 199

Al Alawi, Jamil S. K., 301

Al Wahda dam, 133

Al-Asam, S. M., 178

Al-Ayn aquifer, 181

Al-Batin aquifer, 181

Al-Fusail, A. M., 178, 189

Al-Gazi, A., 171

Al-Ibrahim, A. A., 185

Al-Kenson, D. J., 187

Al-Khairu, 'Iz al-Din 'Ali, 56

Al-Mahmoud, M. A., 182

Al-Mansour, K., 182, 187

Al-Mawardi, A. b H., 268, 270

Al-Mugran, A., 183

Al-Nabban, M. F., 268, 270

Al-Rahbi, 'A., 268, 270

Al-Shaybani, M., 269

Al-Sufy, A. M., 183

Al-Tokhais, A. S., 186

Al-Wahda Dam, 17

Al-Weshah, R., 31

Algeria:
    agriculture/irrigation in, 5, 12,
        151-52, 153, 156, 168;
    allocation/reallocation in, 5, 151-52;
    arable land in, 286, 289;
    conservation in, 164, 165;
    and demographics, 148, 289;
    and desalination, 221;
    development programs in, 156;
    domestic use in, 152;
    and economic issues, 150-51;
    and the future agenda, 286, 289;
    government policies in, 23, 166;
    hydroelectric power in, 152;
    industrial use in, 156;
    information networks in, 23;
    infrastructure in, 156;
    institutional framework in, 168, 169;
    and an integrated approach to water
        management, 23;
    land area of, 286;
    municipal use in, 156;
    planning in, 156;
    pricing policies in, 166, 168;
    sources of water for, 74, 150, 151,
        156, 164;
    wastewater/reuse, in, 158, 165;
    water quality in, 165.
    See also Maghreb countries

Alheritière, Dominique, 46

Allan, J. A., 68, 71, 74

Allocation/reallocation:
    and climate, 244;
    and costs/cost recovery, 26, 222;
    current trends in, 3-12;
    and demand management, 87, 94, 98;
    and demograghics, 83;
    and efficiency, 88-89;
    and the evolution of water strategies, 91;
    and the future agenda, 308, 312;
    and government policies, 83;
    and institutions, 99-100;
    overview of, 66, 91;
    and politics, 83, 89, 99-100;
    and pricing, 222;
    and priorities, 23, 84;
    and supply and demand, 1, 2-12, 254,
        255, 308, 312.
    See also Cooperative arrangements;
    Law; *specific country, region, river, or
    type of use*

American Society of International Law, 46

Amman, Jordan, 21, 26, 31, 76, 82, 142

Amran aquifer, 189

Amu Darya basin, 43

Anderson, E., 82

Anderson, R. L., 273

Antecedent rights, 271

Anti-Lebanon Mountains, 130

Anti-Taurus Mountains, 128, 130

AOAD. *See* Arab Organization for Agricultural Development

Appreciable harm, 47, 48, 49, 53, 263-65

Aquifers, 235, 297. *See also specific aquifer*

Arab Authority for Agricultural Investment and Development (AAAID), 116

Arab Fund for Economic and Social Development, 4, 116

Arab League of States, 45, 57, 102, 116, 137

Arab Mashrek:
agriculture/irrigation in the, 124, 125-26, 127, 142, 143;
allocation/reallocation in the, 140-41, 142;
climate in the, 121, 123, 130-32, 140-41;
and conservation, 142, 143-44;
and cooperative arrangements, 133, 141, 142, 144;
demographics in the, 127, 141, 142;
droughts in the, 121, 123, 143-44;
food security in, 138;
geography of the, 127-32;
history of the, 123-27;
industrial use in the, 142;
and information systems, 141;
international water in the, 133;
and organizations, 143-44;
overview of the, 121-23;
and politics, 127, 135-39, 143;
rainfall in the, 121, 123, 130, 131-32, 140-41, 144;
supply and demand in the, 127-35, 136, 143-44;
and transfers, 141-42;
urbanization in the, 126-27;
and water balance, 121.
See also Euphrates River; Iraq; Jordan; Lebanon; Syria; Tigris River

Arab Organization for Agricultural Development (AOAD), 8-9, 102, 116

Arab world:
arable land in, 286, 289;
and demographics, 289;
development in the, 312;
and economic issues, 312;
evolution of water strategies in the, 91-94;
and the future agenda, 285-316;
industry in the, 312;
overview of the, 1-2, 65-70, 89-99, 286;
rainfall in the, 286, 297;
and rural use, 289;
and supply and demand, 297-98, 301, 305, 310-12;
and urbanization, 289;
water crisis in the, 310-12.
See also specific country or region

Arabian Peninsula:
and agriculture/irrigation, 171, 173, 178, 180, 181, 182, 185, 190, 193-94, 197, 201-2;
and climate, 174-75;
and conservation, 171, 173, 190, 194, 197-98, 199, 202;
and cooperative arrangements, 173, 181-82, 202;
and costs, 201;
and the delivery system, 199;
demand in, 184- 92;
and demand management, 197-99;
and demographics, 171, 184, 185;
and desalination, 171, 173, 182-83, 185, 190, 191- 92, 194-96, 198, 201, 202;
and domestic use, 173, 178, 180- 81, 182, 185, 190, 198, 199, 202;
and economic issues, 190- 91;
and the environment, 192-95;
and floods, 178;
and food self-sufficiency/imports, 173, 185, 194;
geography of the, 175-76;
and government policies, 173, 185, 197, 199, 200- 201, 202;
and industry/industrial use, 183, 185, 190, 194, 202;
and institutional arrangements, 173, 201, 202;
and new water, 190, 195-97;
overview of the, 171-74, 190-91, 201-2;
and per capita consumption, 198;

and planning, 173, 200, 202;
and pricing policies, 173, 198-99,
    201, 202;
problems on the, 171, 173;
rainfall on the, 174-75, 176;
and rural use, 193;
sources of water for the, 171, 173,
    174-84, 185, 190-91, 193, 195-97,
    199, 201, 202;
and transfers, 195, 196- 97, 202;
and urban use/urbanization, 171, 173,
    183, 190, 198, 202;
and wastewater/reuse, 173, 175,
    183-84, 190-91, 193, 195, 198,
    201;
and water quality, 173, 180, 183,
    190, 192-93;
water temperature on the, 180.
See also specific country

Arabian Shelf, 175, 179

Arabian Shield, 175

Arable land, 6, 8. *See also specific country*

Argentina, 135-36, 264

Aruma aquifer, 179, 181, 187, 188

Aswan Dam/Aswan High Dam, 12, 16,
    79, 84, 280;
    and cooperative arrangements, 50-51,
    54;
    and Nile River development, 102,
    104, 105, 107, 108, 109, 111, 112,
    113, 115

Ataturk dam, 57, 139

Atbara River, 103, 105, 109, 115, 116

Atlas Mountains, 149, 150

'Auda, 'Abd al-Malik, 53

Australia, 6, 26

Authman, M. N., 178

Automated water management control
    systems, 18

BAAC (British Arabian Advisory Com-
    pany), 178

Bahr El-Ghazal, 105, 111

Bahr El-Jebel, 105, 111

Bahrain:
    agriculture/irrigation in, 4, 19, 182,
        184, 185;
    and conservation, 199;
    and cooperative arrangements, 181;
    and demand management, 199;
    and demographics, 289;
    and desalination, 183, 187, 191, 308;
    domestic use in, 182, 191;
    and the environment, 192-93, 194;
    and food self- sufficiency, 185;
    and the future agenda, 289, 308;
    and government policies, 185;
    and industry/industrial use, 191, 194;
    and new water, 196;
    sources of water for, 176, 179, 180,
        181, 182, 187, 196, 267;
    and supply and demand, 185, 187,
        190, 289, 308;
    and transfers, 196;
    wastewater/reuse in, 184, 187;
    and water quality, 192-93

Bakour, Yahia, 20, 25, 29, 130-31, 133,
    141

Balikh River, 55, 130, 140

Bangladesh, 135

Banias River, 59, 71, 77, 97

Barada River, 133

Barberis, J., 264, 267

Barghouti, Shawki M., 11, 29, 30

Bari, Zohurul, 56

Baumol, zz, 47

Bazzaz, Fakhri A., 249

Beaumont, Peter, 124, 127

Bekaa Valley, 133, 138

Bellagio Draft Treaty (1988), 267

Ben-Adam, Y., 268, 270

Berkoff, Jeremy, 17

Beschorner, Natasha, 57, 58

Bilder, R., 266

Birecik dam, 56

Biswas, Asit K., 44, 276

Biyadh aquifer. *See* Wasia-Biyadh aquifer

Blue Nile: annual discharge of the, 105; and Nile River development, 103, 108, 109, 110, 111, 115, 116; and transboundary water, 45-46, 50, 52-53, 105, 265

Border irrigation, 11

Bos, E., 5

Bourne, C. B., 264

Boutayeb, N., 160

Brackish water, 3, 195-96, 199, 211, 216, 220. *See also* Desalination; *specific country or region*

Braidwood, Robert J., 124

Brazil, 135-36

Brooks, David B., 130

Brown, L. R., 136

Burami aquifer, 181

Burdon, D. J., 179

Burundi, 49

Bushnak, A. A., 183, 187-88, 192

Buwaib aquifer, 179

California, 31, 34, 82, 249, 309

Cameroon, 224

Canada, 6

Capital costs, 24, 27, 226-27, 229, 230, 238, 240

Caponera, Dante A., 262

Carbon dioxide, 243, 244-46, 248-52

Cayor Canal, 205

Central Region: actions needed in the, 118-19; agriculture in the, 102, 104, 118;

and conservation, 104, 107, 111, 112; and cooperative arrangements, 117, 118-19; countries included in the, 103; and demographics, 102, 103, 117, 118; droughts in the, 117, 118; and economic issues, 104, 118; and the environment, 102, 117, 118-19; and food, 102-3, 110, 116; geography of the, 101-2; and government policies, 119; and industrialization, 112; and information systems, 119; institutional structures in the, 118, 119; overview of the, 101-5, 117-19; and politics, 117; rainfall in the, 107, 118; sources of water for the, 103-5, 117; and transboundary water, 117; and urbanization, 102, 112, 117, 118. *See also* Nile River; *specific country*

Century Water Scheme, 51, 56

Cess, R. D., 245, 246, 250

Cetin, Hikmet, 55

Chad, 267

Chase Investment Bank, 56

Chile, 30

China, 30, 205

Chitale, M. A., 29

Choucri, Nazli, 258, 275

Climate, 70, 243-52, 254. *See also* Rainfall

Cloud-seeding, 30, 36

Collins, Robert, 50, 51, 52-53, 74

Colonialism, 43, 45, 49, 50, 56

Columbia River, 33-34

Community of co-riparian states, 46-47, 262, 263

Competition index, 6

Conflict: basic principles concerning, 262-72;

causes of, 260- 61, 273-75, 277-82;
consequences of, 282;
and the future agenda, 312-13;
idea of, 260-62;
and the law, 262-72, 282- 83;
and modeling, 261-62;
paramount question about, 282-84;
and power, 276-82;
recommendations concerning, 282-84;
and scarcity, 253-55;
and security, 255-59, 273-74;
and the superordinate principle, 275-76, 284;
and supply and demand, 312-13;
and treaties, 272, 274, 282-83;
types of, 260, 282.
See also War; *specific country or region*

Congo Free State, 49, 108

Conservation:
    and agriculture/irrigation, 28-30, 36, 143;
    and allocation/reallocation, 5;
    and cooperative arrangements, 34, 54;
    and cost recovery, 24, 25, 26, 27, 36;
    cross- generational, 26;
    and demand management, 87;
    and desalination, 203, 234;
    and efficiency, 199;
    and erosion, 164-65, 170;
    examples of, 5, 28, 142;
    and the future agenda, 297, 298, 301, 308, 314;
    and government policies, 24, 30, 36-37;
    importance of, 36-37;
    and industrial use, 314;
    and national/regional organizations, 143-44;
    and non- conventional sources of water, 298;
    overview of, 67;
    and pricing policies, 30, 198-99;
    soil, 164-65, 170;
    and supply and demand, 297, 298, 301, 308, 314;
    and technology, 28, 199;
    and urban use, 36, 314;
    and wastewater, 234;
    and water management, 5, 24, 25, 26, 27, 28-30, 36-37;
    and water quality, 142, 170.
    See also specific country or method

Consumption. *See* Per capita water availability/consumption; Usage; *type of use*

Conway, D., 70

Cooperative arrangements:
    and allocation/reallocation, 33, 34, 35;
    and appreciable harm, 47, 48, 53;
    asymmetry of, 42-43;
    bases for, 283-84;
    benefits of, 61-62;
    and climate, 250;
    and colonialism, 43, 45, 49, 50, 56;
    and community of basin interests, 46-47;
    and conservation, 34, 54;
    and costs, 62;
    enforcement of, 44;
    and the environment, 33, 35;
    and the Fashoda syndrome, 44, 45;
    and financing, 44, 55-56;
    and food security, 58, 61;
    and the future agenda, 314;
    and good faith, 47, 48-49;
    and government policies, 61;
    and the Harmon doctrine, 46;
    and the Helsinki Rules, 47, 48-49;
    incentives for, 283, 284;
    and induced solutions, 44, 56;
    and information systems, 23-24, 32-33, 35, 36, 48, 52, 53, 57, 284;
    and infrastructure, 61-62;
    and institutional arrangements, 33, 284;
    and an integrated approach to water management, 24;
    and international law, 40-41, 46-49;
    and involuntary/hegemon solutions, 43, 46, 58-59, 60, 61;
    and the legal framework for bargaining, 46-49;
    and limited territorial sovereignty, 46, 47;
    and mediation, 283;
    and military power, 43, 44-45;
    and multi-good bargaining, 43, 51-52, 55, 58;
    and new water, 74;
    and opportunity costs, 62;
    and politics, 1, 40-46, 55;
    and reasonable use, 47;

and regional management, 33-34;
and rights, 56, 57;
as a solution to water management
  problems, 23-24;
and superfairness, 47;
and the superordinate principle, 275-76;
and supply management, 34;
and technology, 1, 34-35, 284;
and transfers, 56, 61;
voluntary, 42;
and war, 60;
and water management, 1, 18, 23-24,
  32-35, 36.
See also International law; Islamic law;
Law; *specific country, region, or river*

Cory, H. T., 51

Coser, L. A., 260

Cost recovery, 24-28, 36, 83. *See also*
  specific country or type of cost

Costs:
  and accounting methods, 221;
  and allocation/reallocation, 222;
  and climate, 250;
  and cooperative arrangements, 62;
  and the delivery of water, 89, 230-31,
    233;
  and demand management, 89;
  and efficiency, 224;
  and fractured rock zones, 82;
  and the future agenda, 301, 310-11,
    314, 315-16;
  of groundwater, 74;
  and infrastructure, 224;
  lack of data about, 225;
  of new water, 75-76, 77-78, 84;
  and pipelines, 75-76, 205, 226, 240-41;
  and pricing policies, 222, 224, 315-16;
  and supply and demand, 13, 83-84,
    301, 310-11, 314, 315-16;
  of supply methods, 222, 224-26;
  and transfers, 142, 196, 197, 205,
    206, 225;
  and wastewater/reuse, 94, 231, 233;
  of water provision, 221-22.
  See also Cost recovery; Desalination:
  costs of; Financing; *specific type of use;*
  type of cost

Cotgrove, S., 254

Country Report, 186

Cranston, M., 255

Cretaceous aquifer, 179

Crops. *See* Agriculture/irrigation

Cyprus, 28-29, 77

Dabbagh, Taysir A., 203, 224

Dammam aquifer, 179, 180, 181, 182,
  187, 188

Dams: earth, 35. *See also specific dam*

Dan River, 59, 71

Danish, S., 187

Darghouth, M., 23, 26, 29

David, Peter, 256

Dead Sea, 34-35, 82, 199

Delivery of water, 2, 87, 89, 230-31,
  233. *See also* Leakages;
  type of system

Dellapenna, J., 265, 266, 267

Demand management:
  and agriculture/irrigation, 97;
  and allocation/reallocation, 2, 87, 94,
    98;
  as an alternative to new water, 83;
  and cause of demand, 87;
  and conservation, 87;
  and costs, 89;
  and the delivery of water, 87, 199;
  and demographics, 87;
  and efficiency, 87, 89, 94;
  and the environment, 97;
  and the evolution of water strategies, 92;
  and food, 86-87, 95-96;
  and the future agenda, 94, 297, 305;
  and information systems, 159;
  and infrastructure, 94;
  and the national economies, 95-99;
  and new water, 83, 86-87, 98;
  overview of, 67, 86-89;
  and politics, 87, 94, 98-99;
  and pricing policies, 89;

and reuse, 94, 95, 98;
supply management vs., 19, 94;
and technology, 67, 83, 94;
and wastewater/reuse, 67, 94, 95, 98.
See also specific country or region

Demirel, Suleyman, 57

Demographics:
    and allocation/reallocation, 4-5, 6, 83;
    and arable land, 6, 8;
    and climate, 243, 244, 248;
    and conflict, 260, 274, 275;
    and demand management, 87;
    and desalination, 220;
    and the evolution of water strategies,
        92, 93-94;
    and the future agenda, 289, 305;
    and new water, 83;
    overview of, 66, 71;
    and security, 257;
    and supply and demand, 2, 4-5, 6, 13,
        35, 65, 66, 71, 254, 255, 289, 305;
    and technology, 91;
    and transboundary water, 41;
    and water management, 1, 2, 4- 5, 6,
        8, 13, 35.
See also specific country or region

Desalination:
    advantages/disadvantages of, 206,
        234-35;
    and agriculture/irrigation, 301;
    and the annual designed capacity,
        183;
    and brackish water, 68, 78, 195-96,
        211, 216, 220;
    and climate, 248;
    and conservation, 203, 234;
    construction of plants for, 190, 191-
        92, 194, 195, 202;
    costs of, 17, 31-32, 78, 84, 98, 158,
        191-92, 195-96, 199, 201, 203,
        204, 206, 220, 221-22, 224-31,
        233-34, 238, 240- 41, 301, 310-11,
        314;
    and the delivery of water, 230-31, 233;
    and demographics, 220;
    and developing countries, 204, 239-40;
    and domestic use, 98, 182, 185, 191,
        198, 220;

and economies of scale, 220, 229;
and electrical production/energy, 84,
    206, 220, 225, 227, 240, 310-11;
and the environment, 78, 194-95,
    234-35, 237;
and feed water, 216, 220, 221;
and financing, 192, 206, 238-39;
and the future agenda, 235-36, 241,
    297, 301, 308, 310-11, 314;
and government policies, 199, 237-38;
and groundwater, 78;
and industrial use, 78, 98, 183, 185,
    191, 220, 314;
and information systems, 196, 237;
and institutional arrangements, 236-
    39;
and international organizations, 236-
    39;
life span of plants for, 227, 228-29;
modeling approach to, 231, 233-34;
neglect of, 204-5;
as new water, 78;
as a non-conventional water source,
    30, 31, 36-37;
number of plants for, 183;
and O and M costs, 78, 192;
overview about, 68, 73, 203-4, 239-40;
and planning, 201;
and the private sector, 183, 191, 192,
    204, 238;
processes for, 206-15, 216, 220-21;
and research, 203, 204, 236-39;
and sea water, 78, 208, 211, 216,
    221, 310-11;
statistics about, 19, 31, 32, 68;
and strategic vulnerability, 78, 206;
and supply and demand, 205-6, 235,
    236, 239, 241, 297, 301, 308, 310-
    11, 314;
and technology, 203, 204, 206, 220,
    236, 237, 239;
and training of personnel, 204, 205;
transfers compared with, 205-6;
and urban use, 78, 98, 183, 220, 314;
and wastewater/reuse, 78, 198, 220,
    231, 234, 240;
and water management, 17, 19, 31-
    32, 36-37;
and world volumes of desalination
water, 216.

See also specific country; specific country or region

Deudeney, D., 259

Development/developing countries, 20, 25, 27, 156-58, 204, 239-40, 312. *See also specific country*

Dhofer Mountains, 175, 197

Diama reservoir, 151, 157-58

Disi aquifer. *See* Qa Disi aquifer

Disposal costs, 231

Diyala River, 125-26, 127-28

Djibouti, 103, 309

Domestic use:
   and allocation/reallocation, 2, 4-5, 83, 270-71;
   and costs, 76;
   and desalination, 98, 182, 185, 191, 198, 220;
   and the future agenda, 286, 305;
   and Islamic law, 270-71;
   and supply and demand, 1, 4, 286, 305;
   and water quality, 142.
See also specific country or region

Dougherty, J. E., 256

Draa River, 158

Dracup, J. A., 204

Drainage. *See* Wastewater/reuse

Drip irrigation, 11, 20, 29

Droughts, 19, 115, 117, 162, 246, 250. *See also specific country or region*

Dubai, 191, 193

East Ghor Canal, 17, 133

Economic issues:
   and agriculture/irrigation, 2, 10;
   and conflict, 260;
   and the economic value of water, 18, 88, 91;
   and the future agenda, 303, 312, 314;
   overview of water and, 66, 68, 70-84;
   and security, 258;

and transboundary water, 43;
and water management, 12-37.
See also National economies; *specific country, region, or issue*

ED. *See* Electrodialysis process

Edgell, H. J., 179, 181

Edmunds, W. M., 74

Efficiency, 67, 87, 88-89, 94, 142, 199, 224

Egypt:
   allocation/reallocation in, 5, 6, 9, 10, 13-14, 50-51, 53-54, 104, 108-9, 111;
   and appreciable harm, 264-65;
   and climate, 132, 248;
   competition index in, 6;
   and conflict, 264-65, 278, 280, 313;
   conservation in, 12, 28, 29-30, 54, 107, 112-13;
   and cooperative arrangements, 34, 42, 43, 45-46, 47, 49-54, 56, 61, 74, 80, 107-9, 110, 276, 277;
   cost recovery in, 25;
   and demand management, 89, 95;
   demographics in, 13-14, 113, 280, 289;
   development in, 13, 107-14;
   domestic use in, 71;
   economic issues in, 68, 104, 280, 303;
   and the environment, 15-16, 112, 118, 280;
   floods in, 107, 112;
   and food, 9, 30, 61, 87, 95, 113, 289;
   and the future agenda, 111, 286, 289, 303, 308, 311, 313-14;
   geography of, 71;
   government policies in, 10, 13-16, 23, 30, 61, 113-14;
   and institutional arrangements, 114;
   and an integrated approach to water management, 23;
   land in, 13, 14, 286;
   and new water, 74, 80, 82, 95;
   non-conventional water sources in, 30-31;
   ownership in, 14;
   politics in, 68, 105, 109;
   private sector in, 30;

rights of, 49-50, 51, 54;
socio-economic problems in, 14-15;
sources of water for, 2, 13, 30-31, 65-
    66, 74, 80, 82, 95, 105, 113, 267;
and storage, 107, 108, 109, 111, 112;
and supply and demand, 71, 80, 84,
    87, 111, 255, 286, 289, 303, 308,
    311, 313-14;
and transfers, 59, 61, 102, 205;
urban use in, 71;
wastewater/reuse in, 14, 79, 80, 82,
    89, 95, 112, 113, 114, 118;
water quality in, 113, 114.
See also Egypt— agriculture/irrigation
in; Nile River

Egypt—agriculture/irrigation in:
and climate, 248;
and conservation, 28, 29-30, 112-13;
and cooperative arrangements, 51, 110;
and cost recovery, 25;
and crops, 9, 10, 14, 16, 29-30, 112,
    113, 313-14;
and demand management, 89;
and demographics, 13-14, 112, 113;
and drainage, 12, 15;
and efficiency, 9, 14, 20, 89, 113-14;
and the environment, 15-16, 112-13;
expansion of, 5, 108, 109, 112;
and the future agenda, 289, 311,
    313-14;
and government policies, 9, 10, 12,
    13-16, 61, 113-14;
historical aspects of, 107;
problems facing, 14-15;
and technology, 14;
and wastewater/reuse, 15, 79, 80, 89,
    95;
and water quality, 15

Ehleringer, J. R., 249

El Ain, 78

El-Saie, zz, 239

El-Zawahry, A. E., 178

Elahi, A., 21

Electrical production: and desalination,
    220, 225, 227, 240

Electrodialysis process, 195, 208, 211

Encyclopedia of Islam, 270

Energy, 206, 250, 310-11

Environment:
and agriculture/irrigation, 11-12;
and climate, 250;
and conflict, 260, 261, 274, 275;
and cooperative arrangements, 33, 35;
and cost recovery, 26;
and demand management, 97;
and desalination, 78, 194-95, 234-35,
    237;
and the future agenda, 316;
overview of, 66, 67, 74;
and the Rio de Janeiro Conference
    (1992), 74;
and security, 257, 258-59;
and supply management, 19;
and technology, 15-16, 35.
See also specific country or region

Equatorial Lakes, 105, 107, 108, 111

Equitable utilization, 263-65, 266

Eritrea, 42, 49

Erosion, 164-65, 170

Ethiopia:
agriculture/irrigation in, 54, 110;
and climate, 248;
and conflict, 265, 280;
and cooperative arrangements, 108,
    110;
geography of, 71;
supply and demand in, 71, 73, 91;
and transboundary water, 42, 43, 45-
    46, 49, 50, 52-53, 54, 108, 110, 265

Ethiopian plateau, 103, 105

Euphrates River:
and allocation/reallocation, 5, 139;
and conflict, 312, 313;
and conservation, 12;
and demand management, 97;
development of the, 134, 137;
flow of the, 134;
and the future agenda, 312, 313;
and the history of the Arab Mashrek,
    124, 126;
information system about the, 23, 57,
    139;

length of the, 137;
and politics, 2-3;
sources of the, 124, 126, 128, 130, 133;
and supply and demand, 3, 65, 73, 91, 312, 313;
and supply management, 19;
as transboundary water, 41, 42, 43, 44, 45, 54-59, 61, 133-34, 136, 137-38, 139-40, 277;
and wastewater, 79

Evaporation, 29, 31, 73, 82, 105, 121, 144, 305

Fajer, E. D., 249

Falkenmark, Malin, 301

FAO, 110, 181

Fashoda syndrome, 44, 45

Fields, C. B., 249

Financing:
    and cooperative arrangements, 44, 55-56;
    and desalination, 192, 206, 238-39;
    and infrastructure, 96-97;
    and transfers, 205;
    and water management, 21.
    See also Cost recovery; Costs; *specific country or project*

First-in-use-first-in-right, 263, 271

Floods, 11, 197, 246, 250

Food imports, 91, 92-94, 99-100. *See also* Food self- sufficiency; *specific country or region*

Food security, 153, 250, 257. *See also* Food imports; Food self-sufficiency; *specific country or region*

Food self-sufficiency:
    as aim of agriculture, 8-9, 30;
    and climate, 244;
    and cooperative arrangements, 58, 61;
    and demand management, 86-87, 95-96;
    and the evolution of water strategies, 92-94;

and food security, 257;
and the future agenda, 310;
and government policies, 30;
overview of, 66, 68, 70;
and a water crisis, 310.
    See also specific country or region

Fossil water, 243-44, 250. *See also specific source of water*

Fox, I.K., 135

Fractured rock zones, 82, 314

France, 47, 56

Frederick, K., 31

Frederiksen, H. D., 23, 25, 26, 27

Frey, Frederick, 43, 260, 261, 262, 274, 276, 277

Friendship and Good Neighborliness Treaty (1926), 56

Furrow irrigation, 11

Future agenda:
    areas of concern for the, 313-16;
    and conflict, 312-13;
    and costs, 301, 310-11, 314, 315-16;
    and demographics, 289, 305;
    and development, 312;
    and economic issues, 303, 312, 314;
    and energy, 310-11;
    and the environment, 316;
    and food, 310;
    and government policies, 310-13, 315-16;
    and infrastructure, 285, 312;
    and institutional arrangements, 315;
    and modeling, 301, 303, 305, 308-10;
    overview of the, 285-86, 289;
    and planning, 312;
    and politics, 310, 315;
    and pricing policies, 301, 310, 311, 315-16;
    and the private sector, 315;
    and sources of water, 297-98, 301;
    and studies/investigations, 313-14;
    and urbanization, 305;
    and a water crisis, 310-12;
    and water quality, 312.
    See also specific country, use, or source of water

GAP. *See* Southeast Anatolia Project (Turkey)

Garber, Andrae, 132

Garon River, 197

Garretson, Albert Henry, 46, 262

Gaza. *See* West Bank and Gaza

GCC. *See* Gulf Cooperation Council countries

Geographic Information Systems (GIS), 33

Geography, 71. *See also specific country or region*

Geophysical explorations, 314

Geopolitics. *See* Politics

Gezira project, 45, 80, 82, 108-9, 115

GIS. *See* Geographic Information Systems

Glater, J., 204

Gleick, Peter, 256, 259, 260

Global warming, 118, 244-46, 247, 248

Godana, Bonaya Adhi, 46, 49

Golan Heights, 59, 60, 77, 130, 140, 280

Goldstone, J., 274

Gomez, C., 301

Good faith, 47, 48-49

Good neighborliness, 263, 267

Gordon, H. B., 247

Government policies:
and agriculture/irrigation, 9-10, 12-13, 25, 26, 30, 161;
and allocation/reallocation, 10, 83;
and conservation, 24, 30, 36-37;
and cooperative arrangements, 61, 283, 284;
and cost recovery, 24, 25, 26, 27;
and desalination, 199, 237-38;
and food self-sufficiency, 30;
and the future agenda, 310-13, 315-16;

and municipal use, 36;
overview of, 66, 68, 70;
and pricing policies, 315-16;
and supply and demand, 310-13, 315-16;
and water management, 9-10, 12-18, 23-24, 25, 26, 27, 36-37.
See also Politics; *specific country or region*

Great Britain:
as a colonial power, 43, 45, 49, 50, 56, 65-66, 107-8, 109, 115;
and cooperative arrangements, 43, 45, 49, 50, 56, 107-8, 109, 115;
and Nile River development, 107-8, 109, 115

Great Man-Made River, 75-76, 102, 205

Greater Zab River, 127-28

Greenhouse gases, 243, 244-46, 247, 248, 250

Groundwater, 3, 23, 35, 70, 74-75, 78, 82, 143-44, 267, 308. *See also specific country or region*

Gruen, George E., 77

Guariso, Giorgio, 52

Guiers Lake, 158

Guinea, 150

Gulf Cooperation Council countries, 216, 221, 236, 237

Gulf States, 5, 79-80, 82, 132, 196. *See also specific state*

Gulf War, 54, 55, 71, 78, 277, 280

Gurr, Ted R., 44, 254

Haarsma, R. J., 247

Hajar Mountains, 175

Hamadah aquifer, 181

Hannoyer, Jean, 58

Hansen, J., 246

Hardin, Russell, 54

Harmon Doctrine, 46, 52-53, 58, 262

Hasbani River, 59, 71, 130, 133

Hayton, R. D., 267

Helsinki Rules (1966), 47, 48-49, 266-67

Homer-Dixon, T., 256, 258, 261, 275, 282

Houghton, J. T., 244, 245, 246

Howe, Gordon R., 124

Howell, P. P.., 74

Hydroelectric power. *See* Energy; *specific country or region*

Hydrological explorations, 314

Hydrometeorological Survey of the Upper Nile and Lake Victoria Basin, 52

Ice melt, 247

Icebergs, towing, 197

ILA. *See* International Law Association

ILC. *See* International Law Commission

Imported food. *See* Food imports; Food self-sufficiency; *specific country or region*

Incentives. *See* Government policies

India, 16, 135, 276

Induced solutions, 44, 56

Indus River, 276

Industrial use:
and allocation/reallocation, 4-5, 84;
and conservation, 314;
and costs, 76;
and desalination, 78, 98, 183, 185, 191, 220, 314;
and the future agenda, 286, 305, 309, 314;
overview of, 67;
and supply and demand, 1, 77, 286, 305, 309, 314;

and wastewater, 80;
and water quality, 142.
See also specific country or region

Infiltration gallery systems, 188

Information systems:
and cooperative arrangements, 23-24, 32- 33, 35, 36, 48, 52, 53, 57, 284;
and demand management, 159;
and desalination, 196, 237;
and groundwater, 70, 74;
and the law, 267;
and planning, 200;
as a solution to water management problems, 23-24.
See also specific country, region, or river

Infrastructure, 61-62, 94, 96-97, 224, 285, 312. *See also specific country or region*

Institut de Droit International, 264

Institute for Biospheric Research, 249

Institutional arrangements, 33, 67, 89, 99-100, 236-39, 284, 315. *See also* Private sector; *specific country or region*

Intergovernmental Panel of Climate Change (IPCC), 245

International Atomic Energy Agency, 235

International Committee for the Hydrology of the Rhine Basin, 250

International law, 33, 40-41, 46-49, 136, 272, 274

International Law Association (ILA), 47, 48, 266

International Law Commission (ILC), 45, 47, 48, 57, 263-64, 266-67

International organizations, 97, 236-39. *See also specific organization*

International rivers:
definition of, 57, 136;
number of, 286.
See also Transboundary water; *specific river*

International water course:
   definition of, 48.
   See also Transboundary water; *specific river*

Iran, 136, 197

Iraq:
   acquired rights of, 139;
   agriculture/irrigation in, 5, 12, 97, 134, 137, 140, 289;
   allocation/reallocation in, 5, 8, 244;
   and conservation, 12;
   and cooperative arrangements, 42, 43, 44, 45, 47, 54-59, 61, 134, 138, 139, 140, 181, 277;
   demand management in, 97-98;
   demographics in, 134;
   and desalination, 221;
   and economic issues, 303;
   and the environment, 74, 97;
   and food, 12, 134;
   and the future agenda, 286, 289, 303, 313;
   government policies in, 74;
   history of, 124;
   infrastructure in, 71;
   land in, 8, 286, 289;
   and new water, 196;
   planning in, 140;
   politics in, 74;
   sources of water for, 134, 137-38, 139, 140, 179, 181, 196;
   and supply and demand, 134, 254, 255, 286, 289, 303, 313;
   supply management in, 97-98;
   and transboundary water, 42, 43, 44, 45, 47, 54-59, 61, 134;
   and transfers, 140, 196;
   urban use in, 71, 97;
   and wars/conflicts, 134, 135-36, 139, 280, 313;
   and wastewater, 74, 79.
   See also Arab Mashrek; Euphrates River; Tigris River

Irrigation. *See* Agriculture/irrigation

Islamic law, 51, 268-72

Israel:
   agriculture/irrigation in, 59, 60, 95-96;
   and conflict, 264, 277-78, 279-80, 282, 283;
   and the consequences of scarcity, 255;
   and cooperative arrangements, 34-35, 41-42, 43, 44, 45, 59-61, 97, 133, 276;
   and demand management, 95-96;
   desalination in, 60;
   domestic use in, 59, 60;
   and food self-sufficiency, 95-96;
   geography of the, 130;
   industrial use in, 59;
   rainfall in, 132;
   sources of water for, 17, 77, 130, 133, 134-35, 138, 140, 267;
   and transboundary water, 41-42, 43, 44, 45, 59-61, 134-35, 136, 138;
   and transfers, 59, 61, 71, 77, 134-35, 205;
   wastewater/reuse in, 60, 95-96

Italy, 108

Jaghjagh River, 133

Jauf aquifer, 179, 187

Jebali, Ali, 301

Jebel Alawi River, 130

Jebel Awlia dam, 42, 50, 108

Jebel Sheikh, 71

Jellali, Mohamed, 301

Jezirah plateau, 127, 130

Jilh aquifer, 179

Johnston (Eric) Plan, 276

Jonglei Canal, 51, 52, 53, 54, 74, 95, 102, 111, 116

Jordan:
   agriculture/irrigation in, 5, 9, 10, 12, 13, 16, 17, 18, 20, 26, 29, 43, 59, 77, 96, 133;
   allocation/reallocation in, 5, 6, 8, 9, 10, 18, 23-24, 96;
   competition index in, 6;
   and conflict, 277-78;

conservation in, 29, 60, 96;
and cooperative arrangements, 18, 34-35, 41, 42-43, 44, 59-61, 97, 133, 181, 276;
cost recovery in, 26, 96;
delivery system in, 96;
and demand management, 94, 95, 96;
demographics of, 71;
and desalination, 17, 78;
domestic use in, 20, 26, 59, 71, 76, 96;
and financing, 96-97;
and food self-sufficiency, 87, 95;
geography of, 71;
government policies in, 10, 16-18, 23-24;
history of, 124;
industrial use in, 59, 76;
information networks in, 24;
and an integrated approach to water management, 23-24;
land in, 8;
and new water, 78, 82, 196;
non-conventional water sources in, 31;
and pipeline experiences, 76-77;
politics in, 96- 97;
pricing policies in, 96;
sources of water for, 2, 13, 16-18, 31, 76-77, 78, 82, 132, 134-35, 138, 179, 181, 196, 267;
supply and demand in, 71, 84, 87, 132, 255;
and supply management, 20;
and transboundary water, 41, 42-43, 44, 59- 61, 138;
and transfers, 71, 76, 134-35, 196;
urban use/urbanization in, 71, 96;
wastewater in, 78, 82;
water quality in, 17.
See also Arab Mashrek; Jordan River; Jordan Valley

Jordan River:
and agriculture/irrigation, 95;
and conflict, 96, 274, 277-78, 280, 282, 283, 312;
and cooperative arrangements, 32, 276;
and demand management, 95;
and the future agenda, 312;
and geography, 130;
information system concerning the, 23;
management of the, 138;
and pipelines, 77;
and politics, 2-3;
and supply and demand, 71;
and supply management, 19;
and transboundary water, 41, 42, 45, 59-61;
tributaries of the, 130;
and water quality, 95

Jordan Valley, 12, 16-17, 18, 26, 29, 35, 77, 82, 96

Juba River, 103, 312

Jurassic aquifer, 179

Kally, Elisha, 43

Karun River, 127

Keban dam, 56, 57

Kedri, Sadek A., 155, 157

Keenan, J. D., 31, 32

Kenya, 49, 110

Khabur River, 55, 58, 128, 130, 133, 140

Khadam, M.., 30

Khashin El Girba Dam, 115

Khazen, A., 161

Khouri, J., 178, 189

Khouri, N., 23, 205, 231

Khuff aquifer, 179

Khulab aquifer, 181

Kimball, B. A.., 248

Kirk, E. J., 258

Kohlan aquifer, 189

Kolars, John, 54, 127, 128, 130, 131, 137, 254

Kufrah aquifers, 82

Kurds, 43, 55

Kuwait:
agriculture/irrigation in, 19, 182, 184, 187;
and allocation/reallocation, 4;
and cooperative arrangements, 45, 181;
and demographics, 289;
and desalination, 183, 187, 191, 203, 216, 227, 308;
domestic use in, 182, 187;
and the environment, 193, 194;
and the future agenda, 289, 308;
industry in, 187, 194;
and new water, 196;
non-conventional water sources in, 31, 32;
rainfall in, 132;
sources of water for, 2, 31, 32, 176, 179, 180, 181, 182, 196;
and supply and demand, 187, 190, 289, 308;
and transboundary water, 45;
and transfers, 196, 205;
wastewater/reuse in, 184, 187, 193

Kuwait Fund, 221

Lake Albert, 53

Lake Assad, 76

Lake Lanoux arbitration (1957), 47

Lake Manzala, 113

Lake Nasser/Nubia, 80

Lake Tana, 108

Lake Tiberius, 133

Lake Victoria, 52, 53

Law:
as a bargaining ploy, 272;
and conflict, 262-72, 282-83;
and information systems, 267.
See also International law; Islamic law; Treaties

Le Marquand, D. G., 135

Leakages, 30, 192-93, 199, 230, 233, 297

Lebanon, 132-33, 138, 279-80, 283, 286, 289, 303. *See also* Arab Mashrek; Jordan River

Legal issues. *See* International law; Islamic law; Law

Leitner, G. F., 226, 235

Lemarquand, D. G., 277

Lesser Zab River, 127-28

Libya:
agriculture/irrigation in, 5, 76, 98, 151-52;
allocation/reallocation in, 5, 8, 76, 151-52;
conservation in, 165;
and costs, 75-76;
and desalination, 157, 158, 221, 308;
development programs in, 157;
and the future agenda, 286, 308;
and geography, 71;
and the Great Man-Made River, 75-76, 102, 205;
industrial use in, 158;
institutional framework in, 168;
land in, 8, 286;
new water in, 75-76, 77, 78, 98;
pipeline experience in, 75-76;
politics in, 76;
rainfall in, 123;
and reclamation, 157;
sources of water for, 74, 75-76, 77, 78, 98, 150, 267;
and supply and demand, 152, 154, 254, 286, 308;
and transfers, 98, 157, 205;
urban use in, 71;
wastewater/reuse in, 158, 165.
See also Maghreb countries

Limited territorial sovereignty, 46, 47, 262, 263

Lincoln, D. E., 250

Litani River, 59, 130, 133, 138

Liyah aquifer, 181

Lloyd, J. W., 74, 268

Lonergan, Stephen C., 130, 259

Lowermilk, W. C., 35

Lowi, Miriam, 44, 60

Lubchenco, J., 244

Lutz, E., 67

Maas, A., 273

McCaffrey, S., 264, 267

McClelland, Elizabeth, 54

Machar marshes, 111

MacNeil, Jim, 118

Maghreb countries:
  agriculture/irrigation in the, 147, 151-
    52, 153, 160, 165, 167-68;
  allocation/reallocation in the, 151-
    52, 160, 163;
  climate in the, 148-50, 151, 153;
  conservation in the, 161, 166, 168,
    169, 170;
  cost recovery in, 168;
  and demand management, 98, 158,
    159-61;
  and demographics, 147, 148, 153,
    154, 169-70;
  development in the, 154, 156-58;
  domestic use in the, 71, 151, 152,
    166;
  droughts in the, 151, 152, 154, 160,
    161-63, 165;
  and economic issues, 150-54, 166-68;
  and the environment, 164- 66;
  floods in the, 151, 158;
  and food security, 153;
  government policies in the, 147-48,
    166;
  hydroelectric power in the, 152, 153-
    54, 160;
  industry in the, 147, 152;
  and information systems, 159;
  infrastructure in the, 98, 170;
  and institutional arrangements, 152-
    54, 166-69, 170;
  and new water, 77, 78, 158-59;
  overview of the, 147-48, 169- 70;
  and planning, 154, 155-56;
  and pricing policies, 166;
  rural use in the, 170;
  sources of water for the, 77, 78, 150-
    51, 152-54, 158-59;
  standard of living in the, 148;
  storage reservoirs in the, 153, 170;
  supply and demand in the, 71, 147,
    148, 150, 154, 158;
  and technology, 154;
  and transfers, 153;
  urban use in the, 71, 151, 153, 160;
  urbanization in the, 147, 148, 153,
    165;
  water quality in the, 160, 164, 165-
    66, 168, 170.
  See also Algeria; Libya; Mauritania;
    Morocco; Tunisia

Majradah Canal, 205

Maktari, A.M.A., 268

Malaysia, 205, 224

Mali, 150, 158

Mallat, Chibli, 46

Managil extension, 115

Manantali reservoir, 151, 157-58

Maqarin Dam, 278

Mar del Plata Water Conference (1977),
  53

Marginal costs, 26

Mather, J. R., 144

Matson, Ruth, 43, 262, 274, 277

Mauritania, 149, 150, 151, 157-58, 168,
  286, 303

MAW (Ministry of Agriculture and
  Water, Saudi Arabia), 178, 179, 181

MED. *See* Multi-effect distillation

Medusa Bag project, 77

Melen Project, 205

Merguellil River, 158

Meroe site, 73

Mesopotamia, 127, 130

Mexico, 30, 46

Meybeck, Michel, 112

Michael, Michael, 46, 47

Mikhail, Wakil, 58

Military power, 43, 44-45. *See also* War

Minjur-Druma aquifer, 179, 180, 187

Mitchell, William A., 130, 131, 137

Modeling:
   and climate, 245, 246, 247, 250;
   and conflict, 261- 62;
   and desalination, 231, 233-34;
   and the future agenda, 301, 303, 305,
      308-10;
   and power, 262, 278-80;
   and supply and demand, 301, 303,
      305, 308-10

Mohamed, A. K., 178, 189

Morocco:
   agriculture/irrigation in, 5, 12, 20, 23,
      26, 29, 30, 98, 151-52, 153, 155,
      157, 160, 163, 167, 289;
   allocation/reallocation in, 5, 8, 151-
      52, 155, 160, 163;
   conservation in, 29, 30, 163;
   cost recovery in, 26;
   and demand management, 94, 98,
      160-61;
   and demographics, 148, 289;
   development programs in, 157;
   distribution system in, 163;
   domestic use in, 20, 152, 163, 167;
   droughts in, 162, 163;
   and economic issues, 150-51;
   and the environment, 164;
   and the future agenda, 286, 289, 303;
   government policies in, 23-24, 166;
   hydroelectric power in, 152, 163;
   information networks in, 23-24;
   infrastructure in, 157, 163;
   institutional framework in, 168, 169;
   and an integrated approach to water
      management, 23;
   land in, 8, 286, 289;
   planning in, 155-56;
   and pricing policies, 166- 67;
   and rain technology, 158;
   rainfall in, 162;
   rural use in, 157, 163;
   sources of water for, 158;
   and supply and demand, 286, 289,
      303;
   and supply management, 20;
   and transfers, 157;
   urban use/urbanization in, 157, 167.
      See also Maghreb countries

MSF. *See* Multi-stage flash distillation

Multi-effect distillation (MED), 206,
   208, 211

Multi-good bargaining, 43, 51-52, 55, 58

Multi-stage flash distillation (MSF), 32,
   183, 206-8, 216, 220, 221, 227, 229,
   238, 241

Munasinghe, M., 67

Municipal use. *See* Urban use; *specific
   country or region*

Muzafer, S., 276

Naff, Thomas, 43, 262, 269, 274, 276,
   277

Najran aquifer, 181

Nakashima, M., 25

National economies, 10, 13, 95-99

Nationalism, 256, 259, 260, 264

Neogene aquifer, 179, 180, 181, 182,
   187, 188

Netherlands, 224

New water, 73-82, 83, 84, 86-87, 91, 94,
   98, 301, 314. *See also* Desalination;
   specific country or region

Night temperatures, 246-47, 250

Nile Basin Organization, 53

Nile Projects Commission Report, 51

Nile River:
   administration of the, 109;
   and agriculture/irrigation, 107, 109,
      110;
   and allocation/reallocation, 108-9,
      110, 115;
   and climate, 248;

and conflict/war, 112, 264-65, 312, 313;
and conservation, 12, 104, 107, 108, 111, 112;
and cooperative arrangements, 107-8, 109;
development of the, 107-17;
and the environment, 111, 112-17;
and evaporation, 105, 107;
and the future agenda, 312, 313;
and information systems, 23, 52, 53;
natural characteristics of the, 105-7;
and new water, 73;
and politics, 2-3, 105, 107-11;
and rights, 49-50, 51;
and supply and demand, 3, 91, 312, 313;
and transboundary water, 42, 43, 44, 45-46, 49-54, 56, 61, 65-66;
and transfers, 59, 61, 205;
and water quality, 114;
yield of the, 104

Nile River Agreement (1929), 49-50, 65-66, 108-9, 115

Nile River Agreement (1959), 50-51, 104, 109-10, 111, 115, 272, 276, 277

Nile River Agreement (1975), 51

Non-conventional water sources, 30-32, 36-37, 298. *See also specific source*

Norby, R. J., 249

North, A., 74

North Africa:
agriculture/irrigation in, 10, 67, 89;
allocation in, 10, 66;
conservation in, 67;
demand management in, 67;
and demographics, 66;
and desalination, 68;
economic issues in, 66, 68;
and the environment, 66, 67;
food security in, 66, 68, 70;
government policies in, 10, 66, 68, 70;
industrial use in, 67;
institutional arrangements in, 67;
overview of, 65-70;
ownership of resources in, 67;

politics in, 68, 89-99;
supply and demand in, 66, 86-89, 91;
technology in, 67;
urban use in, 67;
wastewater/reuse in, 67, 68.
See also Arab world; *specific country*

Northeastern African aquifer, 267

Nubian aquifer, 114, 286, 312

O and M costs (operation and maintenance costs):
and cost recovery, 24, 25, 26, 27, 28;
and desalination, 78, 192, 226, 227, 229, 235, 238;
in the Jordan Valley, 17

Oechel, W. C., 246, 247

OED. *See* World Bank

Okidi, Charles Odidi, 49

Oman:
agriculture/irrigation in, 184, 185, 188;
climate in, 174, 175;
and cooperative arrangements, 181;
dams in, 178;
demand in, 185, 188, 308-9;
and demand management, 199;
and desalination, 183, 188, 191;
and domestic use, 188, 191;
and the environment, 192;
and floods, 188;
and food, 185;
and the future agenda, 308-9;
and government policies, 185;
and industrial use, 191, 199;
and new water, 196, 197;
sources of water for, 176, 178, 179, 181, 182, 188, 196, 197, 308-9;
and transfers, 196;
and urban use, 308-9;
wastewater/reuse in, 184, 188;
and water quality, 192

Operation and maintenance costs. *See* O and M costs

Ophuls, William, 254

Opportunity costs, 26, 27, 62, 221, 311, 316

Organization for Development of the Senegal Valley, 157-58

Organizations:
international, 97, 236-39;
national/regional, 143-44, 315.
See also specific organization

ORMVA. *See* Regional Agricultural Development Authorities

Orontes River, 41, 42, 130, 133, 138

Owen Falls dam, 42, 53

Ownership, 40, 41, 67, 270

Pakistan, 197, 276, 312

Palestine:
agriculture/irrigation in, 95-96;
allocation/reallocation in, 95;
and conflict, 282;
and cooperative arrangements, 61;
and demand management, 95-96;
and food self-sufficiency, 95-96;
and new water, 78, 82;
and pipeline experiences, 77;
sources of water for, 61, 77, 78, 82, 130, 134-35, 267;
transfers to, 61;
wastewater/reuse in, 82, 95-96.
See also West Bank and Gaza

Pallas, P. P., 82

Parnall, Theodore, 47

Peace Pipeline. *See* Turkey: Peace Pipeline of

Pearce, D., 67

Per capita water availability/consumption, 1, 2, 4, 87, 101, 254, 289, 308-9

Permanent Joint Technical Commission (PJTC), 51

Pfaltzgraff, R. L., 256

Pipelines, 75-77, 196-97, 199, 205, 225, 226, 240-41, 301

Pittock, A. B., 247

Planning:
and climate, 250;
and the future agenda, 312;
and information systems, 200;
integrated river basin, 26, 33- 34;
need for contingency, 19;
and supply and demand, 312;
and transfers, 205.
See also specific country or region

Ploss, Irwin, 60

Politics:
and allocation/reallocation, 83, 89, 99-100;
and cooperative arrangements, 1, 40-46, 55;
and cost recovery, 27;
and demand management, 87, 94, 98-99;
and financing, 96- 97;
and food imports, 99-100;
and the future agenda, 310, 315;
and infrastructure, 96-97;
and institutional arrangements, 315;
overview of, 68, 70-71, 73;
and supply and demand, 2-3, 66, 84, 310, 315;
and transboundary water, 40-46, 55, 135-39;
and transfers, 142;
and treaties, 272.
See also Government policies; *specific country or region*

Ponting, Clive, 101

Population. *See* Demographics

Porter, H., 249

Postel, Sandra, 254, 301

Power:
asymmetry of, 277-78, 280;
and conflict, 276-82;
and cooperative arrangements, 43, 44-45;
matrix model of, 262, 278-80;
military, 43, 44-45

Pricing policies:
and agriculture/irrigation, 11, 36;
and allocation/reallocation, 10, 11, 222;
and brackish water, 199;
and conservation, 30, 198-99;
and costs, 222, 224, 315- 16;
and the delivery of water, 89;
and demand management, 89;
and the future agenda, 297, 301, 310, 311, 315-16;
and government policies, 315-16;
and supply and demand, 83-84, 89, 297, 301, 310, 311, 315-16;
and supply management, 19;
and technology, 11;
and a water crisis, 310.
See also Cost recovery

Private sector, 26, 36, 315;
and desalination, 183, 191, 192, 204, 238

Production costs, 233

Productivity, 6, 9, 29

Protocol on Matters Pertaining to Economic Cooperation between the Republic of Turkey and the Syrian Arab Republic (1987), 58

Pyne, R.D.G., 231

Qa Disi aquifer, 76, 82, 132, 267

Qatar:
agriculture/irrigation in, 4, 19, 182, 184, 188;
and cooperative arrangements, 181;
and the delivery of water, 199;
demand for, 187-88;
and demand management, 198, 199;
and desalination, 183, 187, 188, 191, 308;
domestic use in, 188;
and the environment, 192, 194;
and the future agenda, 308, 309;
and government policies, 199;
and industry, 188, 194;
and new water, 196, 197;
and pricing policies, 198, 199;

sources of water for, 176, 179, 181, 182, 187-88, 196, 197;
and supply and demand, 190, 308, 309;
and transfers, 196, 197;
wastewater/reuse in, 184;
and water quality, 192

Queiq River, 133

Rady, M. A., 14, 20, 25, 29-30, 31, 112, 113-14

Rahad II project, 111

Rain technology, 158

Rainfall:
and agriculture/irrigation, 286;
and climate, 245, 246, 247, 250;
and the future agenda, 286, 297;
information about, 70;
map of, 286;
and water supply, 70, 71.
See also specific country or region

Rainwater harvesting. *See* Water harvesting

Rajagopalan, V., 23

Ramsar Convention, 316

Ras al'Ain aquifer, 42

Reasonable use, 47, 264

Reclamation Act (U.S., 1902), 25

Reclamation. *See specific country or region*

Reed, E. C., 230

Regional Agricultural Development Authorities (ORMVA), 26

Regional cooperation. *See* Cooperative arrangements

Regional information clearinghouse (RICH), 141

Regional management, 33-34, 94

Regional organizations, 143-44, 315

Replacement costs, 221

Returns to water principle, 88

Reuse. *See* Wastewater/reuse; *specific country or region*

Reverse osmosis (RO), 32, 183, 195, 208, 211, 216, 220, 221, 228, 229, 241

Rhine River, 250

RICH. *See* Regional information clearinghouse

Rights. *See specific right, river, or country*

Rio de Janeiro Conference (1992), 74

Rio Grande River, 46

Riyadh, Saudi Arabia, 30, 132, 181-82, 186, 193

R'kiz Lake, 158

RO. *See* Reverse osmosis

Rogers, Peter, 47, 98, 261, 313

Romm, J. J., 256

Roseires Dam, 50, 111, 115, 116

Rosensweig and Perry, 249

Rubenstein, Jonathan, 60

Rural use, 289, 305. *See also specific country or region*

Rus aquifer, 187

Rwanda, 49

Sadukham et al., 226, 235

Sahara desert, 149

Saiq aquifer, 188

Sajour River, 133

Sajur River, 130

Sakaka aquifer, 179, 187

Salameh, Elias, 96, 132

Salim, Omar M., 155, 157

Salination, 11, 12, 16, 68, 112, 113, 118. *See also* Desalination; *specific country or region*

Saq Aquifer, 132, 179, 180, 181, 187

Sarat Mountains, 175

Saudi Arabia:
   agriculture/irrigation in, 5, 19, 98, 181, 182, 184, 185, 186-87, 193;
   and allocation/reallocation, 4, 5, 8;
   climate in, 174, 175;
   and conservation, 186, 199;
   and cooperative arrangements, 181-82;
   demand in, 185, 186-87;
   and demand management, 198, 199;
   and desalination, 183, 186, 191, 216, 308;
   domestic use in, 182, 184, 186, 190, 191, 199;
   and the environment, 192-93;
   and floods, 178;
   and food self-sufficiency, 87;
   and the future agenda, 286, 308;
   geography of, 175;
   and government policies, 193;
   industrial use in, 184, 191;
   land in, 8, 286;
   and new water, 98, 196, 197;
   non-conventional water sources in, 31, 32;
   and pricing policies, 198;
   and the Qa Disi Aquifer, 132;
   sources of water for, 2, 31, 32, 98, 141-42, 176, 178, 179, 180, 181-82, 186-87, 190, 193, 196, 197, 267;
   and supply and demand, 87, 190, 286, 308;
   and transfers, 98, 141-42, 196;
   urban use in, 186;
   wastewater/reuse in, 184, 186;
   and water quality, 192-93

Savage, C., 77-78

Scarcity, 2-12, 253-55. *See also* Supply and demand

Schneider, S. H., 245, 250

Scientists, need for, 283-84

Sea water. *See* Desalination

Sea-level rise, 118, 247-48

Seckler, David, 313

Security, 255-59, 273-74. *See also* Food security

Sen, Amartya, 68

Senegal, 150, 158, 205

Senegal River, 149, 150, 151, 157-58

Sennar dam, 45, 49-50, 108, 115

Setit River, 111, 116

Sewage. *See* Wastewater/reuse

Shahin, M., 178

Sharia, 268-69, 270, 271

Shatt al-Arab River, 127, 131, 205, 248

Silting, 11-12, 164, 170

Silver, R. S., 203

Singapore, 205, 224

Sinn River, 133

Sirhan aquifer, 181

Six Day War, 280

Snow cover, 247

Sobat River, 52, 103, 105, 111

Somalia, 103, 104, 117, 286, 303, 309

Souss River, 158

Southeast Anatolia Project (Turkey), 44, 54-55, 58, 59, 73, 139

Soviet Union, 44, 57, 280

Space holder contracts, 34

Spain, 47

Sprinkler irrigation, 11, 29, 160, 161

Standard of living, 87, 184, 185, 186, 202, 275, 286. *See also specific country or region*

Starr, Joyce, 205

Storage. *See specific country, region, or reservoir/dam*

Submerged tube process, 206

Subsidies. *See* Government policies

Sudan:
agriculture/irrigation in the, 5, 9, 16, 21, 24, 29, 30, 80, 82, 108, 109, 110, 111, 114-17, 289;
allocation/reallocation in the, 4, 5, 9, 104, 108-9, 111, 115;
and climate, 107, 114, 248;
and conflict/war, 117, 280, 313;
conservation in the, 29, 30, 116;
and cooperative arrangements, 24, 45, 46, 47, 49-54, 61, 74, 80, 107-9, 110, 115, 276, 277;
costs/cost recovery in the, 21, 25;
demographics in the, 116, 117, 289;
domestic use in the, 117;
drought in, 117;
and economic issues, 303;
and the environment, 111, 116, 117;
and financing, 117;
and food, 30, 102-3, 115-16;
and the future agenda, 286, 289, 303, 308, 313;
geography of, 71;
government policies in the, 24, 30;
and an integrated approach to water management, 24;
land in the, 286, 289;
and new water, 73, 74;
politics in the, 105, 109, 116;
productivity in the, 116-17;
sources of water for the, 73, 74, 105, 114-17, 267;
and supply and demand, 71, 286, 289, 303, 308, 313;
and transboundary water, 45, 46, 47, 49-54;
wastewater/reuse in, 80, 82.
See also Jonglie Canal

Sudd swamp region, 74, 105, 107, 108, 111, 112, 116. *See also* Jonglei Canal

Superfairness, 47

Superordinate principle, 275-77, 284

Supply and demand:
and allocation, 2-12;
categorization of, 83-84;
and costs, 83-84, 301, 310-11, 314, 315-16;

and demographics, 2, 4-5, 6, 13, 35, 65, 66, 71, 289, 305;
and development, 312;
and economic issues, 303, 312, 314;
and the evolution of water strategies, 91-94;
and modeling, 301, 303, 305, 308-10;
and a water crisis, 310-12;
and water management, 2-12, 35;
and water quality, 35, 312.
See also Conservation; Cooperative arrangements; Demand management; Future agenda; New water; Scarcity; Supply management; *specific country, region, source, or use*

Supply management, 19-21, 34, 94

Surface water, 23, 70-71, 91. *See also specific country or source of surface water*

Switzerland, 224

Syr Darya basin, 43

Syria:
agriculture/irrigation in, 9, 10, 20-21, 25, 29, 61, 97, 133, 134, 140;
allocation/reallocation in, 6, 8, 9, 10, 244;
competition index in, 6;
and conflict/war, 97, 135-36, 139, 277-78, 279-80, 313;
conservation in, 29, 97-98;
and cooperative arrangements, 41-43, 44, 45, 54-59, 60, 61, 97, 133, 134, 138, 139, 140, 181, 276, 277;
cost recovery in, 25;
demand management in, 97-98;
demographics in, 133, 140;
and development of the Euphrates River, 137;
domestic use in, 71, 133, 140;
and the environment, 97, 140;
and food, 58, 61;
and the future agenda, 289, 303, 313;
geography of, 71;
government policies in, 10, 24, 61;
industrial use in, 133, 138;
and an integrated approach to water management, 24;
land in, 8, 289;

and new water, 196;
planning in, 140;
precipitation in, 131;
sources of water for, 31, 133-34, 137-38, 139, 140, 179, 181, 196;
and supply and demand, 71, 133-34, 255, 289, 303, 313;
supply management in, 97-98;
and transboundary water, 41-43, 44, 45, 54-59, 60, 61;
and transfers, 196;
urban use in, 71, 97;
wastewater/reuse in, 79, 140.
See also Arab Mashrek; Euphrates River; Jordan River

Tabqa (Ath-Thawra) Dam, 45, 57, 58, 134, 137

Tabuk aquifer, 179, 180, 181, 187

Tal, Abraham, 43

Tanzania, 49, 110

Taqa aquifer, 188

Technical experts, 283-84

Technology:
and agriculture/irrigation, 10-11, 13, 14, 28;
and allocation/reallocation, 10-11;
and climate, 250;
and conflict, 275;
and conservation, 28, 199;
and cooperative arrangements, 1, 34-35, 284;
and demand management, 67, 83, 94;
and demographics, 91;
and desalination, 203, 204, 206, 220, 236, 237, 239;
as a double-edged sword, 11;
and the environment, 15-16, 35;
and the evolution of water strategies, 91;
and food imports, 91;
and groundwater, 74;
need for increasing availability of, 66;
and pricing policies, 11;
production, 10-11;
and transfers, 205;
and wastewater/reuse, 94

Teclaff, Ludwik A., 46, 108, 262, 267

Teerink, J. R., 25

Tharthar Canal and Reservoir, 128, 140

Thornthwaite, C. W.., 144

Tigris River:
  and allocation/reallocation, 5;
  and conflict, 312, 313;
  and conservation, 12;
  and demand management, 97;
  development of the, 134, 137;
  flow of the, 134;
  and the future agenda, 312, 313;
  and the history of the Arab Mashrek, 124, 126;
  information system concerning the, 23;
  and new water, 73;
  and the Peace Pipeline, 141-42;
  and politics, 2-3;
  and renewable water resources, 3;
  sources of the, 127-28;
  and supply and demand, 65, 91, 312, 313;
  and supply management, 19;
  as transboundary water, 42, 43, 54-55, 56, 58-59, 133-34, 137-38, 140;
  and transfers, 140, 141- 42;
  volume of the, 137;
  and wastewater, 79

Transactions costs, 26

Transboundary water:
  access to, 40;
  as characteristic of Arab world, 286;
  definition of, 48, 57;
  and economic issues, 43;
  and established patterns of use, 41;
  and the Fashoda syndrome, 44, 45;
  and international law, 40-41, 46-49;
  justifications for use of, 47-48;
  and limited territorial sovereignty, 46, 47;
  and military power, 43, 44-45;
  ownership of, 40, 41;
  and politics, 40-46, 135-39;
  and the public domain, 40;
  and reasonable use, 47;
  rights to, 40, 41;
  and war, 44-46.
  See also Conflict; Cooperative arrangements; International river; *specific river, country, or region*

Transfers:
  advantages/disadvantages of, 205-6;
  and agriculture/irrigation, 315;
  and cooperative arrangements, 34-35, 56, 59, 61;
  and costs, 142, 196, 197, 205, 206, 225;
  desalination compared with, 205-6;
  examples of, 205;
  and financing, 205;
  and the future agenda, 301, 308, 315;
  and planning, 205;
  and politics, 142;
  and strategic vulnerability, 206;
  and supply and demand, 301, 308, 315;
  and supply management, 20;
  and technology, 205.
  See also Jonglie Canal; Pipelines; *specific country, region, or water source*

Treaties, 272, 274, 282-83. *See also specific treaty*

Treaty of Lausanne (1923), 139

Tripoli Declaration (AOAD), 102

Tuijl, W. V., 26, 29, 30

Tunisia:
  agriculture/irrigation in, 5, 12, 29, 30, 151-52, 153, 156-57;
  allocation/reallocation in, 5, 8, 24, 151-52;
  conservation in, 29, 30, 165;
  cost recovery in, 26;
  demand management in, 94, 98, 155, 161;
  and demographics, 148;
  and desalination, 161;
  development programs in, 156-57;
  droughts in, 162;
  and economic issues, 151;
  government policies in, 24, 166;
  hydroelectric power in, 152;
  industrial use in, 156-57;

information networks in, 23, 24;
infrastructures in, 157, 158;
institutional framework in, 168, 169;
land in, 8, 289;
planning in, 155, 161;
and pricing policies, 166;
rural use in, 156-57;
sources of water for, 74, 150, 155, 158;
and transfers, 205;
urban use/urbanization in, 156-57,
  165;
wastewater/reuse in, 158, 165;
water quality in, 165.
See also Maghreb countries

Turkey:
agriculture/irrigation in, 54, 140, 244;
climate in, 123, 131, 132, 247;
and cooperative arrangements, 34, 42,
  43, 44, 45, 47, 54-59, 60, 61,
  134, 138, 139-40, 277;
and development of the Tigris River,
  137;
and economic issues, 312;
and new water, 73;
Peace Pipeline of, 77, 141-42, 196-
  97, 205, 226, 240-41;
planning in, 140;
as a source of water, 42, 43, 44, 45,
  47, 54-59, 60, 61, 121, 128, 130,
  134-40, 280;
sources of water for, 73, 139-40;
and supply and demand, 77, 91;
and transfers, 61, 134-35, 141-42,
  205;
and war/conflict, 139, 278, 279-80.
See also Euphrates River

Turner, K. T., 67

Tvedt, Terje, 133, 134, 254

Uganda, 42, 49, 50, 53, 110

Ukayli, M. A., 178

Um er-Radhuma aquifer, 179, 180, 181,
  182, 187, 188, 189

Undugu Group of Nile States, 52

United Arab Emirates:
agriculture/irrigation in the, 182, 184,
  185, 188;
climate in the, 174;
and cooperative arrangements, 181;
dams in the, 178;
demand in the, 185, 188;
and desalination, 183, 188, 191, 308;
domestic use in the, 188, 191;
and economic issues, 303;
and the environment, 192;
and food, 185;
and the future agenda, 303, 308-9;
government policies in the, 185, 188;
and industrial use, 191;
and new water, 196, 197;
sources of water for the, 141- 42, 176,
  178, 179, 181, 182, 188, 196, 197,
  267;
and supply and demand, 190, 303,
  308-9;
and transfers, 141-42, 196, 197;
urban use/urbanization in the, 188,
  308-9;
wastewater/reuse in the, 184, 188;
and water quality, 192

United Nations, 48, 74, 264, 276, 303.
See also International Law Commission

United Nations Development Program,
  24, 110

United States:
and agriculture/irrigation, 6, 301;
and cooperative arrangements, 33-34,
  44, 46;
and cost recovery, 26;
and demographics, 289;
and desalination, 220;
energy in the, 311;
and the Harmon doctrine, 46;
as a mediator, 44, 46, 276;
and supply and demand, 301;
and transboundary water, 44, 46;
and transfers, 196

United States Agency for International
  Development (USAID), 44, 113

United States Office of Technology Assessment, 204

Universities, 204, 238

Uqba, A. K., 178

Urban use:
and allocation/reallocation, 84, 98;
and conservation, 36, 314;
and cost recovery, 25, 83;
and the delivery of water, 89;
and desalination, 78, 98, 183, 220, 314;
and the future agenda, 301, 305, 308-9, 314;
and government policies, 36;
and non-conventional water sources, 31;
overview of, 67;
and supply and demand, 301, 305, 308-9, 314;
and wastewater/reuse, 80, 95, 98.
See also Urbanization; *specific country or region*

Urbanization, 2, 77, 83, 112, 289, 305. *See also* Urban use; *specific country or region*

'Urf, 268-69, 271

Usage:
and demographics, 289;
equitable/reasonable, 47, 263-65, 266;
and pricing policies, 89;
as a reason for scarcity, 254;
and urbanization, 289.
See also Supply and demand

Users: role of, 36

Utton, Albert E., 46, 47, 267

Vesilind, P. J., 244

Vienna Conference (1815), 57

Wade, Neil, 226, 227, 235

Wadi Shebelli, 103

Wagnick, K., 183, 235, 236

Wajid aquifer, 179, 180, 181, 189

War, 44-46, 60, 78, 277. *See also* Conflict; Military power; *specific war*

Wasia-Biyadh aquifer, 179, 180, 181-82, 187, 188, 189

Wastewater/reuse:
and agriculture/irrigation, 16, 18, 231;
and conservation, 234;
and cooperative arrangements, 33;
and costs, 231, 233;
and demand management, 67, 94, 95, 98;
and desalination, 78, 198, 220, 231, 234, 240;
and the environment, 16, 33;
as a non-conventional water source, 30- 31;
public opinion about treated, 80;
as a source of new water, 78-82, 297;
and technology, 94;
and urban use, 80, 95, 98;
and water quality, 79.
See also specific country, or region

Water: economic value of, 18, 88;
importance of, 67, 273, 310;
returns to, 88

Water banks, 34, 36

Water crisis, 310-12

Water harvesting, 18, 30, 31, 36, 143

Water imports. *See* Transfers

Water management:
actions for enhancing, 36-37;
and allocation/reallocation, 3-12;
conclusions about, 35-37;
and conservation, 5, 24, 25, 26, 27, 28-30, 36-37;
and cooperative arrangements, 1, 18, 23-24, 32-35, 36;
and demographics, 1, 2, 4-5, 6, 8, 13, 35;
and economic issues, 12-37;
and financing/cost recovery, 21, 24-28, 36;
and government policies, 9-10, 12-18, 23-24, 25, 26, 27, 36-37;
importance of, 37;

integrated approach to, 23-24, 36, 37;
and non-conventional water sources, 30-32, 36-37;
overview of, 1-2, 89-99;
and solutions to water problems, 2, 23-30;
and supply and demand, 2-12, 35.
See also specific country, region, or type of management

Water quality:
and agriculture/irrigation, 142;
and conservation, 142, 170;
and domestic use, 142;
and the environment, 16;
and the future agenda, 312;
and industrial use, 142;
and the law, 264;
and leakages, 192-93;
and supply and demand, 13, 35, 254, 312;
and wastewater/reuse, 79.
See also specific country or region

Water Resources Center (WRC), 114

Waterbury, John, 45, 49, 50, 52, 53, 56, 59

Waterlogging, 11, 12, 16, 112, 113, 118, 194

Watershed management, 35

Weather modification, 197

West Bank and Gaza:
agriculture/irrigation in, 20;
and the consequences of scarcity, 255;
and food self-sufficiency, 95;
geography of the, 130;
groundwater in the, 132;
and transboundary water, 41-42, 43, 59, 60, 61, 138.
See also Palestine

West Bank mountain aquifer. *See* Yarqon-Taninim aquifer

White Nile, 42, 52, 105, 107, 108, 109, 110

Whittington, Dale, 52, 54

Whyte, A.V.T., 254

Wihda dam, 42, 60, 61

Wild, J., 98

Wilkinson, J. C., 268

Willcocks, William, 108

Williams, W., 249

Wittfogel, Karl A., 126

World Bank:
and cooperative arrangements, 44, 49, 53, 55;
and cost recovery, 27-28;
Egyptian studies by the, 14, 16, 113;
and the Euphrates system, 56, 57;
and financing issues, 21, 113;
GNP studies of the, 104;
and information systems, 23-24;
Jordanian studies by the, 17, 21;
and transboundary water, 44, 49, 53, 55;
and Uganda, 53;
water resource base studies of the, 2;
WSS projects of the, 27

World Court, 276

World Health Organization, 222

World Resources Institute, 2, 3

Wright, E. P., 74, 82

Xie, M., 20, 29, 30

Yarmuk River:
and agriculture/irrigation, 97;
and the geography of the Arab Mashrek, 130;
and pipeline experiences, 77;
as a source for Syria, 140;
and supply and demand, 71;
surface water availability of the, 132;
as transboundary water, 41, 42, 43, 44, 59, 60, 61, 133, 138;
and transfers, 71

Yarqon-Taninim (Western) Aquifer, 130, 267

Yemen:
  agriculture/irrigation in, 5, 12, 20,
    189;
  and allocation/reallocation, 5, 6;
  competition index in, 6;
  and cooperative arrangements, 181;
  dams in, 178;demand in, 189;
  and desalination, 183, 189, 191;
  domestic use in, 71, 184, 189;
  and floods, 189;
  industrial use in, 184, 189;
  land area of, 286;
  and new water, 196, 197;
  non-conventional water sources in, 31;
  rainfall in, 174, 175, 189, 191;
  sources of water for, 31, 176, 178,
    179, 181, 182, 189, 196, 197;

and supply and demand, 71, 289;
technology in, 189;
urban use/urbanization in, 71, 181;
wastewater/reuse in, 184

Zagros Mountains, 127

Zagros River, 131

Zaire, 42, 49, 52

Zaki, E., 21, 24, 25, 30

Zeroud River, 158

Ziz River, 158

Zubaidi, zz, 237